THE FACE OF DECLINE

To the people of the anthracite region who shared their stories
and helped us understand their lives

THE FACE OF DECLINE

THE PENNSYLVANIA ANTHRACITE REGION IN THE TWENTIETH CENTURY

Thomas Dublin
and
Walter Licht

CORNELL UNIVERSITY PRESS
Ithaca and London

Title page: Anxious wives of miners, Knox Mine disaster, Port Griffith, January 1959. Photograph by George Harvan.

Cornell University Press gratefully acknowledges receipt of grants from the University of Pennsylvania and the State University of New York at Binghamton which aided in the production of this book.

First published 2005 by Cornell University Press
First printing, Cornell Paperbacks, 2005

Printed in the United States of America

Library of Congress Cataloging-in-Publication Data

Dublin, Thomas, 1946–
  The face of decline : the Pennsylvania anthracite region in the twentieth century / Thomas Dublin and Walter Licht.
    p. cm.
  Includes bibliographical references and index.
  ISBN-13: 978-0-8014-3469-3 (cloth : alk. paper)
  ISBN-10: 8014-3469-6 (cloth : alk. paper)
  ISBN-13: 978-0-8014-8473-5 (pbk. : alk. paper)
  ISBN-10: 0-8014-8473-1 (pbk. : alk. paper)
  1. Anthracite coal industry—Pennsylvania—History—20th century. 2. Coal miners—Pennsylvania—History—20th century. 3. Pennslyvania—Economic conditions—20th century. I. Licht, Walter, 1946– II. Title.

HD9547.P4D83 2005
338.2'725'097480904—dx22

2005015831

Cornell University Press strives to use environmentally responsible suppliers and materials to the fullest extent possible in the publishing of its books. Such materials include vegetable-based, low-VOC inks and acid-free papers that are recycled, totally chlorine-free, or partly composed of nonwood fibers. For further information, visit our website at www.cornellpress.cornell.edu

Cloth printing   10 9 8 7 6 5 4 3 2 1
Paperback printing   10 9 8 7 6 5 4 3 2 1

Contents

Illustrations

THE FACE OF DECLINE

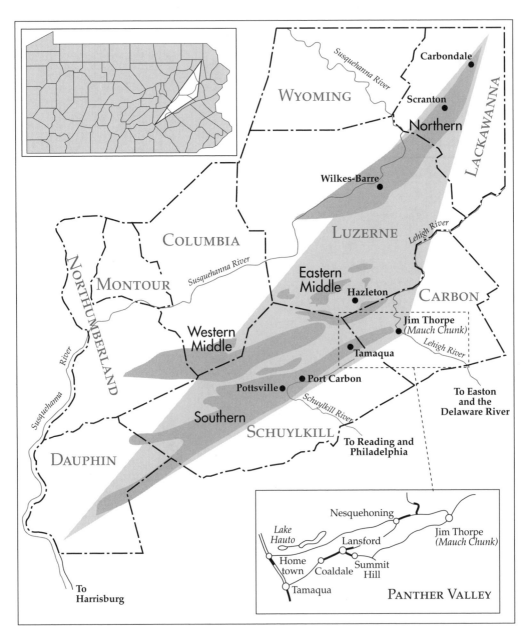

Pennsylvania anthracite region, 2004 (with coal fields shaded). Adapted courtesy of Stackpole Books.

Introduction

MIKE SABRON DIED ON JANU-
ary 19, 2003, at St. Luke's Miners
Memorial Hospital in Coaldale, Pennsylvania, just a few miles from his home in Summit Hill. We had the good fortune to interview
Mike, a former anthracite coal miner, during the
course of our study on the roots and consequences of sustained economic decline, a process all too familiar to regions and cities across
the United States at the turn of the twenty-first
century. Mike lived and worked in the Pennsylvania anthracite region, an area that first prospered on the mining of coal and then collapsed
economically beginning in the 1920s with
changes in the fuel marketplace. This region's
history provides an opportunity to explore the
causes of economic disintegration and the responses of capital, labor, communities, and government. Mike experienced the full brunt of the
closing of the mines and the ensuing regional
crisis, and his story speaks to the personal dimensions of economic decline.

Mike was part of a dying breed of men, literally—a member of the last generation of Pennsylvania anthracite mineworkers. Mike began
working in 1931 at the age of eighteen for
Lehigh Navigation Coal (LNC), a company operating in the Panther Valley, in the southern
portion of the region. He progressed through a
series of jobs at LNC—first in the bagging plant,
then driving mules, loading coal, and finally,
working as a contract miner beginning in 1947.
As demand for anthracite coal declined after
1950, the mines worked only two or three days a
week, and Mike and some buddies took painting
jobs to supplement their shrinking earnings. In
April 1954, however, LNC closed all its mines in
the Panther Valley. Unemployed, and with a
family to support, Mike joined a group of displaced miners and began commuting on a

weekly basis to Linden, New Jersey, to work in a
General Motors plant. He hated the regimen of
the auto assembly line and jumped at the first
opportunity to return to mining when one of the
valley's collieries reopened in October. Mike
worked almost continuously until the last underground mine in the area closed for good in 1972.
After a short period of working at temporary
jobs, Mike retired, soon getting involved in local
efforts to preserve the history of anthracite mining. Mike Sabron's story—and scores of others
we gathered from his fellow former miners,
their wives, and children—tell of crisis, coping,
resilience, and love of place.[1]

The "place" of Mike Sabron's life, the Pennsylvania anthracite coal region, contributed
mightily to the rise of the United States to world
economic supremacy. The region occupies a triangular area in northeastern Pennsylvania that
begins near the Susquehanna River in Dauphin
County north of Harrisburg at its southwestern
point, extends in a north-northeasterly direction
through Wilkes-Barre and Scranton in the
Wyoming and Lackawanna Valleys to just north
of Carbondale, then southward again to Jim
Thorpe (formerly Mauch Chunk) along the
Lehigh River, and, finally, back in a southwest-
erly direction through Carbon and Schuylkill
Counties. Under the surface of this near-five
hundred–square-mile area lay 95 percent of the
nation's supply of anthracite coal, a fuel that
powered U.S. industrialization, a remarkably efficient source of heat to generate steam, smelt
iron ore, and warm the homes of the growing
urban population of a modernizing nation. In
1890, this tiny area provided more than 16 percent of the nation's energy consumption.[2]

Coal played a singular role in the industrial
transformation of Western Europe and the
United States in the nineteenth century. Two

Mike Sabron while working at Lanscoal, ca. 1970.
Photograph by George Harvan.

Mike Sabron, Lanscoal miner, Lansford, fall 1996.
Photograph by George Harvan.

forms of coal—bituminous and anthracite—became the major sources of energy that displaced waterpower and fueled the steam engines and locomotives that changed the character and pace of modern life. Bituminous, or soft coal, is far more common than anthracite, more amenable to mechanization, and continues to play a substantial role in the U.S. economy today as a fuel for generating electricity. Bituminous coal is found across a wide swath of Appalachia stretching from northern Alabama to western Pennsylvania, through Ohio, Indiana, and Illinois, and in Wyoming, Colorado, and Utah, while commercial anthracite deposits are limited almost exclusively to the tiny, triangular region in northeastern Pennsylvania.[3]

The emergence of anthracite mining in the nineteenth century wrought dramatic economic, environmental, and social changes in northeastern Pennsylvania. With the discovery and extraction of coal, the area bustled with entrepreneurial activity, rapid canal and railroad building, and corporate growth and the deepening involvement of finance capital. The population exploded as successive waves of European immi-

grants arrived. At the century's beginning, only three or four residents per square mile lived in the sparsely settled, undeveloped region. By 1900 population density averaged 293 people per square mile and area mines were producing fully 57 million tons of anthracite coal annually. What had been an utter wilderness became the anthracite coal region.[4]

The expansion of mining required vast labor power. Immigrants came first from England, Wales, Scotland, Ireland, and Germany, and then from southern, central, and eastern Europe—all of whom made the region rich in ethnic institutions and folklore. These newcomers soon encountered perilous and exploitative working conditions and participated in repeated and legendary labor battles, conflict that remained characteristic of the anthracite region into the middle of the twentieth century.[5]

The history of the anthracite region in the period of its ascendancy is alluring and has attracted generations of chroniclers and scholars.[6] World War I, however, marked the industry's peak. Starting in the 1920s, anthracite coal lost its market to much less expensive and more convenient fossil fuels, particularly oil and gas, but also bituminous coal, and the region entered an enduring economic decline. The area's contribution to and presence in the national experience began to fade. This book explores the Pennsylvania anthracite region during these difficult times. Extensive sections of U.S. cities and counties that once brimmed with industry now lay fallow. For Americans of the twenty-first century, the long-term economic collapse and stagnation of areas are an unavoidable chapter in the history of the nation. The Pennsylvania anthracite region, an early example, illuminates the dynamics of decline: the roles of capital, labor, and the state and the impact on and responses of communities, families, and individuals.

The issue of economic decline forced its way into public view and policy discussions in the early 1980s as Americans witnessed major changes in the national economy. The anxieties of the time were well founded. The country had just experienced a troubling decade marked by hyperinflation; the highest unemployment levels since the Great Depression of the 1930s; soaring interest rates; the falling value of the dollar; unprecedented losses in productivity and manufacturing jobs; the fiscal insolvency of major corporations and federal, state, and municipal governments; huge balance of payments deficits as well-produced and competitively priced goods from Japan, West Germany, and other nations flooded the U.S. marketplace; and an end to the easy access to world resources that Americans had enjoyed for decades. Pundits and scholars alike predicted a permanent eclipse of the American economy with the United States going the way of former great empires.[7]

A return to better economic times in the mid-1980s shelved the doomsday talk, and optimism returned, at least for a period. A deep recession in the early 1990s revived prophecies of an end to the American dream; such terms as "deindustrialization" and "downsizing" appeared regularly in academic works and newspaper headlines alike.[8] But, just as the ink was drying on the latest lamentations, the U.S. economy surged forward in a heady five-year period of growth with an exuberant stock market. This bubble too would burst, and, in the first years of the twenty-first century, uncertainty characterized economic activity in the United States.[9]

The uneven and confusing record of the U.S. economy in the last decades of the twentieth century offers a warning. A long-term historical perspective is necessary to judge whether an economy is experiencing deep collapse, momentary fluctuation, or structural change. Diagnoses and prognostications of the moment are quickly dated. In this respect, the Pennsylvania anthracite region provides an opportunity to examine economic decline over the long term.

There are good reasons for exploring economic disintegration within a narrow compass. While pronouncing on the rise or fall of nations attracts attention and sells books, national economies are too large and complicated to support easy generalizations. Sectors or regions of economies are surer units for analysis. The matter is also relative. The British have long bemoaned their country's descent from a once preeminent position in the world economy— blaming either the traditionalism and risk aversion of their business classes or the protectiveness and recalcitrance of their unions.[10] Yet, Great Britain remains an affluent nation by any measure. In addition, the causes of long-term economic collapse are uncertain and open to debate. The shortsightedness of American managers and finance capitalists, inflexible mass-production technologies, high wages, sclerotic union rules, globalization, capital mobility, and the fading American work ethic have all been cited in the recent, intermittent discussions of America's fall from grace.[11] While a focus on the anthracite region may not conclusively address all of these issues, an examination of the region's decline over an eighty-year period offers insights that prove helpful in analyzing parallel phenomena of more recent origin.

The economic collapse of the anthracite region of northeastern Pennsylvania is unmistakable. Vacant storefronts are common throughout the downtowns of the area's cities and towns— and have been for decades. Elderly retirees, who subsist largely on government transfer payments, including social security pension benefits and black-lung compensation payments, populate anthracite communities disproportionately.[12] The natural landscape is scarred. Mammoth open pits where coal has been excavated by huge mechanical shovels; abandoned breakers and underground mines; man-made mountains of culm, the refuse of the mines, dotted with the hearty but stunted birches that are all that manages to survive on this rugged landscape; and even occasional smoke rising from the earth where underground coal fires still burn: All contribute to the region's look of desolation.[13]

Sociodemographic statistics confirm the visual impression. Population decline is a sure sign of economic stagnation. In 1930, close to 1.2 million people lived in the anthracite region; 836,000 live there today. The area's population loss amounts to almost 30 percent in the past seventy years—a period in which the overall population of the United States increased by 130 percent (see table 1 in appendix 1). A shift in the age structure of a region is also an indicator of economic decline as young adults are forced to seek employment elsewhere. Between 1950 and 1990, the proportion of the region's population under thirty years of age dropped from 55 percent to 40 percent, while the share above the age of sixty-five grew from 5 percent to 17 percent (see table 7 in appendix 1).[14] While an increase in life expectancy in the United States has contributed to an overall increase in the proportion of the elderly, the graying of the anthracite region has been distinctly stronger than for the nation as a whole, and the decline in the proportion of young wage earners has been much steeper than national figures.

Unemployment numbers, of course, tell the story of economic dislocation. From the 1930s to the 1960s, unemployment rates for the anthracite region ran two to three times higher than national figures; since 1970, unemployment rates have been either slightly higher or at par with the nation's, but this leveling has occurred after severe job and population losses had taken their toll. With the sustained economic collapse of the area, income levels have also fallen far below national averages. Median family and per capita income figures for the region in recent decades have been 15 to 19 percent lower than for the country as a whole (see table 2 in appendix 1).

The signs and measures of crisis are clear. So are the causes. The search for explanations for the long-term decline of the Pennsylvania anthracite region does not have to reach far. The growth and prosperity of the area before 1920 rested on the mining of anthracite coal. The erosion of the market for anthracite triggered and sustained the economic disintegration of the region; the failure effectively to replace mining with other enterprise—a subject to be treated in chapters 5 and 8—has contributed to the region's persisting distress. The actual reduction in anthracite production has been momentous. Pennsylvania anthracite miners extracted 100 million tons of coal in 1917, the crowning year. Production plummeted decade by decade thereafter; by 2000, the few operating strip mines in the region yielded less than four million tons of anthracite. Employment in the mines dropped precipitously with the falling demand for anthracite. At the apex before World War I, 181,000 mineworkers labored in the coalfields of the region (providing direct support to more than 700,000 residents). By 2000, the industry employed fewer than 1,000 workers (see table 3 in appendix 1).[15]

The Pennsylvania anthracite region shares a fate with other areas in the nation and the world. The economic decline of coal-mining communities has been an international phenomenon across the twentieth century. In the United States, overall coal production fell dramatically from 1920 through the 1960s (see table 3 in appendix 1). Notable growth in output occurred thereafter—but well below rates of increase for gas and oil production and consumption—largely through highly mechanized, open-pit mining of bituminous coal in western states.[16] Consequently, aggregate employment of coal miners dropped steadily throughout the century; census enumerators counted almost 785,000 coal miners in 1920 and only 72,000 in 2000. Anthracite took the greatest hit, but declines were also substantial in bituminous (or soft) coal areas of western Pennsylvania, Ohio, Illinois, West Virginia, Kentucky, Tennessee, and Alabama, leaving mining communities there in as economically depressed circumstances as those in the Pennsylvania anthracite region.

Historic coal-producing areas—once thriving, now stagnating—can be traced on a world map. Coal-mining regions in Great Britain, Belgium, France, Germany, Poland, and Ukraine provide notable cases of economic decline that echo the experience of Pennsylvania anthracite (see table 8 in appendix 1).[17] Coal's decline in Western Europe, though, has differed from U.S. cases in an important respect. In many of these countries during the first half of the twentieth century, Social Democratic governments assumed ownership and administration of coal mines in reaction to declining markets and the plight of workers. Nationalization did not stem the crisis in the demand for coal, and these governments have been forced into phased closings of the mines. Such decisions took place amid national public and parliamentary debates and with strong trade union–led protests that garnered national public attention and support. Government leaders responded with comprehensive programs for relief, relocation, redevelopment, and land reclamation. In Pennsylvania and elsewhere in the United States, governmental responses to the economic decline of coal-producing communities have been scattered and piecemeal in comparison. Private ownership of the mines, the role of finance capital in the liquidation of assets, the federal political system with reliance on localism, and ineffectual trade unionism militated against the kind of national assistance policies that have figured so prominently in the decline of coal mining in Western Europe and benefited miners there in ways unimagined by their Pennsylvania anthracite counterparts.

Cross-national comparisons point to an important theme of this book. Economic decline is not a uniform, abstract process. Just as scholars have recently emphasized the unevenness of capitalist economic development—industrialization, for example, unfolding in varying ways with varying consequences in different regions and sectors of the economy—unevenness marks long-term economic collapse as well.[18] Political context matters, as is evident in international comparisons. But even within the Pennsylvania anthracite region itself, the decline occurred in staggered ways, and the impact and responses varied across time, from community to community, and among groups—men and women and old and young faced economic decline variously.

The general conclusion here matches the results of contemporary studies of deindustrialization in the United States. Bleak findings abound. In communities recently battered by plant closings, workers remain unemployed for long periods of time, ultimately finding lower-paying jobs with fewer benefits and less security. Yet the experience of the loss of manufacturing jobs varies significantly among workers by age, sex, race, education, skill level, and location, and displaced workers and their families successfully cope through personal and collective means. Human resilience is evident in the midst of catastrophic economic loss.[19]

Mike Sabron and his neighbors across the Pennsylvania anthracite region have long practiced the individual and collective strategies manifest today in areas recently beset by lasting job loss. For example, whole communities during the Depression rose to demand that employment in the mines be spread equitably among all workers. At the same time, men in southern areas of the region, usually working in family groups, dug into hillsides and "bootlegged" coal on their own to help make ends meet. After World War II, the decline accelerated, and married women secured employment outside their homes. They assisted each other in maintaining family and community life, as their unemployed husbands shared information about employment opportunities outside the region and often commuted weekly, boarding together at distant job sites. Younger unemployed miners responded to the crisis by migrating from the region, often employing networks of family and friends, revealing patterns of chain migration that social scientists have described for transatlantic immigration.[20]

The Pennsylvania anthracite region with its extended experience provides an opportunity to understand sustained economic decline as a complicated, evolving process and fine-grained subject. This is not a facile story of gloom. The second half of the region's history—its economic decline and both community and personal responses to it—is as intricate and important as its more-honored first.

◆

The book begins with the development of the anthracite region in the nineteenth century. Chapter 1 examines anthracite's rise, treating the geological history of the anthracite region, early discoveries of coal, entrepreneurial transportation developments, the consolidation of corporate ownership, the structuring of a cartel by financial capital, population expansion and patterning, the working and social lives of the first generations of anthracite miners and their families, and the intense labor organizing and conflict that marked the last third of the nineteenth century, culminating in the emergence of the United Mine Workers of America (UMWA). Over the course of the century, key institutions of organized capital and labor emerged that would subsequently play significant roles in the region's decline.

Chapter 2 examines the Pennsylvania anthracite region at its apogee and in its initial decline, the first three decades of the twentieth century. The era begins with the great Anthracite Strike of 1902, a battle of national significance that witnessed the intervention of such leading figures as President Theodore Roosevelt and the investment banker J. P. Morgan. The settlement of the strike introduced federal government–administered arbitration procedures that established a new system of labor relations in the industry. Anthracite coal workers and their families enjoyed more certain economic times. The communities they lived in now had bustling downtowns with department stores and cultural institutions (and the fine neighborhoods of the established elites). The first decades of the twentieth century also witnessed the incorporation of southern and eastern European immigrants into anthracite communities, whose churches, ethnic businesses, and fraternal organizations contributed to the institutional mix. At the industry's peak during World War I, the social and economic order in the region seemed sound. Prosperity, however, was short-lived, and in the 1920s the first signs appeared that all was not well. The anthracite coal market collapsed, miners faced unsteady employment prospects, and combative labor conflict resumed.

While economic difficulties in the anthracite region predated the onset of the Great Depression, the national economic collapse stimulated the first significant collective responses to decline. The Depression decade reveals the interplay of corporate, union, and governmental forces that would shape the anthracite region throughout its extended decline as well as the independent actions of mineworkers as they coped with the economic crisis. The institutional accommodation reached during the 1902 strike disintegrated. In reaction to the now-deepening crisis, anthracite miners engaged in unique protest and survival activities in the 1930s, including dramatic equalization of work campaigns, insurgencies within the UMWA, and coal bootlegging. These movements provide the focus for chapter 3. The federal government, with President Franklin Delano Roosevelt's New Deal, became increasingly involved, but anthracite mineworkers, not officials in Washington, took the lead in developing creative responses to the economic crisis.

World War II offered a reprieve from the desperation of the Depression and provides the starting point for chapter 4. The war mobilization greatly stimulated the demand for coal and increased production raised hopes that the industry had overcome two decades of decline. The high hopes of the war years were quickly dashed as the bottom fell out of the anthracite market shortly after the war's end. Chapter 4 describes the precipitous second stage of the decline of the regional economy in the postwar decades, focusing on strategies developed by both mine management and the UWMA in the post-World War II period. Case studies of three firms, Lehigh Coal & Navigation, the Glen Alden Company, and Philadelphia & Reading Coal & Iron, reveal efforts to hold on, but also decisions leading to liquidation and diversification of assets, and the financial manipulations of a new breed of outside investors that depleted the region's productive capacity. In these difficult times, local and national leaders of the UMWA, whose primary base of support and focus historically had been in bituminous coal areas, abdicated their responsibility to defend the jobs and prospects of anthracite mineworkers, choosing instead, often in corrupt ways, to defend the international union and their interests as labor bureaucrats.

The next three chapters examine responses to the sustained decline at the grassroots level of communities, families, and individuals. Local chambers of commerce, and local, state, and federal governments offered and implemented varied ideas for the redevelopment of the anthracite region. They included simple community fundraising drives, as well as more elaborate state and federal programs, all aimed at retaining existing businesses and attracting new ones to the area.

Chapter 5 presents both a history and an assessment of these redevelopment efforts. The persistent attempts to reverse the decline challenge any stereotypical view of economically depressed areas. In fact, many redevelopment projects succeeded. However, in the absence of overall planning, communities within the Pennsylvania anthracite region competed with each other to lure enterprise, making for both winners and losers. A study of the fates of 1,250 firms that relocated or expanded operations between 1940 and 1987 sheds light on the strengths and limitations of redevelopment strategies. Oral history interviews of area residents and migrants reveal fault lines in these efforts and the ways that working men and women came to resent the sacrifices they were called on to make in the campaigns.

Chapter 6 treats the personal experiences of older men and women and relations between husbands and wives as the regional economy declined sharply in the post–World War II decades. Drawing on contemporary census records and oral histories, the chapter weaves a story of resilience in the face of economic loss. Men lost a familiar world and faced challenges in securing new employment. Women's lives changed dramatically as homemakers and breadwinners, as many secured employment in area garment factories. Census data highlight shifting occupational and earning horizons for men and women, illuminate coping strategies such as commuting and out-migration, and reveal the impact of economic decline on patterns of work and retirement of older residents of the area. Testimony of area residents and outmigrants reveals both persistence and change in family dynamics and gender norms. This account thus brings together quantitative data from decennial censuses, qualitative perspectives from oral history, and contemporary testimony to construct the adult experience of regional economic decline.

Chapter 7 looks at the children. In earlier times, the sons and daughters of anthracite miners had contributed significantly to family income. Family ties and expectations kept them in the area. After World War II, the personal and career prospects for young men and women turned bleak. What did they do? How young adults variously perceived and faced the crisis—and who stayed, left, and left and returned—are questions explored in this chapter through use of interviews and a questionnaire survey of young men and women who graduated from area high schools between 1946 and 1960, the period of anthracite's most steep decline. Migration proved the dominant response of the younger generation, but even as the great majority of sons and daughters left the region, their values and beliefs connected them with their parents' generation.

Chapter 8 examines the legacies of sustained economic decline. This book begins with the geological history of the region and ends with a look at the impact of both development and decline on the physical landscape. The material environment of the region has been assaulted with only modest efforts to date at reclamation. Official responses to the social dislocations have also been wanting. Chapter 8 speaks to institutional failure—of the failure of corporate, union, and governmental leaders to shape a meaningful response to economic decline. Compared to the more socially conscious initiatives enacted in Western European nations in the face of coal's worldwide decline in the second half of the twentieth century, the road taken in the Pennsylvania anthracite region appears particularly wanting. In this respect, residents of the area share much in common with U.S. industrial workers who have experienced job loss with the massive plant closings of the last quarter century. Yet contrasting with the institutional failures and personal disruptions is the great resilience of the region's men and women who persisted in the face of economic decline and crafted for themselves a sense of place and identity that made their staying worthwhile. They committed themselves to preserving their own and their predecessors' contributions to the region's and nation's history. The book ends with the efforts of Mike Sabron and his Panther Valley neighbors to create a local anthracite mu-seum to honor their collective legacy.

We engage the experiences of economic de-cline from both regional and local perspectives. A wide lens encompasses the entire Pennsylva-nia anthracite region. For finer understandings, the focus narrows to a group of small mining communities in the Panther Valley, an area at the eastern end of the region's southern coal-fields stretching west from Jim Thorpe through Lansford to Tamaqua. These different perspec-tives, both broad and deep, permit us to under-stand changing lives in a changing world, reveal-ing a multilayered and complex human story.

CHAPTER 1

Creating the Anthracite Region
From Prehistoric Times to 1900

THREE HUNDRED AND FIFTY million years ago—well before the age of the dinosaur—what is current-day Pennsylvania was a flat, steamy swampland of lush vegetation at sea level.[1] Dying trees and plants settled to the bottom of the shallow marshes, slowly decomposing in the low-oxygen environment with entangled remains of dead animals and fish to form thick, spongy mats of organic material. Poor people throughout human history have waded into bogs to gather one common end product of that decomposition, peat; once dried it can be lit and serve as a cheap fuel. Eons ago in Pennsylvania, alternately receding and advancing waters brought heavy deposits of sand and mud that covered and submerged the beds of peat. Over millions of years, the process repeated itself, leaving successive underground veins of peat, some deep below the surface, separated by layers of sandstone, shale, and other sedimentary rock formations.

Nature then acted as alchemist: The pressure cooker of the earth transmuted the decomposed vegetation of Paleozoic Pennsylvania into hard, shiny black coal. The weight of overlying sediments compacted the peat to a fraction of its original thickness and forced out moisture and gases, the latter produced as heat from the core of the planet baked off volatile chemicals. Hardened carbon remained from a metamorphosis tens of millions of years in the making. The deeper the layers of peat and the greater and

longer the earth's pressure and heat, the denser and purer the carbon formation. At the top levels, a woody, brown coal, or lignite, formed, a substance not far removed from dried peat. Farther below the surface, lodes of a dull black, compressed but still brittle fuel emerged, bituminous coal.

Had not the earth cataclysmically shifted three hundred million years ago, horizontal strata of lignite and bituminous coals would have predominated underground in Pennsylvania, east and west. Tectonic plate movements and collisions in the east at the time first rumbled and folded the land in on itself, burying layers of soft coal farther below the surface, and then bulging the earth's crust upward to form mountains—what became the northern section of the Appalachian range. Greater heat and pressure transformed bituminous coal into a glistening, ebony, almost pure carbon rock: anthracite coal. Farther to the west, massive veins of bituminous coal remained as a prime resource for the future. There had once been considerably more coal in Pennsylvania than survived into the historic period, but the forces of erosion, particularly glaciation, wore away the outermost, exposed layers of those deposits leaving about ninety-eight billion tons of coal, estimated to be the original reserve that existed when mining of bituminous and anthracite coal began in the late eighteenth century.[2]

Geological history and topography molded

the lives and fortunes of the eventual beneficiaries. Anthracite coal formed in a discrete, 484-square-mile area of eastern Pennsylvania; this concentration became the basis and stimulus for regional economic development. Three distinct coalfields running northeast to southwest formed, shaping settlement and marketing patterns: (1) a northern field through Lackawanna and Luzerne counties, sliced in the southwest by the Susquehanna River through the so-called Wyoming Valley; (2) a two-part eastern and western middle field through Luzerne, Schuylkill, and Northumberland counties; and (3) a long southern basin stretching through Carbon, Schuylkill, and Dauphin counties.[3] (See anthracite region map in introduction.) The placement of the anthracite veins in these respective fields—more horizontal in the north, more pitched in the south—affected methods of extraction, land and mine ownership, management practices, employment opportunities, and labor relations.[4] The locations of these distinct fields in relation to rivers and major urban centers of the Middle Atlantic states eventually influenced both the development of mining and the marketing of coal in each area. Anthracite's great efficiency as an energy source—and its high market value—derived from its dense, pure carbon content, from its high-pressure origins well below the surface. Yet, its location in relatively deep, twisting and narrow veins interlaced with rock strata also proved a liability. Less amenable to mechanization, requiring intense labor to mine and process, anthracite ultimately lost in the energy marketplace to less expensive bituminous coal and other fossil fuels.

The commercial exploitation of anthracite commenced in the early nineteenth century. Entrepreneurially minded and inventive individuals first successfully tackled the problem of transporting newly discovered lodes of anthracite to population centers; in the process they built an intricate regional network of canals and then railroads and an urban infrastructure. Mining boomed with growing market accessibility and demand, but overproduction quickly generated great fluctuations in prices and profits. Outside financiers moved to order the industry through consolidations in ownership and, in this process, formed a powerful cartel of interests across the coalfields. Tens of thousands of

mineworkers drawn to labor in the region's mines faced grave uncertainties as well—irregular employment, inadequate incomes, capricious management, and perilous working conditions—and they began collectively to organize in kind. In a century-long whirlwind of institutional development, human interventions—on the area's unique geological and topographical base—further created the Pennsylvania anthracite region.

◆

The first human settlers in the anthracite region arrived more than ten thousand years ago, their forebears having migrated from East Asia across the Bering Straits and the North American continent.[5] The mountainous, densely forested core of the area proved forbidding to Native American groups. They named the dark hills *Towamensing*, the wild place, and only refugees of tribal warfare ventured into the interior in search of sanctuary. In the centuries before European colonization, various indigenous peoples formed communities around the perimeter of the region. The Lenape lived in small bands along the Delaware River to the east and south, hunting, fishing, and growing beans and squash. A much more formidable nation inhabited the northern Wyoming Valley, the Susquehannocks, who were rivals of the powerful Iroquois confederation. They established large settled communities with a developed agriculture. Surviving evidence suggests that Native Americans of eastern Pennsylvania first exploited the area's coal deposits at the time of European settlement.

With European contact, the Native American groups at the periphery of the anthracite region became enmeshed in crippling trading relations, land disputes and concessions, alliances with the rival imperial powers, Great Britain and France, and related intertribal conflicts. The Lenape lost their domain, through successive, mostly fraudulent land purchases by the Pennsylvania colonial government. Their traditional economy gave way to trade in furs for imported European goods. Driven farther north, they found themselves ostensibly befriended by Moravian missionaries and tribes allied with the French and then embattled with Iroquoians and white settlers from Connecticut who also staked claims to

the Wyoming Valley. Various raiding parties reduced their numbers. Communicable diseases
carried by Europeans—smallpox, measles, and
influenza—further decimated the Lenape and
other Native Americans of eastern Pennsylvania
in the late seventeenth century.[6]

The erasure of a Native American presence
did not lead to immediate settlement of Euro-
Americans. German and Scottish immigrants
began to farm in the southern and eastern outskirts of the anthracite region in the 1730s and
1740s. A Moravian leader, Count Nicholas Lud-
wig von Zinzendorf, conducted a first survey of
the interior and found it so uninhabited and bar-
ren that he named it St. Anthony's Wilderness.
In 1762, a hundred settlers from Connecticut
arrived in the Wyoming Valley under the auspices of the Susquehanna Company; the land
development company claimed rights to the
area based on a charter granted to Connecticut
by King Charles II a century earlier. The settlement remained disputed and precarious: Indian
attacks decimated the settlement in 1763 and
1778; Pennsylvanians' counterclaims led to two
wars. The disputes were finally resolved in the
1780s with the acceptance of Pennsylvania's jurisdiction over the area, but with individual land
claims of the Connecticut settlers sustained.[7] In
the 1820s, travelers heading to Wilkes-Barre
and other places established by the Susquehanna Company still found sparsely settled communities.[8] The discovery of anthracite coal
would change all this.

There are many claimants to first discovery.
Legend has it that Necho Allen, a hunter, lit a
campfire in 1790 at the base of Broad Mountain
near present-day Pottsville and awakened to find
the seeming rock ledge that he slept under
aglow and on fire.[9] His traditional claim to the
discovery of anthracite is disputed by earlier developments. Native Americans were earlier dis-
coverers. Nanticoke Indians along the Susquehanna River near present-day Wilkes-Barre
were mining and utilizing "stone coal" in the
1760s. In 1766 a delegation of Nanticokes complained to representatives of the colony's proprietors in Philadelphia that whites had been mining coal from their workings. Other surviving
records make it clear that knowledge of the coal
deposits was common among Connecticut and
Pennsylvania settlers of the Wyoming Valley.[10]

The development of anthracite depended on
new uses. Obediah and Daniel Gore deserve
credit for the early spread of anthracite; in 1769,
the brothers demonstrated the first application
of anthracite in blacksmithing. During the
American Revolution, Matthias Hollenback, a
land speculator, arranged for the shipping of anthracite to Carlisle, Pennsylvania, where the coal
fueled the forges of patriotic gunsmiths.[11] And
then there is Philip Ginder (spelled Ginter in
some accounts), a favorite figure in the popular
history of anthracite. Ginder, a prosperous German immigrant miller, found extensive beds of
coal on a mountain near Summit Hill in 1791
while out either hunting or searching for rock
for millstones (the story varies) and immediately
recognized the commercial potential. He shared
his discovery with Jacob Weiss, an influential
landowner and merchant. Weiss, after ascertain-
ing that the find was "stone coal," joined business friends from Philadelphia in forming the
Lehigh Coal Mine Company in 1792—the first
local enterprise of its kind—to exploit Ginder's
discovery. Ginder refused remuneration, asking
Weiss instead for assistance in securing title to
land he had purchased in the area.[12]

Ginder and Weiss warrant a place in history
because their actions led to the first commercial
exploitation of Pennsylvania anthracite. Two
challenges limited initial business success. Anthracite is difficult to light unless regular drafts
of air stoke the burning. Extensive use for home
heating awaited the development of stoves with
appropriate grating and draft systems. Innovations in stove fabrication came quickly in the
early nineteenth century enabling the wider use
of anthracite. Transporting coal from mines to
urban markets proved a much more formidable
challenge. Remarkable entrepreneurial activity
in the first decades of the nineteenth century led
to the construction of a dense network of canals
and eventually railroads that allowed for the
massive shipping of anthracite to Philadelphia,
New York City, and other urban centers of the
northeast. A number of men played essential
roles.

Judge Jesse Fell of Wilkes-Barre usually receives credit for solving the technical problems
involved in utilizing anthracite coal for home
heating, although others no doubt came upon
the solution independently. Having employed

anthracite for two decades in manufacturing iron nails, in 1808 Fell succeeded in burning anthracite in his fireplace utilizing a simply constructed iron grate. Neighbors quickly followed in adopting the grate and demand for anthracite picked up markedly as the market expanded from blacksmiths who had been using anthracite with a forced draft to a much larger number of area residents who could now burn anthracite to heat their homes.[13]

Simultaneous with Fell's success in burning anthracite on an open grate, Abijah and John Smith, of nearby Plymouth, were trying to solve the transportation difficulties to reduce the cost of shipping anthracite to urban centers beyond the region. Owners of coal lands, the Smith brothers constructed flat-bottom arks and transported their coal down the Susquehanna River. Their first shipment in 1807 went unsold, but the next year, following the announcement of Fell's success in burning anthracite, they had some success. By 1812 they were shipping arkloads of coal to Havre de Grace at the mouth of the Susquehanna River and selling their coal through an agent in New York City. These modest successes sparked further efforts to promote anthracite coal and open new markets in Philadelphia and New York.[14]

Jacob Cist naturally took to the new opportunities. Cist grew up among pioneer anthracite coal developers. His father was a major investor in the Lehigh Coal Mine Company, and Jacob Weiss was his uncle. He would marry the daughter of Matthias Hollenback. Cist made his own mark during the War of 1812. Inhabitants of the new nation's urban centers relied for fuel on wood and purchases of Virginia and British bituminous coal. Both were in short supply during the war, and Cist figured that he might reap tidy profits by transporting anthracite to Philadelphia. Conveying coal by wagons from the Wilkes-Barre area, where he lived, proved unmanageable, and Cist fixed on marketing coal mined from the flagging Lehigh Coal Mine Company to the south. Recognizing possibilities for shipping coal to Philadelphia by water, he ordered the construction of small, flat boats that could navigate the shallow, turbulent Lehigh and Delaware rivers. Cist launched advertising campaigns in the Quaker City and surrounding areas to convince residents of the advantages of clean-burning anthracite, especially when used in the new, safe and efficient stoves his company supplied. The War of 1812 ended before Cist could earn his imagined fortune, and he is remembered more as a trailblazer than for his direct business success.[15] He provides a first case, too, of what would be a notable feature of subsequent anthracite development: The movers and shakers of the trade would be the transporters and merchandisers of coal, not the operators of the mines.

Joseph White and Erskine Hazard followed on Cist's lead, realized his grand objectives, and secured anthracite's development. White and Hazard operated a wire works on the Schuylkill River near Philadelphia. They recognized opportunities to profit on the transport of coal from the southern coal basin down the Schuylkill River. However, they failed to secure a charter from the state to form a canal company; opposition from rival interests and concern for their potential monopoly control of Schuylkill River transport doomed their petition.[16] Spurned, White and Hazard turned to Cist's plan to exploit the mine at Summit Hill and utilize the Lehigh River as a gateway. To that end, they signed a long-term lease with the Lehigh Coal Mine Company and lobbied successfully for a charter from the Pennsylvania legislature to improve navigation on the Lehigh River, which effectively granted them a monopoly of shipping and waterpower rights on the river. White and Hazard then oversaw the building of a graded roadway connecting the Summit Hill Mine and the Lehigh River, completed in 1819. The road was paved with stones in 1822, making an excellent surface for the coal wagons. Improvements on the Lehigh permitted downriver navigation by wooden arks loaded with coal and sharply reduced the transportation costs to Philadelphia. By 1825 White and Hazard's Lehigh Coal & Navigation Company (LC&N), incorporated in 1822 and still in existence in the 1970s, was able to ship more than 28,000 tons of anthracite to Philadelphia.[17]

The early success in developing the Philadelphia market led White and Hazard to sponsor further transportation improvements. They improved the roadway to the river by building an ingenious gravity railroad, with coal and mules riding downhill and the mules hauling the empty

Coal mine at Summit Hill, 1821. Richard Richardson, *Memoir of Josiah White* (1873), p. 48.

coal cars back to the mine. Even more important, they improved the river navigation with the construction of a system of dams and locks that permitted down- and upstream transportation on sturdy, reusable barges pulled by mules. The partners sponsored further extensions and improvements of canal transportation on the Lehigh River, which stimulated mine openings throughout the eastern middle and southern fields of the region. The forty-six-mile waterway system they developed, with thirty-six miles of canal and ten miles of navigable river, included a total of fifty-six locks, enabling canal boats to negotiate a descent of 350 feet over its route. By 1840, more than 250,000 tons of anthracite mined in the area reached Philadelphia annually through White's and Hazard's enterprise.[18] Their financial success encouraged others to promote transportation improvements to exploit the rich coal reserves in the other coalfields.

With White and Hazard rebuffed in their original designs for developing the Schuylkill River for transportation, rivals pursued the prospect, at first seeing potential for the shipping of lumber and agricultural products from Schuylkill County, and then, coal. In 1815, a group of German residents in the area received a state charter of incorporation to form the Schuylkill Navigation Company. Prominent Philadelphia merchants and bankers, including the foremost financier of the age, Stephen Girard, invested in the new venture, but shares in the company bought by farmers in Schuylkill County proved as important for the financing of the 108-mile-long canal and river system that stretched from Port Carbon (near Pottsville) to the Quaker City. The highlights of the construction included a 450-foot tunnel bored through a mountain, a great engineering feat for the times, and 120 locks permitting a safe descent of 585 feet for coal barges along its route—all at a cost of $2.2 million, making the Company among the wealthiest private corporations in the nation at that date. The canal opened for business in 1825, and by the early 1840s, the Schuylkill Navigation Company transported more than 500,000 tons of anthracite coal annually to Philadelphia. With a transportation link achieved, fortune seekers rushed to the southern field to purchase coal lands and excavate coal, creating hundreds of small-scale mining operations.[19]

With anthracite of the middle and southern fields now tapped, the last frontier remained the vast reserves of the northern anthracite field. The Susquehanna River provided an outlet for coal from the southwestern end of the northern field, the Wyoming Valley, and efforts were soon afoot to improve that route. Construction on the North Branch Canal began in 1828, and a 55-mile stretch paralleling the river between Nanticoke and Northumberland opened in 1831; three years later a 17-mile extension brought the canal north to Pittston. The North Branch Canal connected via the Susquehanna, Union, and Schuylkill canals to reach the Philadelphia market, while further passage down the Susquehanna River accessed Baltimore.[20]

Parallel to the efforts to link coal lands along the Susquehanna River to markets, there developed a more ambitious plan to connect the northeastern end of the northern field to New York City. Maurice and William Wurtz, dry goods merchants from Philadelphia who owned land in the Lackawanna Valley, saw the potential and the problem. At first, they tried to haul coal by wagon to the upper reaches of the Delaware River and float it slowly downstream on flat boats. They could never compete this way with LC&N operating to the south. The Wurtz brothers then eyed a greater prize, New York City, to the east. Building a canal from the northern field to the Delaware and then to the Hudson River would allow anthracite direct entry to the New York market. Others affirmed their business acumen; within hours of the offering of stock for their Delaware & Hudson Canal Company, the brothers raised $1.5 million for land purchase and construction. Hiring more than 2,500 canal laborers in 1825, the company rushed the building project to completion in four years. The Delaware & Hudson Canal featured an inclined-plane railroad powered by stationary steam engines that transported coal over a 17-mile mountainous stretch to canal docks and a 110-lock canal system that raised and lowered canal boats more than 1,000 feet over its 108-mile course from Honesdale, Pennsylvania, to Rondout, New York, on the Hudson River.[21]

By the 1820s, then, three pioneer groups of transportation entrepreneurs had launched construction of road, river, and canal systems that supplied anthracite coal for the nation's urban and industrial development. These entrepreneurs came, in time, to control both the produc-

tion and transportation of anthracite coal from the three major coalfields. The Schuylkill Navigation Company engineered the massive shipment of coal from the western middle and southern fields to Philadelphia; coal arrived in the Quaker City from the eastern middle and southern fields through the Lehigh Canal and the connecting state-built Delaware Division Canal. Coal from the Lehigh field would also enter the New York City market with the completion of the Morris and Delaware and Raritan canals through New Jersey. Lastly, New York City received great tonnage of anthracite from the vast northern field of the region carried by the Delaware & Hudson Canal.

The three major canal systems opened up coal deposits for exploitation but they were too slow and seasonal to promote the full development of anthracite mining. At midcentury they served the needs of the area's three distinct coalfields, but these water routes were being challenged and soon would be supplanted by a more revolutionary technology—railroads. Eventually it would be the railroads that provided the basis for the full development and consolidation of the region. As the transition from canal to railroad took place, anthracite mining and anthracite communities underwent profound change, but one principle guided development: Those who controlled the transportation of anthracite dominated the industry and the region.

The history of U.S. steam railroading begins in the anthracite region. On August 8, 1829, before a crowd of onlookers in Honesdale, Horatio Allen pulled the throttle on the Stourbridge Lion, a British-built steam locomotive purchased by the Delaware & Hudson Canal Company, making the first test run of a railroad locomotive in the United States. The canal company had laid track and imported the locomotive to facilitate the shipping of coal from the firm's mines to canal loading docks. The original experiment failed in an inauspicious inaugural run—the locomotive proved too heavy for the track and railbed—and company officials decided to begin with horsepower. Still, pressure to move tons of coal efficiently from mines to waterways spurred the building of lighter locomotives that successfully pulled cars over rail feeder lines throughout the anthracite region in the 1830s.[22] Lateral lines also emerged to carry

Chutes loading canal boats on the Lehigh River. *Harper's Weekly*, February 22, 1873, p. 156.

coal cars from various feeder mines to canals, forming a crisscrossing of tracks. Intense (and dense) railroad development thus marked the coal counties of northeastern Pennsylvania at the very dawn of the railway age, but in response to short-haul demands and as an adjunct to the canal system. By the 1840s, a new set of transportation entrepreneurs fixed on railroads for the shipment of coal to urban markets, aiming to bypass the anthracite canals entirely. In the process, they established three major trunk lines—the Lehigh Valley, the Reading, and the Delaware, Lackawanna & Western railroads—that would soon control much of the production, shipping, and merchandising of coal and deeply influence the lives of residents in the region and far beyond.[23]

Asa Packer rose to riches from modest beginnings. He arrived in Mauch Chunk (subsequently renamed Jim Thorpe) in 1833, a twenty-eight-year old carpenter, and immediately saw the profits to be made in applying his skills and

energies to the coal trade. He amassed his initial fortune as coal boat operator, canal boat builder, canal construction contractor, and coal mine operator. He then set out to break the regional monopoly of the Lehigh Coal & Navigation Company by forming a competing rail line. The LC&N responded by warning investors against Packer's risky venture, but Packer marshaled sufficient financial support to buy existing feeder lines and start construction. His Lehigh Valley Railroad (LVRR) opened in 1855, linking Mauch Chunk to Easton on the Delaware River. In the next two decades, Packer, through acquisition and leasing of rail companies and new construction, extended the LVRR northward into the Wyoming Valley and New York State and eastward through New Jersey to the port of New York City, creating a major artery for the transport of anthracite coal to urban and industrial centers. His company also acquired tens of thousands of acres of coal lands. He fought continuing political battles and rate wars with

LC&N, cutting dramatically into the coal traffic of the rival company. Packer also bankrolled the building of iron and steelworks in and around Bethlehem, Pennsylvania. He died in 1879, the wealthiest man in the anthracite region, a stimulator of railroad and industrial development, and patron of cultural and educational institutions in the area. His estate continued to hold a controlling interest in the Lehigh Valley Railroad, which lost much of its dynamism, however, with Packer's passing. In the depression decade of the 1890s, control of the railroad would shift to its financial underwriters, most notably the investment banking house of J. P. Morgan.[24]

An even more powerful railroad corporation emerged to serve the southern and western middle fields, the Philadelphia and Reading. The great promoter of a railroad to tap the vast coal deposits of Schuylkill and Northumberland counties was a scholar, Frederick List. In the 1820s, List, a German immigrant, used the editorial pages of a German-American newspaper in Reading to promote railroad building; List previously had written influential essays and books advocating government sponsorship of industrial capitalist development. List rallied local investor support and secured the backing of Stephen Girard and other Philadelphia financiers for the construction of a railroad connecting Philadelphia and Reading. The 100-mile line opened in 1839. Through acquisitions of feeder lines and further laying of track, List's Philadelphia and Reading Railroad reached northward into the heart of the Schuylkill coal basin and east toward the Lehigh fields. By the mid-1840s, the Reading, as the railroad would be popularly known, formed a transportation network that reached two-thirds of the coal reserves of the entire anthracite region and shipped more than three times the tonnage of its chief competitor, the Schuylkill Navigation Company.[25]

Throughout its history, the Reading generated controversy. Opposition materialized at its very inception. State legislators, facing rising resentment against government-established monopolies, granted a charter of incorporation to the railroad in 1833 that limited the company's operations to shipping, denying it the right to own coal lands or operate mines. The railroad would gain sufficient political leverage by the late 1860s to have the prohibitions rescinded, al-

lowing the Reading to purchase 160,000 acres of prime coal lands. The Reading then became directly involved in the great anthracite labor struggles of the late nineteenth and early twentieth centuries.[26]

The Reading also battled with the Schuylkill Navigation Company. Officers of the canal company mounted a rancorous campaign to discourage investors from backing the Reading, forcing Reading executives to raise capital from New York and British bankers. The canal company also engaged in fierce rate wars in the 1840s, but the railroad still captured the lion's share of coal traffic. The competition between the two rivals eased in 1849, when an agreement left the Reading with two-thirds of the coal traffic but required the road to raise its shipping rates to permit the canal its share of the profits. In 1870, the rivalry formally ended with the Reading leasing and thus controlling the canal. Tonnage handled by the Schuylkill Navigation declined steadily, from more than a million tons in 1869 to just over 70,000 in 1911. The Reading similarly countered competition from other railroads; by the 1890s, through the purchase of stock or lease arrangements, the company controlled the Lehigh Valley Railroad, the Central Railroad of New Jersey, and the Delaware, Lackawanna & Western, attaining a virtual stranglehold on the shipping of anthracite coal. Overextended, the Reading would collapse into bankruptcy during the economic depression of the mid-1890s, and come under the control of its financial guarantor, J. P. Morgan.[27]

Iron manufacturing, not coal alone, gave rise to the Delaware, Lackawanna & Western Railroad (DL&W), which became the major carrier of anthracite from the northern field of the region to New York City. In the 1840s, George Scranton, pooling the financial resources of his extended family and borrowing heavily from New York financiers, built a series of rolling mills that coalesced into the Lackawanna Iron and Coal Company. Scranton gambled on creating the largest mill in the country for the manufacture of iron rails. The company never fulfilled his dreams. Scranton founded his ironworks in the heart of the Lackawanna Valley, an area rich in anthracite but poor in iron ore and limestone for iron production. With the extra expense of transporting iron ore and limestone, the firm could not compete successfully with the high-quality, low-cost rails imported from England. Scranton's bold venture into rail manufacture, however, became the platform for his family's notable sponsorship of railroad construction, urban real estate development, and coal mining.[28]

The importing of raw materials and the marketing of rails required efficient overland transportation. In the 1850s, Scranton gathered partners and funds to build railroads north into New York State to link with the Erie Railroad and east to the Delaware River to connect with rail lines serving the port of New York. Through further construction and purchase and leasing arrangements, a consolidated DL&W emerged, stretching from the Lackawanna and Wyoming valleys of the anthracite region north through upstate New York to the Great Lakes and east through New Jersey to New York City. Intended for the transport of iron rails, the railroad carried enormous tonnage of anthracite coal. The Scrantons, who owned huge tracts of land surrounding their ironworks, also laid out a city that would eventually bear their name, selling lots to other manufacturers, commercial enterprises, and homeowners. From the village of Providence with 1,000 people in 1850, Scranton emerged as a town of 9,000 in 1860 and grew into a diversified urban center with 45,000 resident by the 1880s, making it by far the leading city in the region.[29]

Although the Scrantons, as owners of a major ironworks, played a leading role in the emergence of the DL&W and the early growth of Scranton, the railroad soon came under the control of outside investors. Moses Taylor, of National City Bank, and other New York investors came to dominate the DL&W board of directors, with Taylor providing the vision that transformed the railroad from a minor carrier in the 1850s to a leading anthracite railroad two decades later. Taylor crafted the policy that led the DL&W to acquire coal companies with some 25,000 acres of anthracite coal lands in the 1860s, and negotiated the deals with other rail lines in the 1870s that made the DL&W a major player among the anthracite railroads that dominated coal production and transportation for the next fifty years.[30]

The grasp of financial capital on the transportation companies that drove development raises the complex issue of ownership and control in the anthracite industry. Consolidation, in holdings and the power to set coal prices, was a central element in the rise of anthracite in the nineteenth and early twentieth centuries. Price instability and debt liabilities constituted the prime forces behind concentration. By the turn of the twentieth century, effective control of the anthracite railroads, and hence of the anthracite industry, shifted to outside finance capitalists, who had little interest in vigorous competition and, for the sake of an orderly industry, would even be willing to reach accommodations with organized labor.

The anthracite railroads prospered as varied markets for anthracite coal emerged and expanded across the nineteenth century. Anthracite generated heat with remarkable efficiency and as a newfound source of heat for making iron spurred U.S. industrialization. After 1840, anthracite replaced charcoal in the production of iron and fueled the manufacture of almost half of the crude iron produced in the United States by the 1850s; rails produced in iron rolling mills, in turn, contributed to the rapid growth of railroads. The productive heat of anthracite was also crucial in the rising manufacture of metal machinery, and, more critically, turned water into steam for powering industrial machinery. The application of steam engines—lessening the early reliance on waterpower at isolated river locations—greatly expanded factory production. Anthracite thus fueled U.S. economic development, which, in turn, further increased demand for the hard coal of northeastern Pennsylvania.[31]

However, commercial and home heating provided by far the greatest market for anthracite, and here price volatility had notable impact. Both seasonality and the boom-and-bust cycles of the late nineteenth-century U.S. economy produced great fluctuations in the demand for anthracite, affecting prices, profits, and the ability of anthracite coal producers to meet their debt obligations to such financiers as J. P. Morgan. Controlling production and shipments of coal—in other words, avoiding the oversupply of anthracite to the market—emerged as the solution. The steps taken toward controlling prices varied over time and included consolidating ownership of coalfields, collusive agreements

among shippers, mergers, interlocking corporate directorships, and the squeezing of independent operators.[32]

Railroad companies rushed to purchase coal
lands for economic reasons, but geology, politics,
and law also shaped the process. For example,
the Lehigh Coal & Navigation Company dominated the mining of coal in the eastern middle
and the eastern end of southern coalfields of the
region. And then Packer's Lehigh Valley Railroad moved in to compete with LC&N in its territory and also gained substantial coal holdings
in the Wyoming Valley. Similarly, the Delaware
& Hudson Canal Company and the Delaware,
Lackawanna & Western Railroad cornered ownership of coal lands in the Lackawanna Valley of
the northern field by the 1860s. In both districts,
the original canal companies had received charters from the Commonwealth of Pennsylvania
that provided monopoly rights in shipping and
authority to own and manage mines. The competing railroads, while not so authorized, secured control of mining operations by purchasing the stock of coal companies that could supply
their roads. The advantages possessed by these
transportation giants were compounded by the
requirements that the region's geology imposed
on anthracite mining, which over time called for
the sinking of ever-deeper shafts and the outlay
of more substantial amounts of capital.[33]

In the southern and western middle fields,
the Philadelphia and Reading Railroad secured

overall ownership of coalfields in more gradual
and complicated ways, but it rapidly became
identified as the monopoly player in its area. The
Reading's hold on coal mining and merchandising emerged slowly amid conflict. When both
the Schuylkill Navigation Company and the
Philadelphia and Reading Railroad received corporate charters, state legislators explicitly forbade the transportation companies from owning
coal lands and engaging in mining. Anger at the
monopoly rights awarded earlier to the Lehigh
Coal & Navigation and the Delaware & Hudson
Canal companies and the general antimonopoly
politics of the Jacksonian era figured in the limits placed on coal transporters in the southern
field.[34] In the vacuum created by public authorities, more than a hundred independent mine
operations opened in the 1830s and 1840s in the
area. Geology mattered as well. In the southern
field, coal veins typically lay near the surface;
little capital investment was required and small-
scale enterprises could flourish.[35]

A major transformation occurred in the
1860s, again involving politics, law, and geology.
Railroad companies now lobbied successfully for
legislation allowing them to own coal lands and
engage in mining. The Reading quickly formed
a subsidiary, the Philadelphia and Reading Coal
and Iron Company, and embarked on massive
purchasing of coal lands. With the depletion of
the more easily accessed coal outcroppings in
the southern field, production shifted north

Honey Brook breaker, No. 2, with exiting coal trains. *Harper's Weekly*, September 11, 1869, p. 581.

toward the western middle coalfields of the anthracite region.[36] Successive thick veins of coal existed there, but well below the surface; substantial investments would be necessary to realize this great potential and reap profits from deep mining, and the Reading had the means. By the mid-1870s, the Reading dominated land ownership, coal production, and transportation throughout the western middle and southern fields.[37]

Control of coal lands by railroad companies came latest in the Wyoming Valley of the northern field. A set of strong, independent mining operations emerged there, owned and managed by a local elite identified with the city of Wilkes-Barre. With the ability to ship coal on the Susquehanna River or on the competing Lehigh Valley Railroad, the Delaware, Lackawanna & Western, or the Reading, coal operators did not have to sponsor transportation innovations or bend to outside investors. Coal in the Wyoming Valley also lay much closer to the surface than in the upper Schuylkill coal basin, meaning that Wilkes-Barre–based operators did not need the massive investments to develop their properties that were required in the other fields. The Wilkes-Barre coal operators thus drew on their savings and coal profits to maintain an exceptional independence—at least until the first decade of the twentieth century.[38]

The Wyoming Valley independents eventually succumbed to the biddings of the major railroads—bought out with lucrative stock offers and board-of-director appointments that now connected them with a larger corporate and financial world. The Central Railroad of New Jersey became a principal owner of the Lehigh and Wilkes-Barre Coal Company in the early 1870s. The significant end of independent operations, though, occurred in 1905, with the Lehigh Valley Railroad's purchase of Coxe Brothers Coal Company for $19 million. Coxe Brothers had been the largest independent firm in the region, possessing 5,000 acres of coal lands and mining 1.3 million tons of coal in 1905.[39] With the tide turned in the Wyoming Valley, independent operations became a declining factor in the landscape of the anthracite region.

The ownership of vast coal operations by a handful of railroad companies did not by itself achieve production control or accompanying stability of prices and profits. Competition among the major carriers led to overproduction and oversupplies of coal on the market. The reality hit with the rise of the Reading Railroad to a dominant position in the 1870s amid a severe economic depression. Officials of the Reading tried to meet the crisis by seeking agreements among their fellow shippers to limit production of coal and set shares of the total that each would be allowed to market—an attempt to establish a classic cartel operation. Repeated efforts to reach such pooling agreements failed in the 1870s and in the succeeding decade; some rail companies refused to enter into discussions, while others pulled out of accords in short order.[40] The breakdown of pooling agreements served as a prime motive for the Reading to absorb its competitors through the purchase of stock and leasing arrangements—an initiative that saw the railroad for a brief moment in the 1890s controlling simultaneously the Lehigh Valley Railroad, the Central Railroad of New Jersey, and the Delaware, Lackawanna & Western. Extremely overextended and caught in another economic depression, the Reading lost its bid for monopoly and ended in bankruptcy with subsequent reorganization.[41]

The major railroads finally succeeded in stabilizing production in the region beginning in 1898 through what J. S. Harris, president of the Reading Railroad, referred in testimony before the U.S. Industrial Commission as a "a general understanding."[42] After rounds of acquisitions, underwritten and overseen by J. P. Morgan and other financiers, the anthracite railroads owned substantial shares in each other's lines, had the same major shareholders (most notably, the banking house of Morgan), and shared the same men on their boards of directors (including Morgan appointees).[43] With this community of interest, operating executives of the railroads understood, as J. S. Harris noted, the proper tonnage to be mined and shipped by their own companies—to the mutual benefit of all parties. Analysts at the time, not coincidentally, found remarkable consistency in the proportion of total tonnage each of the major carriers shipped annually and in the market prices of anthracite.[44] Concentration in the ownership of the coal lands

and good economic times facilitated joint decision making at the turn of the twentieth century,
but the interlocking structure of ownership and
control embedded collaboration.

Stabilization in prices and profits also required restraints on the production and marketing of coal by independent operators. To that
end, the major railroad companies employed
various tactics over the course of the late nineteenth and early twentieth centuries: outright
refusal to ship the coal mined by the independents; discriminatory freight rates; long-term
contracts that fixed tonnage and returns; and,
eventually, purchase, as in the case of Coxe
Brothers.[45] Elimination of independent producers was thus a crucial element in the emergence
of the modern anthracite industry at the turn of
the twentieth century.

The major railroad companies achieved a
hold on the anthracite trade in stages, and market forces, geology, politics, law, and the imperatives of finance capital shaped the process.
While a complicated history, the results were
stark. By 1907, after the corporate absorption of
key independent companies and the signing of
exclusive, sometimes perpetual sales contracts
with remaining independents, the seven major
railroad companies controlled fully 91 percent
of all coal produced in the anthracite region.
Concentration had proceeded so rapidly at the
turn of the twentieth century that over a seven-
year period, the output of independent operators declined from 38 percent to 22 percent of
production. And exclusive sales agreements
meant that independents controlled a mere 9
percent of the anthracite wholesale market.
Moreover, the railroad companies held 96 percent of the region's anthracite reserves, with the
two major players owning the lion's share. The
Reading Railroad (with its subsidiary, the Central Railroad of New Jersey) had claim to 60 percent of the coal reserves; the Lehigh Valley Railroad, 18 percent.[46] With such extraordinary
concentration in ownership, production, and
control and a structured corporate community
of interest, the anthracite railroads formed a
powerful cartel—a combine that directly affected the lives of independent coal producers,
anthracite coal consumers, and the army of
mineworkers whose labors generated anthracite's enormous profits.[47]

◆

The rise of Pennsylvania anthracite involved
entrepreneurship and innovation in transportation, not in the production of coal. To be sure,
the era of development and consolidation witnessed advances in mining—with more efficient
explosives for the dislodging of coal and steady
improvements in ventilation—but such examples pale in comparison to the revolutionary
technological and organizational transforma-
tions occurring at the same time in much of the
nation's industrial system.[48] The mining of anthracite coal across the nineteenth century entailed age-old tools and practices: augers, picks,
and shovels and the application of sheer animal
power, in this case, the muscle of mules and
men.

Geological conditions determined strategies
for the extraction of coal and construction of
mines. A form of open-pit mining emerged to
access horizontal veins that ran near and roughly
parallel to the surface. After clearing the surface
overburden, miners basically quarried the exposed anthracite. The Summit Hill Mine,
opened by the pioneer Lehigh Coal Mine Company in 1792, operated on an open-pit basis.
Coal seams that ran near the surfaces of mountains but were sloped required a different
method, the so-called drift technique. If the coal
veins were shallow but did not approach or
break the surface, the miners tunneled horizontally into the mountains and then mined up the
seams. Open-pit and drift techniques represented relatively inexpensive ways of extracting
coal. The mining occurred at or near the surface,
involved short lengths of veins, and did not require elaborate shafts or tunnels, or much water
pumping or ventilation. Before the 1870s, open-
pit and drift mining flourished in small-scale operations in the southern field of the anthracite
region throughout Carbon and Schuylkill counties. Afterward and elsewhere, geology dictated
deeper and more complex mining.[49]

Exploiting the thick veins of hard coal lying
well below the surface required digging vertical
shafts hundreds (sometimes thousands) of feet
into the earth, then tunneling successive levels
of horizontal gangways off of the shafts through
rock and coal seams. Large mining ventures ultimately featured multitiered tunnel systems

sometimes 1,500 to 2,000 feet deep, running for as long as ten miles under the mountains and valleys of the anthracite region. Mining techniques varied depending on the pitch of the veins of coal in the area to be mined. "Room and pillar" mining was common in the northern field, where the coal veins run horizontally. Miners typically opened up a large area, a "room," advancing the coal face, or exposed vein, as they worked, leaving thick "pillars" of coal behind to support the roof of their mining area. In the steeply sloped veins common in the southern and western middle fields of the region, chute or pitch mining predominated. In chute mining, mineworkers drove an opening upward into the coal, lined it with timber, and floored it with sheet iron. Extracted coal fell down into waiting coal cars. Contract miners worked at the coal face above the chute, and a loader or battery man worked a movable bulkhead to fill coal cars that were then hauled on rails by mules to shaft elevators that took the coal to the surface.[50] Over time, electric locomotives (or lokies) took over much of the underground hauling and mules worked only in steeper, less-accessible work areas, transferring coal cars to the lokies at various staging points along underground gangways. Operating deep, labyrinthine mines required special attention and investment for ventilation and the pumping of water out of the mines. As the most accessible coal veins gave out, the capital requirements of anthracite mining increased exponentially, favoring the anthracite railroad conglomerates that expanded in the last third of the nineteenth century.

Large mining operations required battalions of workers with varied skills: rockmen and muckers to excavate shafts and drill underground gangways; timbermen to shore up gangways; skilled miners, assisted by helpers, to detonate and pick at coal seams to free the anthracite; loaders and mule drivers to transport the coal away from the coal face and out of the mines; fire bosses to test for dangerous buildups of carbon monoxide and methane gas; men and boys, who worked above ground in breakers, to crush, clean, and screen the coal into different sizes for various uses; stationary engineers and pump house crews to operate steam engines; machinists to fabricate and repair tools and equipment; carpenters to build and repair wooden structures; locomotive engineers and railway yard hands; and hosts of supervisors to manage the complex of activity.[51]

Mine operators needed workers, and job opportunities beckoned. To the anthracite region during the nineteenth century came tens of thousands of men with their families, most to work directly in and around the mines, others to labor in area manufactories or businesses providing goods and services to members of mining communities. The population of the region accordingly grew fifteenfold from approximately 50,000 residents in 1820 to more than 750,000 by the turn of the twentieth century (see table 4 in appendix 1). By 1900, the region recorded almost 144,000 men working in and about the mines. That number would peak in 1914 at 181,000.[52] The area experienced its greatest rates of population growth in prosperous times for anthracite mining, from the 1820s through 1860s and then in the 1880s. The growth, in turn, slowed during the depression decades of the 1870s and 1890s.

The anthracite region attracted a polyglot people. Native-born Americans and sizable contingents of immigrants flooded the area. The immigrant flow remained steady throughout the nineteenth century, with the foreign-born continuing to represent upwards of one-fourth of the region's population (see tables 5a and 5b in appendix 1); with their American-born children, hyphenated Americans formed the majority, imparting a definite non–Anglo-Saxon Protestant character to local communities. By 1900, the immigrant composition of the region was shifting. Before 1890, immigrants from England, Wales, and Ireland composed more than 70 percent of the foreign-born. The English and Welsh generally held skilled and managerial positions; common day labor remained the province of the Irish. Old-country religious, class, and imperial hostilities between the British and Irish played themselves out in the Pennsylvania anthracite coalfields, transformed in the caldron of the region's intense labor conflict.[53] Newer arrivals, however, began to dominate in the area. As early as 1890, immigrants from southern and eastern Europe had become a presence and began to assert themselves in union organizing efforts. Ten years later, arrivals from Poland and Russia constituted more than one-quarter of the immigrant population, and a multitude of others from Italy, the Balkans, the Austro-Hungarian Empire, and the Baltic formed another fourth of the newcomers. With the shifting ethnic complexion of the anthracite region, the proportion of Irish-born residents declined dramatically. While the share of English and Welsh immigrants dropped from 31 percent to 21 percent between 1880 and 1900, the Irish proportions declined from more than 40 percent to 16 percent over the same period. Replaced by lower-paid immigrant workers from other parts of Europe, Irish and British-born anthracite miners left the area seeking better opportunities in the Midwest and West.[54]

Migrants to the anthracite region—old and new, native and foreign-born—resided in communities along a continuum ranging from cities to small, so-called patch towns. Scranton became the leading urban center of the region, not because of natural geographical advantages but, rather, due to the strategic energies of the Scranton family and like-minded entrepreneurs. The Scrantons and their associates recruited manufacturers and merchants to the city. The coal business remained central, but the city also boasted its ironworks and notable fabricators of machines, engines, railroad equipment, and tools; textile mills; garment factories; powder works; and grain processors. Frank and Charles S. Woolworth opened a string of five-and-dime stores and made the city their headquarters. Scranton kept the door open to opportunity; its most successful businessmen included many immigrants.[55]

Wilkes-Barre rivaled Scranton as a capital city of the anthracite region, but it was a distant second. Scranton typically counted twice the number of residents (in 1900, the population of Scranton stood at 102,026, Wilkes-Barre's at 51,721).[56] Wilkes-Barre also developed in significantly different ways. The city was established in the late eighteenth century as a county seat. The original town leaders formed a long-lasting, closed elite through intermarriage. They helped sponsor transportation improvements, largely to enhance navigation on the Susquehanna River, but never large canal or railroad projects that connected them firmly to external sources of financing. Through purchase of coal lands in and around the city, they became local coal barons

who placed their profits into further coal development and local banks, mercantile establishments, and industry (as well as cultural institutions).[57]

Several cities emerged in the nineteenth century as counterparts to Wilkes-Barre, but had brief histories as urban centers. Pottsville and Mauch Chunk, for example, were of equal size in population to Wilkes-Barre until the late 1860s—with 10,000 to 13,000 residents—and then receded. Pottsville, in the heart of Schuylkill County, emerged as a bustling commercial city with the completion of the Schuylkill Canal and then the Philadelphia and Reading Railroad. Men who had made quick fortunes during the period of independent mining in the southern field established their offices and fine homes in Pottsville. When the Reading Railroad—with its corporate headquarters in Philadelphia—achieved its monopoly hold on mining and transportation in the area, buying out local mine owners and, in some instances, absorbing them into the company's bureaucracy, Pottsville lost its leadership and growth dynamic. Mauch Chunk had a similar fate. The city rose as the business center for the Lehigh Coal & Navigation Company and then the Lehigh Valley Railroad. Asa Packer and fellow men of wealth built Mauch Chunk into a tourist attraction with hotels, a handsome downtown featuring the railroad's headquarters, and a famed row of millionaires' homes. No sooner had the city become the jewel of the anthracite region than Packer and his associates eyed new opportunities for industrial and urban development in nearby Allentown and Bethlehem. As they shifted their interests and offices there, Mauch Chunk's growth and its importance in the region's urban history waned.[58]

The city of Hazleton proved an exception to the early decline of small cities in the anthracite region. Its population reached 12,000 by the late nineteenth century, like Pottsville, but the city continued to mature as a third urban center in the region (though never as brimming as Scranton and Wilkes-Barre). Hazleton was located in the center of the eastern middle coalfield of the region in lower Luzerne Country, and coal mining remained its key business. But, as the first terminus of the Lehigh Valley Railroad, Hazleton housed the railroad's extensive repair shops and yards and a sizable population of railroad workers. Hazleton also served as the market center for neighboring communities. Location and diversified enterprise—not extraordinary entrepreneurial activity—sustained the city's continued growth.[59]

Scores of small towns dotted the landscape of the anthracite region between the greater and lesser urban centers of the area. With 1,000 to 2,000 people by the 1890s, some surged to 5,000 or 6,000 residents at the peak of coal production in the World War I era. Formed directly near mine openings and breakers, these towns, quite contained, stretched for fifteen blocks or so along a main street, eight to ten streets deep.

Typical of the small towns of the region was a row of mining communities spaced through the Panther Valley between Mauch Chunk and Tamaqua at the eastern end of the southern coalfield: Nesquehoning, Lansford, Summit Hill, Coaldale, and Tamaqua. (See inset in anthracite-region map in introduction.) All owed their existence to the Lehigh Coal & Navigation Company. Starting with excavation of outcroppings in Summit Hill, LC&N oversaw the drilling of major shafts and then the tunneling of tiers of horizontal gangways that ran deep underground along the twelve-mile length of the Panther Valley. Towns emerged at key points in LC&N's operations.[60]

Although LC&N dominated the area and the fortunes of the residents—economically and politically—Nesquehoning, Lansford, Summit Hill, and Coaldale were not company towns. Independent retailers, bankers, and other providers of goods and services to mining families established themselves in these communities. The citizens of the towns also founded their own churches, fraternal organizations, and clubs that reflected the diverse ethnic composition of their population. Churches formed the cornerstones of ethnic communities in the anthracite region, providing cultural and personal familiarity for ethnic newcomers.[61] The churches of Coaldale, for example, included the Evangelical Church, St. John's Primitive Methodist, Welsh Congregational, St. Mary's Church (with a largely Irish constituency), St. Mary's Russian Orthodox, St. John's Greek, St. John's Lithuanian, and SS. Cyril and Methodius Church (for the close to four hundred Catholic Slovak families).[62] Coaldale also boasted two newspapers, a successful town football team, volunteer fire brigades, various fraternal orders, and a state hospital (built after considerable community agitation and partially subsidized by LC&N). The residents of Coaldale remained loyal as well to the successive miners' unions that organized in the region.[63]

Support for trade unions would never be countenanced in the scores of classic company towns of the anthracite region—small, contained, and controlled mining "patch" communities. Eckley provides an example. Before the 1850s, Eckley had been a hamlet of a handful of inhabitants who eked out a living cutting shingles (the place was called Shingleton). The Coxe family of Philadelphia owned the land throughout the vicinity. With discoveries of local anthracite deposits, a group of mine operators secured a long-term lease from the Coxes to mine coal. They proceeded to lay out a town, to be named for Eckley Coxe, with all properties—boarding houses, homes, churches, and stores—owned by the company. With a boom in production, Eckley's population swelled to 1,500 by 1870. Private company police maintained order, but the company store remained the key point of control. Required to buy tools, lamps, and powder from the store, without other retail outlets for the purchase of household needs, and owing rent for both lodging and plots of land used to grow foodstuffs, miners found themselves constantly in debt to the company, their wages from toiling in the mines never meeting the accumulated tabs of expenses.[64]

As with other company towns in the anthracite region, Eckley never experienced extended periods of stability as planned. Surface mining gradually replaced underground mining in the early twentieth century, sharply reducing the number of workers as well as residential sections of the community. Management of the mine changed hands frequently through lease arrangements. Rapid population turnover further marked the town as miners and their families mutinied with their feet, and unions eventually made inroads. Still, Eckley remained a company town, though a faded one, as late as 1963, when the Coxe family sold off its final holdings. Five years later the owner and strip-mine operator, George Huss, leased the site for

Lehigh Coal & Navigation Company mine, No. 7 tunnel, ca. 1870. Photograph by Harrison & Morton. Courtesy of the Raymond E. Holland Industrial History Collection.

the filming of *The Molly Maguires* (a popular movie depicting labor violence in the 1870s, starring Sean Connery). Soon after, the Pennsylvania Historical and Museum Commission acquired title to develop the site as a museum depicting the history of a mining patch town.[65]

In the various communities of the Pennsylvania anthracite region, native-born and immigrant arrivals found work, gained an economic foothold, and formed local religious and ethnic institutions. Yet insecurity marked the lives of area residents. The threat and reality of workplace injuries and fatalities were an overwhelming presence in the work and lives of miners. Fluctuations in the market for coal yielded uncertain employment and income with persistent strains on family budgets. And, finally, life expectancy for miners was limited by a combination of industrial accidents and occupational disease. The writer Stephen Crane visited mines in the Scranton area in 1894 and captured the antagonism between nature and human life underground. "Man is in the implacable grasp of na-

ture," he wrote. "It has only to tighten slightly, and he is crushed like a bug." He enumerated mining's dangers: "If a man escapes the gas, the floods, the 'squeezes' of falling rock, the cars shooting through the little tunnels, the precarious elevators, the hundred perils, there usually comes an attack of 'miner's asthma' that slowly racks and shakes him into the grave." Miner's asthma, later known as black lung, sapped the strength of underground miners and often left them incapable of providing for themselves and their families.[66]

Between 1869, when the Commonwealth of Pennsylvania first collected data on workplace fatalities, and 1900, 10,116 mineworkers lost their lives in the mining operations of the anthracite region.[67] Over the thirty-one-year period, 326 anthracite mineworkers died annually on average, between 3 and 4 miners per 1,000 employed. Circumstances did improve over time. In the 1870s, between 5 and 6 mineworkers per 1,000 died regularly each year in accidents; the number declined to 3 per 1,000 by the end of the century. Even at that reduced level, anthracite mining rivaled railroading as the most perilous of occupations.[68] State authorities also compiled information on the number of nonfatal accidents on the job. In the last three decades of the nineteenth century, the Bureau of Mines recorded more than 24,500 injuries, for an average of 846 per year, with close to 9 of every 1,000 anthracite mineworkers injured annually. These figures definitely underreport the dangers of mining. They include only reported accidents, make no distinctions between minor and permanently incapacitating injuries, and do not speak to the long-term respiratory illnesses that miners incurred through inhalation of coal dust.[69]

Anthracite miners lost their lives and limbs in both major calamities and daily mishaps. Cave-ins, roof-falls, underground explosions and fires caused by buildups of methane gas, flooding of passageways, asphyxiation, and mine car accidents all took their toll. Notable disasters received the greatest attention. Smoke and gases from an uncontrollable shaft fire suffocated 110 men and boys on September 6, 1869, in a mine at Avondale near Wilkes-Barre. On December 18, 1885, twenty-six miners drowned in a mine in Nanticoke. Seventeen men died in another mine flood in Jeansville near Hazleton on February 4, 1891.[70] Major tragedies attracted attention and generated calls for safety measures, but anthracite miners also died or were seriously injured in the normal course of day-to-day operations.

After years of pressure from miners, Pennsylvania passed its first significant mine safety legislation in April 1869. Limited in its application to Schuylkill County, the law set standards for mine ventilation and appointed a mine inspector to visit mines across the county and recommend corrective measures. Even had the law's reach been extended into neighboring Luzerne County, it would not have prevented the Avondale disaster, which occurred five months later when fire and smoke from a burning breaker at the top of the mine's shaft blocked the single exit from the mine. The explosion and fire did not burn the miners, but the resulting smoke and gas asphyxiated 108 working underground and the first two rescuers who attempted to search for survivors.[71]

The Avondale tragedy increased the call for safety legislation, and a more extensive law resulted in 1870, applied this time to the entire anthracite region. The law required two outlets for all existing and future mines and set more specific standards for adequate ventilation. Breakers, a major contributing cause in the Avondale disaster, were no longer to be permitted at the top of a mine shaft. The law provided for the appointment of five mine inspectors, set their qualifications, and noted the procedures by which they would be selected. The law also regulated hoisting facilities, mandated inspections of boilers, required a wash shanty for miners, and prohibited the employment of boys under the age of twelve in the mines.[72]

Pennsylvania passed the nation's first legislation regulating safety in underground mining, and, over the course of the last three decades of the nineteenth century, the Commonwealth updated its laws to address changing concerns. A thorough revision of the safety laws for the anthracite region in 1885 extended and codified the earlier legislation. In addition to raising ventilation standards, a series of provisions established the duties of operators and supervisory personnel related to mine safety and set enforcement provisions. The law required the testing and certification of mine foremen, a provision extended in 1891 to assistant foremen. Finally, the Commonwealth gradually increased the number of mine inspectors from the initial five in 1870 to fifteen in 1903 to a peak twenty-five in 1916. While the appointment of additional inspectors seems progressive, the growth in numbers represented slow and uneven improvement. In 1870, the Pennsylvania anthracite region had one mine inspector assigned for each 7,000 mineworkers and for every 2.8 million tons of coal produced. By 1903 each state official safeguarded the working conditions of 10,000 men for every 5 million tons of coal processed; by 1916 the number of workers per inspector did drop to 6,400, constituting one inspector for every 3.5 million tons of coal produced. Over the forty-year period, inspection lagged behind the expansion of output.[73]

Probably the most important piece of safety legislation in the period was not framed as such. An 1889 law providing for the certification of miners to be employed in anthracite required

The Avondale
Colliery disaster—
bringing out the
dead. *Harper's*
Weekly, September
25, 1869, cover.

that miners be certified as competent by a board of skilled miners and have at least two years of practical experience as mine laborers. Miners consistently argued that their safety depended on the competency of those who worked with them in the mines and that considerations of safety required that all miners meet standards of competence. Mine operators fought this requirement tenaciously, but eventually the courts upheld the miners' argument and the state provisions. The legislation protected miners on a day-to-day basis and blocked mine operators from importing hundreds of nonunion, out-of-state, or immigrant miners to serve as strikebreakers.[74]

Despite the improvements in safety regulations, anthracite mines remained dangerous places to work, far more dangerous, for instance, than mines in Great Britain. While fatalities in British mines averaged about 1.5 per 1,000 miners between 1890 and 1900, comparable rates in Pennsylvania's anthracite mines were more than double. Anthracite miners were dying to heat the homes of the urbanizing Northeast and to support the bottom line of the region's railroads and coal operators. The editor of an English coal trade journal commented on U.S. conditions in the mines, noting that they displayed "a general disregard for life that would never be tolerated" in Great Britain. A recent historian examining mine safety attributed the greater safety in English mines to greater mechanization of the work process and stricter state regulation.[75]

Families of miners injured or killed on the

job had to rely principally on their own resources, the kindness of neighbors, and the ethnic fraternal societies they formed and joined.
Major firms typically provided relief funds, underwritten in part by an endowment from the
company and in part by workers' contributions
set on a sliding scale according to their wages.
The funds were invariably voluntary with death
benefits ranging between thirty and fifty dollars
and modest payments to widows or to injured
workers. The poorest workers often could not
afford to participate, and the payments did little
to relieve the immense need that followed an accidental injury or death. The funds were established unilaterally by the companies, and as
shortfalls developed, benefits or required contributions could be revised by the companies at
will. They offered precious little in the way of security for workers in a dangerous occupation.[76]

Nor were there other institutional sources of
support for families of killed or injured miners.
The cost of private insurance was prohibitive,
and Pennsylvania only enacted its first workmen's
compensation law in 1915. Moreover, courts
generally dismissed suits brought by families
charging employer negligence in the case of accidents with decrees that miners, in accepting employment, accepted fully the risks of their work.[77]
Medical assistance for the injured was also limited. Company physicians attended to the
wounded (with rudimentary instruments) at accident sites and homes; these services were usually paid for by the miners through monthly deductions from their pay.[78] Pressure mounted for
improved care and hospital facilities, and both
state and private authorities responded. Miners'
hospitals typically began with fundraising among
working people, but anthracite companies and
railroads often donated the land, while the Commonwealth made substantial appropriations.
Miners' Hospital in Ashland was such a private-
public project and first opened under public
trustees in 1883. The Coxe family, leading independent anthracite operators, contributed to a
hospital in Hazleton that opened with state support in 1891. Finally, LC&N matched workers'
contributions for the establishment of the Coaldale Hospital, which opened in 1910. Still, the
sudden whistle from a mine following an incident signaled anguish and penury for members
of anthracite communities.[79]

Fears of the loss of primary breadwinners
through accidental injuries and deaths weighed
heavily. For families of miners, such anxieties
certainly did not ease concerns about making
ends meet in seeming normal times. Wages and
systems of payment for anthracite workers varied. Miners at the coal face were paid on a
car/tonnage basis, working, in effect, as independent operators, in control of their schedules and the pace of work and responsible for
the costs of tools, gunpowder, and helpers.[80] The
rest of the mine labor force—loaders, drivers,
muckers, laborers, carpenters, machinists, door
boys, and slatepickers—worked under direct
company supervision and received hourly or
daily wages. Salaried foremen and other managers were paid by the month.[81] No matter the
position or method of payment, employment
and income remained unsteady. Seasonal demand for coal, retrenchments following periods
of overproduction and falling prices and profits,
the impact of the business cycle, injuries and ill
health, and periodic disasters all contributed to
the uncertainty of work. Pennsylvania anthracite
miners surely prospered in times of high demand for coal, as did merchants and other pur-
veyors of goods and services in their communities. Yet the vicissitudes of the coal trade
required special coping strategies on the part of
miners and their families, including child labor.

The economic circumstances and dilemmas
of families of anthracite miners became grist for
a number of public investigations in the late
nineteenth and early twentieth centuries and
they provide telling details. In 1879, officials
of the Department of Internal Affairs of the
Commonwealth of Pennsylvania conducted a
statewide study of the budgets of workers—one
of the first of its kind in the United States—that
included fifty anthracite miners.[82] The investigation took place at the end of a decade marked by
severe economic depression and labor unrest.
Hardly a scientific survey—how the sample of
workers was chosen is unclear and the collected
information is incomplete—the study is revealing, nonetheless.

Investigators gathered information on yearly
incomes of the miners surveyed and expenditures of their families (including rent, fuel, groceries, clothing, education, recreation, and other
expenses). When asked simply whether their

earnings covered expenses, thirty-eight replied, "No," twelve, "Yes." Sixteen of the former supplied complete data allowing for calculations of the shortfall between expenses and income. These miners' earnings averaged $334 in 1879; their spending totaled, on average, $459. As primary breadwinners, their incomes thus covered but 73 percent of family living expenses.[83]

Anthracite miners faced a huge obstacle closing the substantial gap between their earnings and family expenditures: underemployment. In 1876, officials of the Pennsylvania Bureau of Mines began collecting data on days worked per year. For the period 1876–1900, mine workers in the anthracite region worked 186 days a year on average, 62 percent of what would be considered a full year's employment (or 300 days).[84] In 1879, the year of the budget study, an upturn in the economy encouraged greater production, and days worked by anthracite miners rose on average to 209. Had the miners surveyed in the study been steadily employed, not just 70 percent of full time, they would have earned $477 and covered their families' living expenses. Of course, had their employers paid them higher wages, the miners would have been able to make ends meet even with slack times in employment. Without sufficient wages or steady employment, the mineworkers surveyed in 1879 faced a grievous shortfall in income; unfortunately, the budget study offered no clues as to how they coped. Later government investigations would provide a clearer answer; the earnings of older sons and daughters and their contributions to the family economy made up the difference and proved crucial for their families' subsistence.[85]

The precarious economic circumstances of the miners and their families had broader consequences. In nineteenth-century anthracite communities, independent merchants, bankers, realtors, and professionals depended fully on the purchasing power and patronage of the miners. Recognizing the earning potential of older sons and daughters and their role in stabilizing the family (and local) economy, these businessmen organized in local chambers of commerce to entice manufacturing enterprises to their communities.

The need of local commercial interests to provide employment opportunities for second and third breadwinners in families of miners complemented developments unfolding a hundred miles east of the anthracite region, in Paterson, New Jersey. In the 1870s, Paterson emerged as the silk-producing center of the United States, benefiting from the arrival of skilled British silk workers who had been displaced when the free trade policies of the British government and the lifting of tariff protections decimated the British silk industry. No sooner had Paterson claimed the title of "Silk City," than manufacturers there began discussing the desirability of shifting portions of their operations to the Pennsylvania anthracite region. With technological advances in silk throwing and later weaving, manufacturers took advantage of the low-wage and seemingly pliant labor of the children of the coal district, thus lessening their reliance on unionized, skilled male silk workers.[86]

Business and civic leaders in Hazleton led the way in attracting silk firms from Paterson. In 1886, they succeeded in luring one manufacturer with a $90,000 subsidy that they had raised. The same group later formed the Hazleton Improvement Company to buy land and build facilities for other concerns.[87] Communities throughout the anthracite region followed suit, offering additional incentives, including tax exemptions, payment of interest on loans, and free water rights. In 1899, for instance, Bamford Brothers of Paterson established a silk ribbon factory in Wilkes-Barre; in 1904, Lansford landed a branch plant of the Century Throwing Company of Paterson. By the mid-1890s, forty-five Paterson firms had opened branch plants in the anthracite region through such inducements (and some would soon move their entire operations there). The campaign to draw manufacturing companies to the area—thereby boosting the purchasing capacities of families of miners—proved so successful that the *Pottsville Daily Republican* boasted in 1889 that "all our surplus female population is now profitably employed and the town is reaping the benefit."[88] However, events did not necessarily unfold according to plan. Far from forming a docile workforce, young women silk workers in anthracite communities demanded better conditions and joined larger labor protests in the region and industry during the first decades of the twentieth century.

Developments common to twenty-first-century America had their counterparts in anthracite mining communities (and Paterson, New Jersey) in the 1880s and 1890s: international shifts of enterprise, capital flight and deindustrialization, and the resort to low-wage female and child labor in peripheral areas. The campaigns to attract manufacturing firms to the area also intriguingly prefigured mid-twentieth century initiatives at redevelopment in the anthracite region when the mines closed. Local commercial interests—not the anthracite railroads or mining companies—similarly would lead redevelopment efforts, although merchants and others in the 1950s were impelled by the sustained and not cyclical loss of the purchasing power of miners and their families. Once again communities would vie to attract new manufacturing enterprise with costly inducements. Growth in light industrial jobs, especially garment manufacture, provided work for married women. Efforts to achieve economic diversification in the late nineteenth and twentieth centuries had a gendered character: Women benefited occupationally more than men given the kinds of businesses that could be attracted.

The economic insecurity of anthracite miners prompted family and broader community responses. Anthracite miners also acted collectively by organizing unions in an effort to bring greater certainty to their working lives. Just as the actions of railroad owners and managers in creating an effective anthracite cartel led to the consolidation of capital, so too the repeated mobilization and protest of mineworkers contributed ultimately to the emergence of a labor organization that spoke for workers across the region—the United Mine Workers of America (UMWA).

The first recorded strikes in the anthracite region occurred in July 1842 and involved more than 2,000 mineworkers from Minersville and Pottsville who walked off their jobs to protest exorbitant prices charged at company stores. As with many subsequent strikes, local police and militia arrived to protect property and arrest reported leaders, and the walkouts achieved neither improvements in working conditions nor sustained union organization. Payment in scrip and paper notes redeemable only at company stores spurred a second notable wave of strikes in 1849, but again, without any gains.[89]

Unrest in the Pennsylvania anthracite region during the Civil War brought federal troops to the area. Few Irish Catholic immigrant miners volunteered to join the Union army, and some actively resisted conscription when first implemented in 1862. The Irish, who found a haven in the Democratic Party, saw little reason to sacrifice their lives for a war to preserve the Union and challenge slavery promoted by the Republican Party, especially as they engaged in ongoing labor disputes with moralistic Protestant mine owners who were staunch Republicans. In one antidraft incident in Schuylkill County, riots broke out when one thousand Irish miners attempted to block a train transporting inductees; in another, the operator of a Schuylkill colliery was murdered because he had allegedly supplied information about his employees to draft officials. Other episodes convinced federal authorities to move troops to the area to restore order; the soldiers guarded draft offices, but also the mines to quell disruptions of disgruntled Irish workers. While the troops had ostensibly entered the region to uphold the military draft and restore order, they remained and provided military force to back up operators in their labor struggles. This occupation of several anthracite counties represented the first use of federal troops in U.S. history to stem labor unrest.[90]

The early protests of anthracite miners were local, short-lived, and not particularly effective, as differences among the three fields in the region divided mineworkers. The first indicator of change came in 1868, as John Siney, a British immigrant miner, founded the Workingmen's Benevolent Association (WBA), with chapters throughout the region but with a stronghold in the southern field. Like many trade unions in the nineteenth century, the WBA renounced the tactic of the strike, seeking rather harmoniously to negotiate contracts with mine owners that benefited all parties. The WBA justified job actions as a way to curtail production and raise coal prices, thereby permitting higher wages. In this vein, the WBA reached the first collective bargaining agreements in the industry that including a sliding scale of compensation in which wages were pegged to changes in the price of coal, with a minimum set if the price of a ton fell below a certain level. The miners shared in the returns of prosperous times and the risks of downturns but with some protection—thus achieving greater security in their lives.[91]

The WBA experienced success with small-scale operators in the southern field but met stiff resistance elsewhere, principally from Franklin B. Gowen, president of the Philadelphia and Reading Railroad. Determined to stamp out unions in the Reading's territory, Gowen first intervened in July 1870 by increasing anthracite shipping prices while the WBA was out on strike, making it extremely difficult for independent operators who had negotiated settlements with the union to market profitably the coal they were producing. In a second strike in early 1871, Gowen allied with the management of the Lehigh Valley Railroad, and both roads raised shipping rates still higher, effectively blocking any move by independents to settle with the union. Gowen's tactics backed up the most intransigent of the operators and placed the WBA on the defensive.[92]

The WBA ultimately met its demise in the Long Strike of 1875. The strike actually began in December 1874 with a lockout engineered by Gowen. The Reading and its associated railroads agreed not to ship the coal of other operators, forcing more than thirty independent producers to join the suspension of operations. The work stoppage, ostensibly initiated as a response to the declining price of coal, gave operators in the Lehigh and Schuylkill fields a basis for announcing wage cuts. In response the WBA county organizations called for a strike. The two sides faced off. The WBA announced a March 1 deadline, at which point union members would return to work only with an 8 percent wage increase, hoping to gain operator acceptance of the current wage scale, but no operators broke ranks. Battalions of private Coal and Iron Police—authorized by state legislation—and state militia patrolled the coal region, creating an atmosphere of intimidation.[93] The WBA moderated its demands, but still in mid-June representatives of operators refused to meet with the Association. Finally, with no end in sight and with miners having been out of work for more than six months, the WBA called off what had become the Long Strike and miners returned to work at the reduced wages offered by the various operators' groups.[94] The WBA, for all intents and purposes, was dead.

The WBA brought industrial unionism to the Pennsylvania anthracite region, organizing mineworkers across divisions of skill, ethnicity, and location. The union did not survive the Long Strike of 1875, but it did set the stage for the eventual success of the UMWA two decades later. As Anthony Wallace has aptly noted, it was not the WBA's "particular accomplishments" that were important, "but the idea of the large industrial union and the development of a higher consciousness of class, of pan-ethnic solidarity, and of norms of union discipline that served as the basis for more enduring organizations."[95]

Gowen's defeat of the WBA allowed him to clamp down on guerrilla warfare that had percolated in the anthracite region for more than a decade, allegedly involving a clandestine Irish order, the legendary Molly Maguires. Irish Catholic immigrants, who came in great numbers in the mid-nineteenth century to work in the anthracite mines, brought with them traditions of secret revenge against their exploiters—English landlords—and in Pennsylvania they directed their wrath against their Protestant English and Welsh overseers in the mines in a mix of class and ethnic antagonisms. With various murders of superintendents and foremen and bombings and arson, the legend of the Molly Maguires grew in the anthracite region in the early 1870s. Zealous to stamp out the violence, Gowen hired the Pinkerton Detective Agency that worked along with the company's private Coal & Iron Police to infiltrate the group.[96] Pinkerton agent James McParlan ultimately provided testimony that led to the conviction and hanging of twenty men, with Gowen serving as the lead prosecution attorney. Ten alleged conspirators were executed on June 21, 1877, which became known in the anthracite region as the "Day of the Rope." The guilt of the condemned remains clouded more than 125 years after the convictions and executions. Clearly, murders had been committed, but the evidence presented at and after the trials does not appear conclusive to Kevin Kenny, the historian who has most thoroughly explored the matter.[97] In Kenny's view, the trials were patently unjust: Irish Catholics were systematically excluded from the juries, and the investigations and prosecutions were basically private affairs paid for and organized by the anthracite railroads and coal companies. Finally,

THE MARCH TO DEATH.

Molly Maguires: the march to death, Pottsville, June 21, 1877. *Frank Leslie's Illustrated Newspaper*, July 7, 1877.
Courtesy of Kevin Kenny.

it is difficult to determine to what extent McPar-lan's role as an agent provocateur may account for the murders.[98]

After the defeat of the Workingmen's Benevo-lent Association and the hanging of the Mollies, labor organizing stalled in the anthracite region until the entry of the United Mine Workers of America in the mid-1890s. The Knights of Labor—led by Terence Powderly, who also served as mayor of Scranton—made some in-roads among miners in the 1880s, and the decade saw sporadic protests, but the frequency of strikes and the union presence in the region declined dramatically.[99] Events outside the region laid the basis for the rebirth of unionism in anthracite.

The UMWA was formed in Columbus, Ohio, in 1890, the product of a merger of two coal mine unions—a trades assembly affiliated with the Knights of Labor and the National Progres-sive Union of Miners and Mine Laborers. At its founding, the UMWA represented no more than 20,000 bituminous mine workers spread across Illinois, Indiana, Ohio, and western Pennsylvania. Organizers of the UMWA first en-tered the Pennsylvania anthracite region in 1894 to raise funds for bituminous workers then in-volved in a quixotic general strike. A few UMWA activists remained to establish locals in the area.[100] They picked an inauspicious time. Not only had unionism all but disappeared in the region, but a severe economic depression starting in 1893 greatly dimmed prospects for UMWA organizers to build their union. By the turn of the new century, they had little to show—9,000 signed-up members among a po-tential 140,000 recruits.[101]

The UMWA in anthracite would rise phoenix-like in the late-1890s on the heels of grassroots insurgencies of immigrant minework-ers; the union would take advantage of and channel, not lead, a percolating mass movement. Spontaneous protests began surfacing in 1897 in surprising fashion. It was a walkout of young mule drivers that escalated into the famous Lat-timer Massacre of September 1897. When the Lehigh and Wilkes-Barre Coal Company an-nounced a decision to consolidate the company's stables for mules, the teenage drivers went on

Strikers marching to Lattimer, September 10, 1897. MG-273, Charles H. Burg Photographs, Pennsylvania State Archives, Harrisburg.

strike because their trips to and from work were now much longer. Young slatepickers, with their particular grievances, soon joined the walkout; within days, the ranks of the strikers swelled to 2,000 fed by a widely shared antipathy to the company's autocratic superintendent.

On Friday, September 10, 1897, some 400 miners, recent immigrants from southern and eastern Europe, marched with an American flag aloft toward the patch town of Lattimer, hoping to convince miners there to join their walkout. Lehigh and Wilkes-Barre Coal Company executives had received assurances from local law enforcement officers that the protest would be quelled. At 3:45 in the afternoon, Sheriff James L. Martin, with a posse of 150 recently deputized Coal and Iron Police behind him, walked to the head of the marchers' column and ordered them to disperse. He then grabbed the flag held by the lead marcher, and in the scuffle

that ensued the assembled police fired into the crowd. Nineteen unarmed miners lay dead in the road, and thirty-six others lay seriously wounded. Martin and seventy-three of his posse were arrested and charged, but were found not guilty and released. The upshot of the shooting and the trial was an upsurge in membership in UMWA locals in the region.[102]

More strikes followed in 1898 and 1899, in locales that had largely been immune to previous union organizing, namely in the northern anthracite field.[103] The thick, horizontal veins of anthracite in the Wyoming and Lackawanna valleys permitted more stable production, and anthracite mineworkers there generally enjoyed more regular employment and earnings than their counterparts to the south. For not self-evident reasons, labor protest at the turn of the twentieth century centered in the northern field, and UMWA organizers made the greatest inroads there. In these conflicts, the rank and file almost invariably responded to wage cuts or new work rules by striking; district and national UMWA officers followed the workers' lead. In the heat of these struggles, they established District 1, which recruited more than 7,000 of the 9,000 workers who joined the union by January 1900. District 9 in the southern coalfield—where the WBA had reigned a generation earlier—had nearly 1,500 members, and District 7, in the Lehigh field, had only 341.[104] As in so much else in anthracite, UMWA districts followed the contours of the region's three distinct fields.

A strike at the Susquehanna Coal Company in the Wyoming Valley in the summer and fall of 1899 proved a tipping point in labor protest and organizing in the region. The walkout involved skilled miners, and UMWA officials quickly and effectively assumed direction. A classic issue in mine work sparked the confrontation: dockage and topping. Miners blasted and picked coal from veins, and their laborers loaded loosened coal into cars. How were the miners then to be compensated for their labors? Systems varied, but usually payment was by the car. Injustices and daily disputes commonly ensued. Companies supplied different size cars, for example, while keeping payment rates per car the same. Equally unsettling, so-called docking bosses invariably determined the cars to contain less than full contents of anthracite, claiming that slate or other rock constituted significant portions of the cars' contents. The miners would then be "docked" pay accordingly. Docking bosses also insisted that coal be loaded far above the cars' heights for miners to receive full pay (topped, in other words). The capriciousness of these practices drove miners employed by the Susquehanna Coal Company to strike in late July of 1899. The walkout spread through the several mines of the company and ultimately involved more than 4,000 workers. Violence ensued when the company decided to operate the mines, as the wives of the striking miners scolded and harassed the strikebreakers. The solidarity of the strikers forced managers of Susquehanna Coal to negotiate with the representatives of the union. The UMWA secured a contract that did not meet all of their demands, but it established sufficient improvements in standards and procedures for compensation that the union could and did claim a major victory.[105]

The rumblings of discontent in the Pennsylvania anthracite region in the late 1890s and the first foothold achieved by UMWA organizers coincided with the overnight rise of John Mitchell to the national leadership of the union. Mitchell grew up in a bituminous mining community in Illinois and entered the mines at the age of twelve; his father had been killed in a mine accident, and Mitchell had to support his family.[106] Like many young men of his generation, he soon wandered about the country, taking mining jobs in the West, but he ultimately resettled in Illinois, married, and became involved in community affairs. Mitchell first accepted a position as a paid organizer in the UMWA in 1894 at a propitious time. The nascent organization was short in top leadership and opportunities opened for promotion; for personal and professional reasons the top officers had short tenures. Between a paucity of rivals and the force of his personality and beliefs, Mitchell rose swiftly to the presidency of the union. Mitchell exuded a natural concern for the plight of his fellow miners and quickly garnered their loyalty. Although young, he appeared a firm and serious man, one able to achieve gains through reason and just compromise. Early in his career, Mitchell dressed and appeared priestlike, and his look and demeanor appealed to Catholic miners, who mistook him for a cleric, and opened doors with political and business leaders as well.[107] Mitchell also strongly believed in industrywide conferences and agreements that could establish standard labor practices, and, in turn, stability of production and employment across firms. In 1898, he helped attain a regionwide contract with bituminous coal operators that provided an eight-hour day for soft coal miners and union recognition. That historic victory propelled him into the presidency of the UMWA, but also gained him entrance to a world of business and political leaders, including membership in the National Civic Federation. In that organization he came to share the view that the corporate capitalist system could be stabilized through cooperation between capital, labor, and the state.[108]

With his national stature, the evident militancy of anthracite mineworkers, and the UMWA's growing presence in northeastern

Pennsylvania, Mitchell set his sights on the anthracite region in the summer of 1900. A major move into anthracite posed enormous risks; the union had little experience in the hard coal trade and faced a much more powerful and organized set of coal operators than in bituminous. On August 23, union delegates met to map out a campaign and set their demands. The demands boldly included abolition of company stores; reduction in the prices of blasting powder charged miners; elimination of the sliding scale of wage payments, an earlier achievement of the WBA but with few remaining adherents; an end to dockage; payment by weight of coal and the setting of 2,240 pounds as the standard ton; fair distribution of coal cars; the right of miners to appoint checkweighmen to verify the weighing of the coal and prevent unfair dockage by companies for slate loaded with the coal; 15 to 20 percent raises in wage rates; and union recognition.[109]

The demands of the UMWA fell on deaf ears, and the union announced a strike date of September 17. On that day, 100,000 anthracite coal workers walked off their jobs—to the surprise of both Mitchell and the operators. The strike constituted the first truly regionwide strike in anthracite and featured the protest organizing of "Mother" Jones, the great labor agitator of her day, and violent confrontations between strikers and strikebreakers—while Mitchell pleaded for restraint. The strike would ultimately end with the intervention of a powerful political figure, Mark Hanna, a Senator from Ohio and kingpin of the Republican Party. Hanna's intercession represented the first of a series of interventions in the first quarter of the twentieth century of national leaders into labor disputes in the anthracite region as the area became the focus of national concern. Fearing negative consequences of the strike on the re-election bid of President William McKinley, Hanna met with the anthracite railroad executives in J. P. Morgan's office and urged settlement of the strike.

The operators refused to negotiate with Mitchell, but they did agree to post 10 percent wage increases and reduce the price of powder. While local union leaders pushed for holding out for a contract with all demands met, Mitchell convinced them to accept the gains and declare a victory. The union then directed the striking miners to return to work on October 29, a day that would be celebrated in the anthracite region for years to come as "Mitchell Day." The union did not win a formal signed agreement or recognition from anthracite operators, but it was a signal victory, which set the stage though for a more monumental confrontation two years later.[110]

The 1900 strike—with organized capital and labor pitted against each other and the intervention of national political leaders—marked the close of a century of rapid transformation. St. Anthony's Wilderness had been re-created. In contrast to the evolutionary pace of the region's geological history, rapid-fire change characterized the nineteenth century. The anthracite canals quickly succumbed to the railroads; surface quarrying of coal gave way to deep underground mining; the early competition of the independent mine operators yielded to corporate consolidations, under the guiding hand of finance capital; town and city building boomed with waves of immigration; miners previously divided by location, skill, and ethnicity coalesced within the UMWA; and violent confrontations between workers and employers increasingly jarred both the region and the nation. After this frenetic century of development, the anthracite region would enter a two-decade period of remarkable prosperity and stability. The order afforded by the anthracite railroad cartel, new labor agreements, and a flourishing national economy benefited anthracite operators and mineworkers alike. From the vantage point of World War I, few would have ventured to predict an imminent and lasting regional economic collapse.

CHAPTER 2

Apogee and Descent
The Anthracite Region in the Early Twentieth Century

ON THE MORNING OF OCTOBER 3, 1902, George Baer, president of the Philadelphia and Reading Coal and Iron Company (and its parent company, the Philadelphia and Reading Railroad), led a delegation of major anthracite coal operators into the Oval Office of the White House. The group had been summoned by President Theodore Roosevelt to meet with John Mitchell, president of the United Mine Workers of America (UMWA), and fellow union leaders to discuss ways to end a strike involving more than 150,000 anthracite mineworkers that had gripped the nation for four-and-a-half months. Although repeatedly beseeched to intercede, Roosevelt had hesitated throughout the summer to use the powers of his office to resolve the dispute. With congressional elections just a month away, reports of growing anarchy in the coalfields, and anxieties rising in urban centers of the Northeast with winter ahead and coal for heating in short supply, Roosevelt heeded the counsel of his political advisors and finally intervened.[1]

The daylong conference in the Oval Office produced no resolution. The anthracite coal operators upheld their absolute authority to run the mines without any interference from the UMWA (their "divine right" as Baer had famously declared in a public letter earlier that summer).[2] Mitchell, in contrast, appeared eminently reasonable as he spoke for resolving the strike through arbitration. Roosevelt's interces-

sion that day failed, but the meeting was historic, marking the first time a president of the United States had brought parties to a labor dispute into the White House to encourage a settlement.

A different kind of intervention by a major political figure would soon bring an end to the great anthracite strike of 1902. On October 11, Elihu Root, Secretary of War and a former corporate lawyer, traveled to New York (with Roosevelt's blessing) to meet with J. P. Morgan to achieve a peace. Root asked Morgan to convince the anthracite operators to agree to arbitration.[3] Within twenty-four hours, the coal company executives bowed to Morgan but insisted that President Roosevelt appoint a five-member commission with a particular composition (a military engineer, a federal court judge, a mining engineer not connected with the coal business, a person familiar with the anthracite coal trade, and an "eminent sociologist"). Mitchell consented to the arrangement after persuading Roosevelt to expand the panel with a representative from organized labor and a Catholic clergyman. On October 16, Roosevelt announced the formation of the Anthracite Coal Strike Commission with a mandate to investigate claims and determine the terms of a settlement. Mitchell convinced UMWA representatives to vote in favor of arbitration, and on October 23 anthracite mineworkers returned to their jobs after maintaining their walkout for 165 days.[4]

The Pennsylvania anthracite region gained

national attention during the great strike of 1902.[5] The solidarity of the anthracite mineworkers, the popularity of Mitchell, the intransigence of the powerful coal operators, and concerns about shortages of coal with winter approaching placed the anthracite region in the spotlight and demanded the involvement of national political and financial figures.[6]

◆

The great strike resulted from the contradictory character of the resolution of the anthracite strike two years earlier. At that time the pressure of national Republican leader Mark Hanna and financier J. P. Morgan had moved the powerful heads of the anthracite railroads to compromise with the UMWA and its young president, John Mitchell, and offer higher wages to striking mineworkers. But they never recognized the union and remained determined to exercise absolute control over their mining and railroad empire. The conflict was bound to be renewed; only the timing was uncertain.

Mitchell emerged from the strike of 1900 with even greater prestige, and union membership in the anthracite region mushroomed to 53,000. However, the victory had not resolved the industry's labor problems, which quickly became clear. The wage increases unilaterally posted by the coal operators held until April 1, 1901. As that date approached, Mitchell desperately worked with Hanna to secure a contract and quell rank-and-file pressure for another strike. Their efforts headed off a confrontation when the anthracite railroad executives did not rescind the wage increases, but rather extended them until April 1, 1902. Mitchell convinced anthracite mineworkers to stay on the job, but he could not control an outburst of unauthorized strikes that erupted in the summer of 1901.[7] The stage was set for the greater battle that would unfold the next year.

As April 1, 1902, approached, Mitchell met frequently with Hanna, leaders of the National Civic Federation, and associates of J. P. Morgan in his effort to bargain with the operators. The demands of the miners had grown since 1900 and now included an eight-hour day. The anthracite railroad executives remained adamant in refusing to be party to a comprehensive, industrywide agreement with the UMWA, or even

to recognize Mitchell by meeting with him. Mitchell used every means to stall and prevent a walkout, but he could not counter rising worker sentiment and a vote of union delegates at a regionwide meeting in Hazleton in mid-May calling for a strike.[8]

In the ensuing months, Mitchell operated on the national stage, presenting a picture of moderation and reasonableness. He kept meeting with political and business leaders to promote possible negotiations, announcing and reiterating his willingness to accept binding arbitration to settle the dispute. To cool the situation, he successfully forestalled a sympathy strike advocated by bituminous miners, seeking rather to create a substantial anthracite relief fund with contributions from the working soft-coal men. He accepted an offer from Father John J. Curran from Wilkes-Barre to mediate the dispute, but the leading anthracite operators rebuffed the priest's efforts.[9]

While Mitchell appeared the peacemaker, striking anthracite miners became increasingly belligerent in defense of their cause. Immigrant strikers intimidated individuals who crossed picket lines and continued to work during the strike. An article in *Straz*, a Polish-language newspaper, expressed the widely shared anger against strikebreakers: "The ranks of Polish scabs in the last few days have lessened considerably. Coercive rocks and a couple of broken ribs have influenced them—for strikers caught some traitors to the workers' cause and so licked them that they will remember it for a long time."[10] In the Panther Valley, striking miners stripped some strikebreakers and paraded them to Coaldale; in nearby Tamaqua, Coaldale strikers "captured" Hungarian strikebreakers and forced them to march to McAdoo some eighteen miles away. According to one account of this incident, "To make the journey as painful as possible to the marched men, pebbles were put in their shoes by the strikers."[11]

Subsequent testimony at hearings of the Anthracite Coal Strike Commission revealed the intensity of antagonism between strikebreakers and strikers. Victims of strike violence testified, recreating the tense confrontations that were a daily occurrence during the strike. "Son of a bitch scab! I kill you!" was a morning greeting for them. And striking miners went beyond

Strikers chasing scabs, 1902 strike. *Everybody's Magazine*, September 1902. Courtesy of Susan Campbell Bartoletti.

cursing out strikebreakers, as one witness re-called: "They were punching us, kicking us, jumping on us and doing everything they could." When strikers could not find the target of their anger, they might leave a clear message for the offender. The testimony of one witness was thus reported: "William Bardner, a mine docking boss, told the commission that a tombstone bearing his name and the words 'scab docking boss' was placed in front of a hotel on the main street of his town."[12] He got the message.

Violence in mining communities led the governor to send contingents of the Pennsylvania National Guard to the Panther Valley and Shenandoah—where the brother of a deputy sheriff was beaten to death on July 30 by an angry mob because the deputy had escorted strikebreakers to work. However, the strength of the striking mineworkers in closing the mines limited the number of physical confrontations with police forces.[13] Still, contemporary reporters emphasized the lawlessness that the strike entailed. A correspondent for *Collier's Weekly* reported in early October: "It is now five months since labor declared war. With each suc-

ceeding day the operators seem to become more determined and the strikers more desperate, and in consequence, riot and lawlessness become more prevalent. Hardly a day passes that acts of violence are not reported."[14] In addition to violence directed toward strikebreakers, the writer highlighted the collecting of surface coal and even digging into underground veins by members of striking families, both to meet their own needs and to sell to local customers: "It is impossible to exaggerate the lawless spirit . . . the utter disregard of the property rights either of corporations or of individuals." Local law authorities would not prosecute the offenders, and in the rare cases in which arrests were made, the defendants "were acquitted by a sympathetic court and the costs of the case were charged to the company from whose property the coal was taken."[15]

Such reports prompted Theodore Roosevelt's intervention in early October to resolve the anthracite coal strike. The commission that he established to achieve an arbitrated settlement began its work on October 27, just days after the anthracite miners returned to work. In the next

President Theodore Roosevelt, Bishop Michael J. Hoban, John Mitchell (left to right), Scranton. *Scranton Times*, February 19, 1958. Courtesy of the Pennsylvania Historical and Museum Commission, Bureau of Historic Sites and Museums, Anthracite Museum Complex, Scranton.

four months, 558 witnesses appeared before the commission, providing testimony and evidence that filled 10,000 pages in 50 typescript volumes. Clarence Darrow, the great civil liberties lawyer and defender of radicals of his age, presented the case for the UMWA.[16]

On March 21, 1903, the Anthracite Coal Strike Commission released its findings and rulings. The commission's award, basically a compromise of the positions that the UMWA and the anthracite operators had argued, provided for a 10 percent wage increase to most mineworkers—half the increase originally demanded; a 10 percent reduction in hours worked per day—from ten to nine hours for workers paid by the day; and the right of miners to appoint checkweighmen and docking bosses (if a majority of workers voted for them and agreed to have the additional costs deducted from their wages). The commission's report condemned certain practices of coal operators—such as company housing and discrimination in the assignment of coal cars—but provided no remedies. The commissioners did not rule in favor of the broad demands of the UMWA, and, most important, on the issue of union recognition. Instead of a system of collective bargaining, the commission created a permanent six-member Board of Conciliation to adjudicate future disputes; a mineworker and operator from each of the three coal districts in the region comprised the Board's membership. In the event of a stalemate, the dispute was to be referred to an impartial umpire appointed by a federal judge. The Anthracite Coal Strike Commission afforded no official standing to the UMWA—its report specifically reproved the closed union shop—but in allowing "an organization" to appoint worker representatives to the new Board of Conciliation, the commission did convey quasi-official status to the union.[17]

The binding recommendations of the Anthracite Coal Strike Commission brought order to industrial relations in anthracite, at least through World War I. In comparison to the tumult of prior decades, the nature of interactions between mine operators and mineworkers can be characterized as peaceful between 1903 and 1920. Yet, the détente established did not constitute an equal partnership, was brittle at best, and the role and presence of the UMWA in anthracite remained precarious.

A key element in the Commission's settlement was the creation of a six-member Anthracite Board of Conciliation that met to resolve grievances between mineworkers and management. The Board's record in the years following the Great Strike of 1902 was mixed; from the perspective of the UMWA and mineworkers, the body had its supporters and its detractors. Immediately after the mineworkers returned to work, the Board received unresolved grievance cases for consideration, eighty-six in 1903 alone. While few of the subsequent judgments redressed miners' complaints of unjust practices, union officers viewed the Board as a salutary development. As Mitchell argued, "the greatest advantage that comes to us from the Board of Conciliation is not the cases we win. . . . The great benefit that has come to the miners . . . has been the cases not referred to it. The existence of the board has deterred unfair foremen and superintendents from imposing on our men, because they knew their actions were subject to review." Misgivings with the ultimate results of the Great Strike of 1902, however, surfaced soon in declining loyalty to the UMWA, membership dropping to roughly 37,000 in mid-1904 and 23,000 at the end of 1907. Growing distrust of the Board of Conciliation became evident as well; in the 1906–1909 period, miners filed a mere twenty-three grievances to the Board.[18]

The settlement implemented by the Anthracite Coal Strike Commission remained in effect for a three-year period, until April 1, 1906. Weeks prior to the expiration of the binding agreement, pressure built for another major strike. The UMWA reiterated its past unfulfilled demands—emphasizing payment by weight, new wage increases, and union recognition (with a union dues checkoff system that required employers to deduct UMWA dues from workers' paychecks and pay them to the union). The operators proposed a continuation of existing arrangements. With membership down—and recent setbacks for the union in bituminous coalfields—Mitchell once again counseled restraint. The parties then agreed to a simple three-year extension of the 1903 settlement.[19]

History repeated itself in 1909, this time with a new national leader of the UMWA, Tom L. Lewis, after Mitchell stepped down from office in 1908. With nothing gained (or lost), union membership in anthracite slipped further to a low of 10,000 in 1911. Disgruntled immigrant mineworkers also began to heed the messages of a rival, more militant labor organization that had entered the anthracite coalfields, the Industrial Workers of the World (IWW).[20]

The UMWA's fortunes improved in 1912. With heightened demand for coal accompanying a booming economy, coal operators could not afford disruptions in production. They agreed to a new four-year arrangement that included a 10 percent increase in wages; this contract provided the first significant gains for anthracite mineworkers since 1903. Still, anthracite coal operators remained firm in refusing to recognize the UMWA as the formal bargaining agent for mineworkers.[21]

With the success of 1912, membership in the union shot up to a peak of 129,000 in 1913. Yet, the UMWA still did not exercise firm control in the region. A rash of local strikes erupted in anthracite between 1913 and 1916 (with close to 80,000 mineworkers involved in the latter year) and union district officers alternately led and followed in these outbreaks. The IWW also had a active hand in the disruptions. With pressure from below, the UMWA was able to negotiate for further improvements in 1916—an eight-hour workday and 3 to 7 percent wage increases—but the settlement did not garner great support among miners (nor did it provide union recognition or institute payment by weight, old demands by now).[22]

Two years later, at a time of war, the UMWA became the official representative for anthracite mineworkers before the U.S. Fuel Commission, an agency established to ensure uninterrupted production of coal for the nation's military mobilization. World War I, however, saw the union definitely lose its grip in anthracite communities. Unauthorized strikes and election challenges to local leaders both disrupted mining and the union's organization in the region. War's end would then bring a return to intense labor conflict, and in the first years of the 1920s the Pennsylvania anthracite region would again be the focus of national attention and concern.[23]

◆

The first two decades of the twentieth century represented a time of growth, maturation, prosperity, and stability in Pennsylvania's anthracite region. Can this golden age be attributed singularly to the détente in labor relations brought on by the settlement of the Anthracite Coal Strike Commission? Less strife and disruption certainly contributed to more regular and increased production, employment, and earnings. Mineworkers did gain from awards in 1903, 1912, and 1916. Yet, other factors were as, if not more, important in the good fortunes of the anthracite region in the period between the great strike of 1902 and U.S. entry into World War I.

Throughout the last quarter of the nineteenth century, anthracite railroad executives tried repeatedly with limited success to reach agreements among themselves as to levels of production and apportioning of market shares to bring stability in prices and profits. Increasingly, the anthracite railroads purchased large holdings of one another's stock, and the emergence of interlocking directorates made possible tacit understandings that had eluded management earlier. The established "community of interest" among the major operators led to a new era of steady and rising prices, and instead of periodic retrenchments, the first two decades of the twentieth century saw ever-expanding production.[24] General economic good times sustained the cartel arrangement. After a twenty-five-year period that witnessed two major depressions, Americans enjoyed a booming economy after 1900. High demand for coal further bolstered prices and production for the companies and work and earnings for anthracite mineworkers.

As the mining of anthracite coal increased from 57 million tons in 1900 to a peak of 100 million in 1917, the number of days worked per year by miners grew significantly.[25] In the 1890s, anthracite mines had typically operated for an average of only 180 days a year. A decade later, that figure had reached almost 230; by 1917 and 1918 the figure approached 290 days annually.[26]

With greater work time and more regular employment, anthracite mineworkers' incomes improved dramatically in the first two decades of the twentieth century. Average annual earnings for adult workers went from $494 in 1901 to $1,480 in 1921. While notable inflation marked these years, wage increases for mineworkers in anthracite adjusted for inflation amounted to almost 40 percent.[27] Gains in real income were particularly significant after 1914, and by the early 1920s, anthracite mineworkers stood at the top of the nation's labor force in terms of real gains in earnings.[28] The informal and formal mobilizations of anthracite miners contributed to their better financial circumstances, but corporate stabilization and a brisk economy—with, simply, more working days—played more decisive roles.

Higher earnings had dramatic impact on the family economy of Pennsylvania anthracite mineworkers. Early budget studies had revealed a severe shortfall between the annual income of adult miners and basic family expenditures. Underemployment and low wages were root causes, and families had to rely on the earnings of children to make ends meet. (The unpaid labor in the home of the wives of miners—in cooking, cleaning, sewing, tending vegetable gardens, and taking in boarders—represented an equally critical subsidy to family means).[29] The exact contributions of children to household income in the late nineteenth century are not rendered in budget studies of the period. Information compiled by census takers, however, indicates the extent of child labor. For example, in 1880 in the city of Scranton, more than 56 percent of children of Irish and Welsh parents between the ages of six and twenty were employed.[30]

A four-year congressional study of woman and child labor published in 1911, which included the silk industry in the Pennsylvania anthracite region, provided actual figures on the place of children's earnings in total family income. Investigators found that a typical son or daughter of a miner who labored in a coal breaker or silk mill in the area contributed 12.5 percent of family income, with the combined contribution of children amounting on average to almost 38 percent of household earnings.[31] The study similarly reported that less than 47 percent of the twelve- and thirteen-year-old children of the families surveyed were in school and just 5 percent of their fourteen- and fifteen-year-old brothers and sisters.[32] Children in anthracite-silk communities thus served as bulwarks of the family economy with their wages covering the shortfalls between fathers' incomes and family expenses and allowing for savings and purchases of occasional luxuries. Their savings account for the fact that 50 percent of the families canvassed in the congressional investigation owned their own homes.[33]

The place of children's earnings in the family economy, and child labor in general, changed dramatically in the second decade of the twentieth century. Law played a key role. In 1905, Pennsylvania lawmakers enacted general legislation that banned the employment of children below the age of fourteen; fourteen- to sixteen-year-old boys and girls could work only if they secured working papers signed by parents and employers. Prior statutes had set sixteen as the minimum age for working inside mines and fourteen for employment about the mines, but proof of age or parental consent had not been necessary. The 1905 law, however, required parents only to swear under oath to their children's ages; without documentary evidence, the intention of the legislation could easily be circumvented. The initial weakness in the law led to lobbying by the National Child Labor Committee, which resulted in a tightening of the Commonwealth's regulation of child labor. The issuing of work permits became the responsibility of school officials in 1910, a group in a better position to determine the ages of students. A combination of stricter standards in the issuing of work permits, better enforcement of compulsory school attendance laws, and statutes requiring employed fourteen- to sixteen-year-olds to enroll in night schools further reduced the hiring of school-age youngsters. These restrictions were strengthened with the passage of the 1915 Pennsylvania Child Labor Law setting fourteen as the minimum working age and requiring working children to have completed the sixth grade.[34]

Legal change played an important role, but the reliance on child labor declined dramatically in the anthracite region in the early twentieth century also because families no longer had to rely on children's earnings as adult mineworkers achieved substantial gains in real income. The rising age at which children began working in coal breakers reflected this broader development. The experience of employees at Lehigh Coal & Navigation in the Panther Valley reveals the decline of child labor over time. LC&N workers, who were first employed as slatepickers at the company in the 1880s, began work on average at 11.3 years of age. In the 1890s, slatepickers typically began at 13.6 years of age, and for the 1910s the comparable age was 15.3. The shift is steady and striking.[35]

The change was regionwide, and census enumerations of mining families in 1920 reveal the new circumstances. In a sample population of almost 550 mining households from a mix of communities—Scranton, the largest city in the region, Lansford, a midsized coal town, and mining patch settlements in Mount Carmel and Coal townships—more than two-thirds of the fathers were sole breadwinners, and only four wives worked outside the home.[36] These families had 1,741 children living at home, and less than one in six were employed. No child under the age of thirteen worked according to census enumerators. Among children of school age (age six to eighteen), less than 14 percent were employed. Only beginning at age fourteen were significant proportions of children working, with 23 percent of fourteen- to fifteen-year-olds and 61 percent of sixteen- to eighteen-year-olds. Children above the age of twenty who continued to live at home did work in substantial numbers; the census listed more than 71 percent of them as gainfully employed. But, overall in 1920, less than one in seven school-age children in families of mineworkers worked to supplement their fathers' earnings.

Child labor had not been eliminated, but figures for representative mining communities in 1920 indicate swift, dramatic change. Just slightly more than a decade earlier, congressional investigators had found almost 38 percent of children of anthracite miners' families at work. Now, that proportion had declined by almost two-thirds, and the vast majority of children attended school.[37] Even among newly arrived immigrant families, a sea change had occurred. In 1880, more than 56 percent of the children of Welsh and Irish immigrant miners in

View of the Ewen breaker of the Pennsylvania Coal Co., S. Pittston, Pa., January 10, 1911. Photograph by Lewis Hine. National Archives, 102–LH-1938, Washington, D.C.

Scranton who were below the age of twenty worked. In 1920, less than 22 percent of the children of new immigrant mineworker families in Scranton were recorded as employed by census enumerators.[38] Protective legislation and conditions of full production and employment combined to erode what had once been a basic fact of life in the anthracite region, child labor.[39]

The first two decades of the twentieth century brought critical changes in the Pennsylvania anthracite region—in business arrangements, labor relations, and the fortunes of anthracite miners and their families—but continuities also marked the era. Residents of the region—their numbers growing from 759,000 in 1900 to 975,000 in 1910 and more than a million in 1920—continued to live in a variety of communities, from cities to mine patch towns.[40]

By 1920, Scranton was a renowned city of 138,000 residents. A magazine article on Pennsylvania in the *National Geographic* in 1919, entitled "The Industrial Titan of America," touted: "Probably no other city of its class in the world is richer than Scranton."[41] The city boasted one of the nation's first electrified urban transit systems, a downtown with tree-lined boulevards, a theater and restaurant district, luxury hotels, and beautiful Courthouse Square in its center, which featured an eight-story skyscraper (home to the board of trade). With five long-distance rails lines bringing 140 freight trains through the city each day, Scranton was well connected to other cities in the mid-Atlantic and New England.[42]

The city continued to lure immigrants with employment opportunities in mining and manufacturing. In 1920, immigrants constituted one-fifth of the city's population, and another two-fifths were native-born children of immigrants. Two-thirds of the foreign-born came from southern, central, and eastern Europe as part of the second great wave of European immigration in the first decades of the twentieth century.[43] Scranton offered a range of employment opportunities; only 29 percent of adult male residents worked in the mines. Men found good jobs in the machine trades and other industries, women in burgeoning textile and garment factories. More than one-fourth of adult women in Scranton worked outside the home, more than two-thirds in the textile and garment sectors—unusually high figures for the anthracite region.[44]

The city did experience one major blow in the early twentieth century. In 1901, the Lackawanna Iron and Steel Company, a founding firm in the city, closed and moved its operations to Buffalo, New York. Businessmen in Scranton, however, remained committed to maintaining the industrial standing of the city, and in 1915 they launched a public campaign that raised more than $1.1 million to attract and subsidize new manufacturing. They then formed the Scranton Industrial Development Company to facilitate their plan for manufacturing growth, another pioneer initiative and agency in redevelopment.[45]

Immigrants also found sound employment opportunities in Lansford in the Panther Valley, a town connected to the history and fortunes of the Lehigh Coal & Navigation Company. Lansford certainly did not have the grandness of Scranton—or its business and industrial might—yet Lansford also experienced boom times during the first two decades of the twentieth century as its population just about doubled, growing from 4,900 to 9,600. By 1920, foreign-born residents composed more than 28 percent of the town's population, with newcomers from southern, central, and eastern Europe constituting five-sixths of the new immigrants. Fully 64 percent of the adult men in Lansford worked in and around the mines—this distinctly was a mining town. One large silk mill offered employment to women, but only 15 percent of adult women worked for wages, and domestic servants actually outnumbered silk workers.[46] Community members could enjoy a rich associational and social life in Lansford. The town had thirteen churches, numerous fraternal and ethnic societies, sports teams, a daily newspaper, *The Evening Record,* and a department store, Bright's, that attracted customers from across the Panther Valley.[47]

Cities and towns accounted for the majority of the population in the anthracite region, but about 30 percent of the region's residents as late as 1920 still lived in communities with populations of less than 2,500.[48] Beyond the boundaries

of more densely settled places were many small clusters of houses, adjacent to single mines. Mount Carmel and Coal townships, both located in Northumberland County, were typical in their scattered concentrations of rural residents. Mount Carmel Township had 5,561 residents in 1920; Coal Township, 17,574.[49] These populations were dispersed in distinct villages and patches across the rugged mining landscape.

Examination of the patch towns found in one portion of Mount Carmel Township and enumerated together in the 1920 census proves illuminating. Altogether this district included the villages of Alaska, Connerville, Beaverdale, Dooleyville, Natalie, and Strong with a total population of 1,326—about 220 residents per village. Alaska and Natalie each had a mine and breaker. Alaska Colliery was typical of the isolated, rural operations in the township. It employed 777 mineworkers, operated for 272 days in 1920, and produced about 300,000 tons of coal. While Scranton and Lansford had experienced renewed immigration since 1900, this portion of Mount Carmel Township was relatively isolated from recent population flows. Only about a sixth of the residents there were foreign-born, and earlier immigrants from northern and western Europe outnumbered second-wave migrants in the township by almost three to one. Southern, central, and eastern European immigrants in the decades before World War I were more attracted to larger cities and towns than to the rural patch towns that dotted the countryside in the anthracite region. The native-born and earlier immigrants from the British Isles tended to predominate in the smaller communities such as those in Mount Carmel and Coal townships.[50]

Other than mining, there was little paid work in these rural communities. More than 80 percent of employed men in these districts of Mount Carmel and Coal townships worked in or around the mines, figures far higher than those for Lansford or Scranton. Furthermore, there were fewer occupational prospects for women in these communities than in urban centers, although women's unpaid work in backyard gardens no doubt contributed more to family living standards in rural patch towns than in the densely settled cities and towns of the region. Natalie was exemplary of the employment pat-

Lehigh Coal & Navigation Company houses, Lansford, ca. 1900. Courtesy of George Harvan.

terns in isolated mining patch towns. In a town of only 286 residents in 1920, 80 of 87 employed men worked in anthracite; the sole working woman was the "janitress" of the local public school. Natalie was a Philadelphia and Reading Coal and Iron patch town, and at the turn of the century the company provided two wagons every Saturday to take Natalie residents into Mount Carmel. According to a *New York Times* account, women returned after shopping on the 9:00 P.M. wagon, while men took the midnight one after drinking.[51]

Outside of the most rural precincts, recent immigration from southern, central, and eastern Europe shaped the history of the Pennsylvania anthracite region in the first decades of the twentieth century. The shift in immigrant origins from the British Isles to southern, central, and eastern Europe entailed both change and continuity in the anthracite region. After 1900, the new immigrants solidified the ethnic communities they began forming in the last decade of the nineteenth century—an ingredient in their active participation in the labor struggles of

the day. Yet, even with their formidable and organized presence in the cities and towns of the area, new immigrant miners remained at the bottom rungs of the occupational ladder.

Across the anthracite region, southern, central, and eastern European immigrant mineworkers founded and joined a vast array of fraternal organizations that both met secular needs, including social support and mutual insurance, and reinforced their strong religious leanings. Local associations in the area grew and merged into long-lasting regional and even national societies. The Greek Catholic Union, for example, united a number of disparate local lodges of Carpatho-Rusyns in 1892; the organization would grow to 333,000 members in 1,700 lodges across the United States by the 1930s. The Pennsylvania Slovak Catholic Union, similarly launched in 1893, had 37,000 members across the state by the late 1920s. The Russian Brotherhood Organization, founded in 1900, also had its origins in the Pennsylvania anthracite region.[52]

Mutual accident, sickness, and death benefits

A mining town, ca. 1902. Peter Roberts, *Anthracite Coal Communities* (1904), facing p. 336.

offered compelling reasons for mineworkers to join ethnic fraternal organizations. The Greek Catholic Union (GCU) instituted insurance plans that were typical of ethnic fraternal practice in the early years. In 1896, the GCU set its death benefit at $500 for male members and $250 for the wives of members. Monthly dues of $0.60 assured members of these benefits, plus a subscription to the monthly Rusin-language newspaper, *Amerikansky Russky Viestnik*. By 1908 dues had edged up only slightly, to $0.75 per month, while the members' death benefit had doubled to $1,000. Death benefits were paid out to members' widows, whether they lived in the United States or had remained in Europe. Publicity aimed at recruiting new members emphasized how the death benefit could provide capital to permit a surviving widow and family to become self-supporting.[53] Given the dangers of anthracite mining, such appeals must have been persuasive.

Ethnic fraternal organizations also played a crucial role in supporting the primary ethnic community institution—the ethnic church. In Shenandoah, a mining town in the southern anthracite field, the Society of St. Casimir raised the initial funds and purchased a city lot for the first Lithuanian Catholic Church in the area, completed in 1874. Shortly thereafter, this church came under Polish-American control, and the Lithuanians organized the St. George Beneficial Society, which financed the Lithuanian St. George Church, dedicated in 1894.[54] Thirty miles to the east of Shenandoah and a year later, Slovak Catholics in Lansford founded a fraternal organization and contributed to the organization of St. Michael the Archangel Roman Catholic Church. Lodges in the GCU played a similar role in many localities, and the Union itself backed a proposal for the appointment of a Greek Catholic bishop to oversee the spiritual affairs of Greek Catholics in the United States. Given the opposition of the Roman Catholic hierarchy to the service in the United States of married Greek Catholic clergy, Greek Catholics found it difficult to transplant their churches. Many Greek Catholic churches found themselves in a struggle with the Church hierarchy, and lodges of the GCU played a major role, in the words of one church historian, "often resist[ing] the encroachments of the local Roman Catholic bishop."[55]

The ethnic churches and the ethnic fraternal organizations grew together between the first substantial immigration of central and eastern Europeans in the 1880s and the closing of the gates to this migration with the passage in 1924 of the Johnson-Reed Act that restricted immigration by establishing national origins quotas. In 1916 the Bureau of the Census conducted a religious census, and the results of that canvass speak to the changing ethnic and religious composition of the anthracite region. Churches in Lackawanna, Luzerne, and Schuylkill counties—the three centers of anthracite mining— reported more than 520,000 church members in 1916, roughly two-thirds of the overall popula-

tion of the three counties. More than 66 percent of these church members were Roman Catholic or Eastern Orthodox, a remarkably high share. Protestants, in contrast, composed only slightly more than 30 percent of churchgoers.[56]

What is particularly striking for this period is the predominance of ethnic Catholic churches in the region. When eastern and southern European immigrants began entering the anthracite region in large numbers after 1880, they found a Catholic Church that was dominated by Irish Americans. Tensions prevailed between established church leaders and the newcomers, to be

St. Thomas Aquinas Church and towering culm bank, Archbald. Photograph (taken from an adjacent culm bank) courtesy of Ed Casey, Archbald, Pa.

sure, but the church leadership was generally responsive to immigrant needs and demands and permitted the organization of ethnic parishes. In turn, there emerged Polish, Lithuanian, Slovak, Slovenian, Italian, Tyrolese, Magyar, Lebanese, and Greek (Byzantine-rite) Catholic churches, for example, in the Scranton diocese. European-born clergy immigrated to the region, received assignments to the new parishes, and conducted services in the native languages of their parishioners.[57]

Conflicts abounded in this process, among ethnic groups and between particular groups and the Roman Catholic hierarchy. Poles and Lithuanians in Shenandoah struggled for control of St. Casimir Church in the 1870s. Two decades later parishioners at St. Mary's of the Assumption Church in Wilkes-Barre divided on whether to continue as a Greek Catholic Church or follow their pastor into the Russian Orthodox fold. Finally, in 1896, a group of Polish Catholics in South Scranton withdrew from their local church and founded a new parish that became the basis of the Polish National Catholic Church in the United States. Street battles and court challenges marked these conflicts. Lay trustees in parishes successfully fought off Church efforts to claim ownership of churches and rectories and control parish financial matters. Parishioners periodically opposed the appointment of a clergyman from a different ethnic group and made life so difficult for the newcomer that the bishop had little choice but to diplomatically transfer him to another parish. Slovak Catholics, Lithuanian Catholics, Polish Catholics, and Greek Catholics were determined to carry on their own particular European and ethnic traditions, either within the Roman Catholic Church or on their own. While there were some skirmishes, for the most part the Catholic Church in the United States offered a pluralistic response that succeeded in incorporating the newcomers into its ranks.[58]

The ethnic fraternal organizations and the ethnic churches also reinforced the class loyalties that emerged among immigrant miners in the anthracite region. The Pennsylvania Slovak Catholic Union (PSCU) was a strong supporter of the United Mine Workers of America and in 1909 provided the UMWA district leader, John Fahy, a platform at its annual convention from which to make a plea for members. Before the 1920s, ethnic lodges routinely expelled members who became strikebreakers. The GCU, whose membership in the anthracite region consisted almost entirely of mineworkers, took such a stance. In 1902 the GCU urged its lodges to raise funds to support the anthracite strike: "There are over a million people affected by the strike and are in need of financial help. A request is being made to the national officers to create a Strike Fund. We ask our fraternal brothers and sisters to establish strike fund collections at lodge meetings, dances, and weddings."[59] Class, ethnicity, and religion reinforced each other and reinforced the sense of solidarity among immigrant mineworkers and their children in the first decades of the twentieth century.

The labor activism of the newest immigrants from southern, central, and eastern Europe and their determination to build—sometimes contentiously—ethnically based community institutions did not translate readily into occupational success and mobility. Older stock miners continued to dominate the upper echelons of anthracite mine work. The survival of what appear to be complete employment records for the coal operations of the Lehigh Coal & Navigation Company between 1917 and the closing of the firm's underground mines in 1954 permits a unique view of the anthracite workforce at the peak of the industry's production, and particularly on the interplay of ethnicity and the employment experience.[60]

By the early 1920s, LC&N had extensive operations covering 8,600 acres in communities across the Panther Valley. The firm's facilities included the following: fourteen underground mines and two large stripping operations, spread over a distance of twelve miles from Nesquehoning in the northeast to Tamaqua in the southwest; an above-ground electric railroad system that transported coal from the mines to breakers in Nesquehoning, Lansford, Coaldale, Rahn, Greenwood, and Tamaqua and two additional collieries outside the valley at Alliance and Cranberry; two coal washeries at Ashton and Hauto; a large outdoor coal storage area at Hauto; a shop complex at Lansford for fabricating and repairing equipment for all of the mines; and sizable holdings of forest lands providing all the timbering required to support the under-

ground passageways in the mines.[61] In 1919, the
first full year after World War I, LC&N produced more than 4.7 million tons of coal, making it the sixth-largest anthracite producer at the
time. The year had been a good one for the
7,121 miners employed by LC&N as the company's various mines and breakers had operated
for 271 days on average.[62]

LC&N's surviving employment records provide a useful portrait of the workers employed
by the company on January 1, 1920. Employees
ranged in age from twelve to seventy-five and
had been working for the company on average
for more than thirteen years. The typical
mineworker was in his mid-thirties and was the
native-born son of an immigrant. Slightly more
than half were U.S.-born while central and eastern Europeans composed more than a third; immigrants from Italy, Great Britain, and Ireland
constituted the remaining 11 percent of the
workforce. The English and Irish had the most
experience of any group in the mines, having
worked on average more than twenty-five years
at the company, and they held the majority of
supervisory jobs. The native-born (likely sons of
immigrants) were the youngest group in the
mines, but still averaged more than fourteen
years of employment before 1920. Southern and
central Europeans were the newcomers in the
mines, averaging eight and twelve years respectively with the company.[63]

In 1920 slightly more than a third of LC&N's
employees worked outside the mines. More
than 350 engineers and firemen kept steam engines running for power in the breakers and underground. Almost 270 machinists fabricated
and repaired numerous pieces of equipment for
the mines. Another 260 boys and men worked as
slatepickers in company breakers, sorting slate
and coal. Although slatepickers comprised less
than 4 percent of the workforce, almost one out
of every three employees in 1919 had begun
working at the company as a slatepicker. General labor was the second most common initial
occupation. Newcomers typically found their
first employment as slatepickers or laborers and
over time moved through a graded series of jobs
as door tenders, mule drivers, loaders, and finally contract miners, the most common occupation in the mines. Almost 2,000 men worked

for the company as contact miners, more than
27 percent of the overall workforce in 1919.

Experience played a part in one's job assignment, but ethnicity was also a determining influence. In 1920, mineworkers from the United
Kingdom still comprised the elite of the workforce at LC&N. Forty-six percent of the British-
born were contract miners, and 17 percent held
supervisory positions. British and native-born
workers dominated in skilled trade positions
around the mines, comprising more than two-
thirds of the carpenters, masons, blacksmiths,
motormen, locomotive engineers and firemen,
machinists, and shopmen. Laborers comprised
less than 22 percent of the English, Welsh, and
Irish mineworkers, the lowest proportion among
ethnic groups. The British had the longest
tenure in the mines—having begun working for
the LC&N before 1890 on average—and high
occupational status accompanied that longevity.

Comparisons with newcomers from central
and eastern Europe are telling, particularly with
regard to managerial appointments. In 1920, 42
percent of them were contract miners and 25
percent laborers; less than 1 percent held supervisory positions at LC&N. Among the recent arrivals, Slavic workers ultimately fared better
than Italians. In 1920, for example, only 4 percent of the Italian mineworkers in the firm were
contract miners, compared to 42 percent of
eastern Europeans. Immigrants from central
Europe also enjoyed greater occupational mobility than did Italians. Their careers at LC&N
averaged more than twenty-three years, and by
the end of the working lives at the company, 27
percent remained unskilled laborers. For the
Italians who stayed, close to 42 percent held
low-skilled positions when they retired; equally
telling, only 2 of the 56 Italians working for the
company in 1920 ever enjoyed the higher status
and income of contract miners. Ethnic structuring of the workforce had been a feature of anthracite mining during its early development
and little had changed by 1920.

Another stark reality of anthracite mining did
not abate during the prosperous first two
decades of the twentieth century. The threat and
actuality of accidental deaths and injuries in and
around the mines still weighed heavily in anthracite communities—in fact, the number and

rate of job-related fatalities rose between 1900 and the end of World War I. The last three decades of the nineteenth century saw the rate of accidental deaths decline from upwards of 6 per 1,000 employees annually to 3 and below, largely through government enforcement of safety regulations. Between 1901 and 1920, 11,621 anthracite mineworkers lost their lives in mining accidents, an average of 581 per year or 3.54 deaths per 1,000 employees annually. Anthracite miners enjoyed steady and full employment and welcome gains in real income in the new century, but as they looked among themselves, the harsh reality was that 1 in 282 died on average each year from injuries sustained on the job.[64]

The notable rise in mine fatalities between 1900 and 1920 was due to a number of factors. Increased production and mining at deeper levels (with greater ventilation problems) took their toll, but new technologies also played a role. The period witnessed significant increases in fatal accidents suffered by surface workers. More dangerous equipment in breakers, electrical hazards, particularly near rail systems, and boiler explosions all made above-ground work more dangerous than before. Downward trends in accidental injuries and deaths in anthracite mining would be reestablished after 1920 with renewed attention to safety standards and education.[65]

The employment records of the Lehigh Coal & Navigation Company detail the continuing perils of mine work in the anthracite region. In the district where the operations of the LC&N were concentrated, twenty-eight mineworkers died on the job in 1919, one death for every 226 employees.[66] Comparing the fatality figures for LC&N with the region as a whole, and 1919 with surrounding years, indicates that 1919 at the LC&N was a particularly dangerous work year. Still, even taking those differences into account and remembering that the typical mineworker employed at that date would be employed almost twenty-seven years at LC&N, the chances were one in eleven that he would suffer a fatal injury while working for the company—a frightening probability.[67]

Fatal accidents, of course, were only the most serious of the mishaps that occurred in the mines. By the 1930s, the Bureau of Mines tabu-

lated work injuries in coal mining and found that nonfatal injuries outnumbered fatal ones by roughly fifty to one. If 9 percent of anthracite miners employed by LC&N were likely to die from a work accident over the course of their careers, virtually all miners could count on being injured at some point during their working lives; many were probably injured several times.[68]

Those who successfully navigated the dangers of mine work still had to cope with black lung when their working days were behind them. *Black lung,* more commonly known as miners' asthma, was a chronic respiratory disease brought on by lengthy exposure to coal dust at the workplace. Breaker boys, miners, and loaders were regularly exposed to fine particles of coal dust as they worked. When the photographer Lewis Hine was sent by the National Child Labor Committee to record the conditions of work for young anthracite mineworkers in 1911, he found the coal dust in the breakers so thick that he could not take clear photographs.[69] While the accumulation of coal dust in one's lungs did not kill immediately, like a roof fall or an explosion, it did eventually sap miners' strength and left many middle-aged miners unable to continue to work underground. "Twice a boy" was a common expression around the mines that spoke to the practice of experienced miners returning to the breaker in old age when they could no longer manage the physical exertion of underground work.[70] Public Health Service studies in the 1920s and 1930s found that roughly 20–30 percent of employed miners suffered serious disabilities due to black lung.[71] LC&N employment records make no references to black lung, but more than 4 percent of mineworkers left the company with a final remark in their records noting illness, while another 19 percent died while on the job from unspecified causes. The remark on Edward Kissner's employment record when he finally retired was typical for many—"no light work," it noted. Kissner, who had begun working at age ten as a slate-picker in 1887, retired as one sixty-five years later. Truly, he'd been "twice a boy."

The number of mineworkers at LC&N in 1920 who died from accidents or unspecified causes while working far exceeded the number who formally retired or received a pension as their careers in the mines came to an end.[72] Company pensions were available in this period, granted on a case-by-case basis by LC&N's Board of Managers, but given the size of the company's workforce, the number of pensioners was extremely modest. In 1920, only forty-five former employees received support from the company, with pensions that averaged $26 a month. Between 1918 and 1922 on average the company added seven new retirees to its rolls each year, and the size of the overall group of pensioners grew from forty to fifty.[73] Neither company aid to accident victims nor pension support provided anthracite miners much security against the perils of work or illness in old age. Government assistance, moreover, remained limited. State miners' hospitals founded in the late nineteenth and early twentieth centuries, for example, refused as a matter of policy to admit men with incurable diseases, specifically miners' asthma. The Commonwealth of Pennsylvania established a workmen's compensation system in 1915, but black lung was not covered.[74] A government-administered pension plan awaited federal legislation in the 1930s. With little institutional support, mineworkers and their families continued to rely heavily in cases of accidental mine injuries and fatalities and old age on aid from the ethnically based mutual benefit societies they built and joined.

Certain facets of life remained constant in the anthracite region—varied communities, ethnic divisions, and perilous work—but the first two decades of the twentieth century also represented a new stage in the area's history in important respects—with the strength of the anthracite cartel, a détente in industrial relations, steady increases in coal production, relatively full, regularized employment, an ending to child labor, gains in real income, and the distinctive presence of the newest immigrants. The great progress of the era was visibly manifest during World War I as the region buzzed with activity.

Production of anthracite coal leaped to all-time highs in 1917 and 1918 with the demands of the wartime mobilization. With the drafting of young adult mineworkers, labor shortages threatened to curtail production. Newspaper articles reported on the decisions of coal companies to hire women in clerical positions and even

briefly to consider employing women in the breakers. The employment of women in and around the mines previously had been taboo as their presence near operations was held to bring bad luck.[75] The Delaware & Hudson Railroad in an unusual move transported sixty Mexican workers to replace conscripted employees (which "created a stir" according to the *Scranton Times*).[76] These management moves reflected the tight labor market in the mining region, which accounted for the fact that President Woodrow Wilson received various petitions from the area asking him to institute deferments for anthracite mineworkers.[77]

The war did not soften the will of anthracite miners to improve the terms and conditions of their employment, and the inadequacy of wage settlements in the face of a rising cost of living divided rank-and-file mineworkers, district leaders, and UMWA officials in Washington. During the wartime emergency, the mines were formally under the authority of the federal fuel administrator, Harry A. Garfield. The UMWA and insurgent groups of anthracite miners appealed to Garfield for a wage increase in September 1918 to counter the skyrocketing cost of living. Some 25,000 workers in District 9 went out on a brief wildcat strike to support their demands, which were met in part by a wage adjustment implemented in October. District 9 mineworkers, dissatisfied with their district leadership, voted out the incumbent president and replaced him with insurgent leader Christ Golden.[78]

It proved difficult, though, to mount a serious labor struggle while U.S. troops overseas were fighting to make the world safe for democracy. UMWA members in Districts 1 and 7 refused to support the insurgent strike in District 9, preferring to bring pressure in less confrontational ways. Patriotic demonstrations, rather than insurgent strikes, were more characteristic displays of mineworkers during the war. Residents of the anthracite communities remained supportive of the war and manifested their loyalty in Liberty Bond and home garden campaigns. Typical was a patriotic rally held on a Sunday evening in February 1918 in the auditorium of the Lansford High School, which attracted a crowd from across the Panther Valley. The local Slovenski League organized the event.[79] With the mines operating fully and community activ-

ity at its height, few in the Pennsylvania anthracite region could imagine at World War I's
end that the beginnings of the long-term economic collapse of the area lay just ahead.

◆

In 1878, William W. Scranton, chief executive of the Lackawanna Iron and Coal Company,
delivered a sober address at a banquet of local
businessmen in the city that bore his family's
name. Scranton predicted that anthracite coal
would soon lose its market to other fuels and
that the city would survive only through industrial diversification. Attending the event were
members of the local board of trade who already
were aware that coke, produced from bituminous coal, was rapidly replacing anthracite in
iron and steel production.[80] Challenges could
easily be raised to Scranton's pessimistic projections. After all, production of anthracite was
ever increasing, and in the next forty years annual production jumped from approximately 25
million tons to 100 million. Scranton could not
have possibly known how right he was given the
state of statistics at the time, but as he spoke the
downward shift in the proportion that anthracite
held in total fuel consumption had begun. Bituminous dominated already in the late nineteenth century, accounting for close to 60 percent of fuel use, with anthracite near 30 percent.
Yet, by the time of anthracite's greatest year in
production, 1917, hard coal provided only about
13 percent of all energy consumed. Bituminous
more than held its own at nearly 69 percent of
energy consumption, but petroleum and natural
gas had already made serious inroads into anthracite's overall share.[81]

Loss in the share of the fuel market did not
portend disaster for the residents of the anthracite region as long as production, employment, and earnings continued to grow. After
World War I, however, a new history commenced. As anthracite's share of fuel consumption continued to decline, production of anthracite also began to plummet. Ten years after
war's end, 1928, annual production had dropped
to approximately 75 million tons; twenty years
after, to 46 million, less than half of peak production.[82]

There are two ways to approach the causes of
the downfall of the anthracite trade. The first is

to concentrate on events of the 1920s, the
turning-point decade. Intense labor strife—
once again attracting national attention and requiring the intervention of national political
leaders—returned, and a series of disruptive
strikes in the 1920s led customers of anthracite
coal to seek out alternative fuels. Yet a perspective that concentrates on precipitating events is
insufficient. A long-range view is necessary as
well. Moreover, the strikes of the 1920s are important less for what they reveal about the
causes of the collapse of anthracite as for the
story they tell of institutional failure, of the failure of organized interests—the government, the
coal operators, and the UMWA—to shape a response to the inevitable decline of production
that would best serve the people of the region.

Four issues are important in understanding
the long-term fate of anthracite. The first involves the quantity and availability of the resource. The anthracite trade did not collapse for
want of coal to mine. A geological study published in 1926 estimated that 21 billion tons of
anthracite had been formed under the surface of
northeastern Pennsylvania eons ago. Of that,
only 4.4 billion had been removed since the inception of mining at the beginning of the nineteenth century.[83] The problem was not the
quantity of anthracite reserves but the location
of the remaining seams. Depletion had occurred
mostly in the northern coalfields where anthracite was most accessible. At the rate coal had
been mined in that area, geologists estimated
that perhaps another forty-eight years of production could be sustained. A slightly more optimistic outlook held for the eastern and western
middle fields, which originally contained the
smallest deposits of anthracite. According to the
study, another 37 and 124 years of mining at recent levels of output would exhaust reserves in
the eastern and western middle fields, respectively. About 50 percent of all of the remaining
hard coal reserves lay below the surface of the
southern coalfield, which at recent rates of production would supply anthracite for another 408
years. Deep, relatively narrow, pitched veins of
coal, however, made its extraction and cleaning
significantly more costly in the southern than in
the other fields.[84]

A second, and related, matter involves the
market for coal. Anthracite had two different

markets: "domestic" and "steam." *Domestic coal* referred to the relatively large pieces of anthracite (so-called lump, egg, stove, chestnut, and pea—metaphors for size) used in home furnace and stove heating, initially in iron and steel production, and also to power locomotives. *Steam coal* (so-called buckwheat) was used in large boilers to heat apartment and office buildings, for generating steam in electric utility plants, and in some chemical and industrial processing.[85]

Domestic coal always dominated the anthracite trade, comprising from 60 to 80 percent of the market, and for good reason.[86] Larger pieces of coal, mined with little difficulty and entailing less breakage or slatepicking, had low costs of production. Domestic coal, at least until the 1920s, also had the least competition (clean-burning anthracite was much preferred in domestic heating to smoke-producing bituminous). Domestic coal fetched a much higher price per ton than steam and cost less to produce; it was the profitable side of the trade and before 1920 there had been little incentive to shift from its production to steam coal.[87]

With domestic coal soon to face stiff competition from cheaper and more convenient fuels for home heating—oil and natural gas—the future of the industry appeared to be in fine buckwheat coal for electric power generation.[88] Yet the long-horizons were dim here. The costs of producing steam coal in the western middle and southern field—with the exhaustion of reserves anticipated elsewhere—would be high, especially with the need for screening and cleaning. More efficiently produced bituminous would always underbid anthracite in the steam coal market. And, in addition, waterpower was becoming a more serious competitor with the increase in hydroelectric power.[89] The anthracite trade was between a rock and a hard place; increasing competition in both home heating and electric power markets and increasing costs of production made the industry's future uncertain.

The difficulty of mechanization represented a third factor in the decline of anthracite. The orientation of coal veins in the anthracite region limited mechanical extraction, particularly in the southern coalfields. As late as the 1940s, only 5 percent of the coal mined in the anthracite region was mechanically cut; in bituminous areas,

90 percent of the extraction of coal involved machines. Similarly, in bituminous, 73 percent of the loading and moving of coal was handled mechanically—in anthracite, only 41 percent.[90] Anthracite mine work remained relatively labor intensive with labor costs representing upwards of 75 percent of total production costs.[91] The geology of the region precluded major substitutions of capital for labor and limited cost savings through mechanization. In the midst of the market decline, enormous mechanical shovels, tearing through mountains, began to replace human labor. During the decade of the 1940s, for example, the proportion of anthracite coal extracted through strip mining grew from 10 to 25 percent. Even with strip mining's cost savings, total production continued to decline and dipped below 30 million tons in 1953.[92]

A final element in the ultimate fate of anthracite involved the geography of its market. Anthracite began as and remained a regional fuel. Upwards of 70 percent of all anthracite was sold and consumed in the Middle Atlantic states in the late nineteenth century and into the twentieth.[93] New England states represented the next greatest market, absorbing 15 percent of the coal mined in northeastern Pennsylvania. Between 5 and 8 percent of anthracite was exported, primarily to Canada. The rest of the coal was shipped westward, but the share of coal sold in states west of Pennsylvania fell from 15 percent to 5 percent between 1915 and 1930. Transportation costs (and weather conditions) shaped the geography of the anthracite market. The farther the distance, the greater the availability of competing fuels, and the warmer the climate, the fewer the customers for hard coal. As the population of the nation moved steadily westward in the twentieth century, and economic development in other regions outpaced that in the industrial Northeast, anthracite's contribution to fuel use was bound to decline.

Things did not bode well for the anthracite trade in the long run. However, the decline was not linear. Various twists and turns occurred over the course of the descent, different events in each decade affecting the process. The Great Depression of the 1930s and World War II (with its revival of production) had distinct impacts. During the 1920s, labor conflict shaped the first stages of anthracite's collapse.

North End, Boston Stripping, Loree Colliery, Larksville, Luzerne County, 1921. Photograph by John Horgan Jr. Courtesy of the Pennsylvania Historical and Museum Commission, Bureau of Historic Sites and Museums, Anthracite Museum Complex, Scranton.

After World War I labor problems began with the failure of the UMWA and anthracite operators to renegotiate the 1918 contract when it expired at the end of March 1920. Seeking to avoid a costly interruption of production, both sides agreed to extend the current contract while negotiating their differences. With negotiations deadlocked, representatives of anthracite locals, meeting at a convention of the union's three anthracite districts in Wilkes-Barre in late May, accepted President Wilson's offer to appoint a federal commission to study the anthracite industry and propose an arbitrated settlement. The 1902 formula would be tried again.[94]

The U.S. Anthracite Coal Commission of 1920 began meeting in Scranton on June 24 and heard testimony from representatives of the operators and the union.[95] About two months later, the Commission made its final report to the president. Anticipating its final recommendation, early in the summer the Commission made a fateful decision not to admit into the hearings union evidence on the monopoly structure of the industry. The UMWA had hired W. Jett Lauck, an economist and independent consultant, to present the case, and Lauck was determined to have full disclosures on the profit making of the anthracite coal operators and particularly on the excessive earnings obtained in railroad shipping.[96] Narrowing the grounds on which their final decision would be based, the Commission refused to consider the UMWA studies and ultimately recommended a 17 percent wage increase for miners, a figure well below a recent 27 percent wage award for bituminous mineworkers and far less than the 60 percent increase the UMWA had first demanded to keep pace with the 70 percent escalation in living costs in the immediate postwar period. The Commission also rejected union demands for a closed shop, dues checkoff, and extra pay for mandatory overtime.[97]

The Commission's settlement was written into an agreement that went into effect in September, but dissident rank-and-file miners refused to accept the arbitration award. Wildcat strikes, opposed by UMWA leadership, broke out across the anthracite region as 100,000 miners went on "vacation" demanding the overturning of the Commission's award and the reopening of wage negotiations with operators. The walkouts were highly effective at first, but without union support (and accompanying strike benefits) and facing unified federal government and operator opposition, miners began returning to work.[98] John L. Lewis now led the UMWA. Lewis would later capture national attention as a defiant labor leader, but in his first years as UMWA president he revealed his basic conservatism by counseling restraint and respectful behavior—in the same manner as his equally notable predecessor, John Mitchell.[99]

By early October 1920, the mines were operating at full tilt and President Wilson called for a reopening of wage negotiations. However, it took the agreement of both sides to revise the Commission's wage settlement, and operators adamantly opposed any change. They were willing to "negotiate," or at least to maintain the appearance of negotiating, but their position did not budge over the two-year contract period, and mineworkers' wages continued to be pegged at the low level set by the presidential Commission.[100]

The anthracite operators clearly won this round of labor conflict. They had raised coal prices in April and added a second price increase in September following the Commission's award. More important, they succeeded in avoiding scrutiny of their collusive arrangements in allocating production quotas among firms and setting high rates for the transportation of anthracite that raised coal prices to consumers and discouraged the efforts of independent firms to compete in the industry. The 1920 Coal Commission maintained practices followed since 1902 that kept federal intervention or regulation of anthracite to a minimum. Miners seethed and consumers suffered while the power of organized operators went unchecked.

In the early 1920s, the anthracite coal executives faced down union pressure and rising public calls for full disclosure of their finances, government regulation of the industry, and even nationalization of the mines.[101] Federal law and judicial rulings aimed at reestablishing competition and lowering prices for consumers also

failed to loosen cartel arrangements. In 1906, Congress had passed the Hepburn Act, legislation adding teeth to the federal government's prosecution of the monopolistic practices of rail carriers engaged in interstate commerce. The Act explicitly prohibited the transporting of coal by railroad companies that had any "interest, direct or indirect" in the mining of coal.[102] In the next decade and a half, the Supreme Court in evolving rulings forced the separation of the anthracite railroads from ownership of coalfields.[103] Initial decisions upheld the constitutionality of the Hepburn Act, forbade direct ownership of coal lands by the railroads, but at the same time undercut the ruling by allowing the carriers to own stock in separate coal mine and coal sales companies. Ultimately the federal government did require the anthracite railroads to dispose of their coal operations, but even in these last instances, railroad owners purchased the lion's share of the newly "independent" firms. This fictitious separation of coal production and transportation made no difference in management operations. The community of interest established earlier among the anthracite railroads remained undisturbed. Thus the formal monopoly power of the anthracite railroads was broken while their effective power in the industry remained unchecked.[104]

The separation in ownership of coal production from shipping and sales harkened back to state incorporation charters of the antebellum period that forbade rail carriers from owning coalfields—from fear of monopolization. The disaggregating of the 1920s—through antitrust legislation and judicial rulings—however, did not lead to a deconcentration in holdings in either the mining or transportation of anthracite. For example, the Delaware, Lackawanna & Western Railroad sold its coal lands to the Glen Alden Coal Company; the Philadelphia & Reading Coal & Iron Company (P&RC&I) received the fields owned by the parent Reading Railroad.[105] Thirty years after the formal separation of transportation and production, the market shares of the leading coal producers remained as concentrated as ever. The Glen Alden and the P&RC&I companies still accounted for more than 25 percent of the anthracite coal mined in the 1950s, and the leading eight mining concerns produced 60 percent of the total.[106] The major coal producers remained in a position to collaborate in preventing market gluts and falling prices.[107] Moreover, a community of interest—with banking interests still providing the glue—persisted among the rail carriers (now divested of their coal lands). They too maintained agreements as to the proportionate shares of coal each would transport to market.[108] The anthracite coal industry came under political and legal attack in the early 1920s, yet consumers continued to face a powerful combine. So did anthracite mineworkers.

Rounds two and three in the labor battles of the 1920s were fought in 1922 and 1923. The settlement imposed by the U.S. Coal Commission of 1920 held until April 1, 1922. Talks began two weeks earlier, and the coal operators immediately proposed a 21.5 percent reduction in wages, claiming hard times for the industry. The UMWA countered with a demand for a 20 percent increase, an eight-hour day, and, once again, union recognition and a dues checkoff system to stabilize the finances of the union. With no grounds for compromise, anthracite mineworkers walked off their jobs on April 1, this time joined by their fellow miners in the bituminous trade, whose contract had also expired. Thus began the first simultaneous strike of anthracite and bituminous workers in U.S. history, a job suspension involving more than 600,000 men.[109]

Once again, general cries arose for action to resolve labor conflict in anthracite. Senator William Borah of Idaho proposed congressional legislation that would have placed ownership of the mines in public hands; others called for greater government regulation of prices, profits, and wages. For the first time, officials in northeastern cities advised localities and consumers to shift to alternative fuel sources for heating. In response to rising concerns, the presidential administration of Warren G. Harding initially announced a hands-off policy.[110]

By July, without an end in sight, Harding had to intervene. The president in the summer of 1922 not only faced the shutdown of the hard and soft coal industries but also a national strike of railroad shop workers that threatened the commerce of the entire country; later this strike would also require his reluctant intervention.[111] To end the dispute in anthracite, Harding of-

fered to create a Federal Coal Commission to investigate the coal trade and arbitrate a settlement. The anthracite coal operators accepted the plan; the UMWA, burned by the most recent commission, refused. John L. Lewis faced less rank-and-file dissidence in this latest imbroglio. With the exception of one major disturbance in the town of Luzerne, where a mob of two hundred strikers stoned firemen who had stayed on the job in the event of spontaneous fires in the mines, and reports of IWW organizers egging on resistance, the strike of summer 1922 passed peacefully in the anthracite region, although violence did mark bituminous coal regions.[112]

The anthracite strike of 1922 dragged on into early September. With rising talk of a government takeover of the mines, both sides bowed to pressure of top Pennsylvania elected officials to reach an accord. On September 11, after a 163-day closure of the mines, anthracite workers returned to their jobs as both parties agreed to maintain the status quo, no wage reductions or increases, until August 31, 1923. On September 22, Congress passed a bill to create a new U.S. Coal Commission to investigate the coal industry and suggest measure to quell disruptions.[113]

Warren G. Harding died in office in early August of 1923, and his successor, Calvin Coolidge, immediately encountered the challenge of another anthracite coal strike at the end of the month.[114] Coolidge also faced competition from political rivals in his party, most notably Gifford Pinchot, the governor of Pennsylvania, and he needed to demonstrate leadership as he assumed the presidency. Coolidge, however, reacted passively, accepting the counsel of advisors merely to empower the U.S. Coal Commission to mediate the dispute as the expiration of the 1922 settlement neared. Pinchot then entered the breach, ostensibly acting on behalf of the president. He suggested grounds for a potential pact between the UMWA and the anthracite coal operators that included a 10 percent wage increase, a uniform eight-hour day, union recognition, and a compromise arrangement on dues checkoff (now the issue most bitterly dividing the parties). His proposal did not prevent a strike that commenced on September 1, 1923, but within a week's time, he achieved an accord with the sides agreeing to Pinchot's recommendations just on wages and hours. With all eyes in the nation once again on the Pennsylvania anthracite region, Pinchot hoped that his intervention would catapult him to the presidential nomination of his party and the presidency. His seeming upstaging of Coolidge, however, lost him support within the Republican Party, and as the anthracite coal operators remained free to raise prices to meet the expense of the new agreement, he lost favor among the public as well.[115]

Pinchot failed to advance his political career, but his intervention did lead to two years of labor peace in the anthracite region during which time anthracite production averaged just over 90 million tons annually.[116] The last great battle of the decade then ensued on September 1, 1925, when the pact engineered by Pinchot expired. Once again, the UMWA entered negotiations with demands for a 10 percent increase in wages and a dues checkoff system. True to form, the coal operators summarily rejected the union's claims. With no bases for bargaining, anthracite mineworkers walked off their jobs, closing the mines for 170 days until a settlement was reached on February 12, 1926. Exhausted, both sides concurred on a five-year contract that left the old wage scale in place, but provided for binding arbitration on wage adjustments if both parties agreed. The agreement included no provisions for dues checkoff.[117]

An observer of the anthracite coal strike of fall and winter 1925–26 must have felt a measure of déjà vu. President Coolidge refused to intervene and remained offstage, leaving a political resolution to congressional leaders and Pennsylvania politicians. Congress considered various measures to require presidential intervention and to establish new commissions and regulatory agencies, but took no action. Pinchot assumed the most active role. In late November, he proposed a blueprint for a five-year contract that included three key provisions: (1) no increases in coal prices; (2) the creation of a Board of Investigation and Award to determine wage scales with the Board having total access to the financial records of the coal operators; and (3) a modified dues checkoff system. The UMWA welcomed Pinchot's plan and the reopening of negotiations. The coal companies voiced immediate and strong opposition. Failing to mediate, Pinchot then pursued legislative action, specifi-

cally submitting a bill to the Pennsylvania legislature to form a state regulatory commission with powers effectively to control production, prices, and employment standards in the anthracite industry. Lobbyists for the anthracite operators ensured that Pinchot's reform measure never left committee. Ultimately, the strike ended not through the interventions of outsiders but because the worn-down parties needed to end the stalemate.

The strike of 1925–26 unfolded without physical confrontations between strikers, strikebreakers, and the police—a departure from previous walkouts. The allegiances of rank-and-file miners in the anthracite region to the union had been shaky from the outset, but in this latest five-month strike, the leadership of the UMWA appeared secure. John L. Lewis at the time did face challenges to his presidency as union members with Communist Party sympathies and connections formed a "Save the Union Movement" to unseat him. Communist organizers had little presence, however, in the anthracite region at least until 1928. They began openly campaigning in the area that year and formed a branch of a rival union to the UMWA, the National Miners Union (NMU), at the same time that they launched another dual union, the National Textile Workers Union among anthracite-region silk workers. The NMU never gained a foothold in anthracite, but individual Communist activists played roles in regional unemployment and union protests of the 1930s.[118]

Violence did not mark the anthracite coal strike of 1925–26, but the walkout saw heightened concerns about disruptions of coal production. This strike occurred during wintertime, and shortages of coal for home heating had immediate impact. Public officials, particularly in New England, held conferences to discuss alternative fuel possibilities and led campaigns to encourage substitutions. The governor of Massachusetts symbolically removed tons of anthracite from his cellar and replaced it with bituminous coal.[119] Others must have followed his example because state authorities in Pennsylvania estimated that during that winter, regular users of anthracite consumed at least 17 million tons of soft coal.[120] Other studies indicated that conversions to oil burners effectively replaced 5 million tons of hard coal. The actual number of these early conversions is not readily determined, but in the immediate five-year period following the 1925–26 strike, oil use in home heating more than doubled in New England and increased by more than 60 percent in the Middle Atlantic states.[121] The decline of anthracite had much deeper roots than the resumption of intense labor conflict to the region in the 1920s, but disruptions in production accompanying the strikes of 1920, 1922, 1923, and 1925–26 certainly contributed to and hastened the process.

During the first quarter of the twentieth century, the eyes of the nation repeatedly focused on the Pennsylvania anthracite region. A broken-record of strikes from 1900 to 1926 drew the attention and required the intervention of national political, civic, and business leaders. The careers of top political figures often turned on their resolving of labor disputes in the anthracite coal industry. The strikes mattered far beyond the five-county region of northeastern Pennsylvania because of their impact on consumers throughout the populous Northeast. Yet the strikes of the 1920s involved more than the purchasers of anthracite; the fates of the citizens of the region were wrapped up in the greater response. In this respect, organized interests failed them.

The anthracite cartel, though altered, remained imperial. Anthracite coal operators bowed grudgingly and ever so slightly to the demands of mineworkers. They refused to reveal the nature of their finances and simply passed on the costs of labor pacts to consumers. They remained obstinate on the matter of union dues checkoff. Not only did they never formally recognize the UMWA, they were not about to ensure that the union had financial resources to operate. Had they compromised on this issue, strikes would have been avoided or at least shortened. Operators also strongly resisted government regulation, although they succeeded in working binding arbitration to their advantage.

The foil to the operators, the UMWA, remained in cautious hands, the leadership concerned principally with the survival of the organization. The union remained a weak party in dealings with anthracite coal executives and never had a certain hold on the allegiance or actions of the ranks of anthracite mineworkers. The union leadership also held to a principle of voluntarism, believing that solid gains were to be made through the collective strength of the membership, not through government intercessions. The UMWA thus opposed state regulation as determinedly as did the coal operators.

Political leaders acted in varied ways. Some adopted hands-off policies. Others sought to use the state either for investigation, mediation, arbitration, or regulation; public takeover of the mines had momentary support. The political foresight or will, though, did not exist to plan for the anthracite region's future—to develop a blueprint that would have anticipated sustainable levels of production and sales of anthracite, appropriate returns to both capital and labor, and perhaps even planning for a phasing-out for mining in the region with effective initiatives at economic diversification and assistance to the displaced.

The Pennsylvania anthracite region received the nation's notice during the first quarter of the twentieth century, but the basic concern was for the consumers of anthracite not for the region itself. Unfortunately, the area would never again draw such national consideration. The strikes of the 1920s highlight the failures of interested parties—anthracite operators, the UMWA, and government—to work together to solve the region's first encounter with economic decline. In the absence of a coordinated response to the initial decline of the anthracite trade, anthracite mineworkers and their community supporters would soon take their fates into their own hands—operating both outside and within existing institutions. They did so as the rest of the nation joined them in facing economic crisis.

CHAPTER 3

The Anthracite Miners' New Deal
The Thirties

ON TUESDAY, OCTOBER 29, 1929, stock prices on the New York Stock Exchange plummeted sharply, capping a five-day panic that constituted the worst stock market crash in U.S. history. A record 16.4 million shares changed hands that day, with the Dow Jones industrial index dropping to 230, down by nearly 100 points or 30 percent in value since the start of the sell-off, and with all indications that the New York financial community no longer had the reserves to buoy up and rescue the market.[1] In the weeks immediately before and after so-called Black Tuesday, stocks declined by more than $26 billion in worth.[2] The great stock market crash of late October 1929 was not the sole cause of the ensuing decade-long general collapse of the U.S. economy, but it did mark the beginning of the Great Depression. From late October of 1929 to the very depths of the contraction in 1933, the gross national product in constant dollars fell by 31 percent, industrial production by 50 percent, and unemployment affected 25 percent of the U.S. workforce, more than 13 million people.[3]

The inadequate response of President Herbert Hoover to the crisis led to the election of Franklin Delano Roosevelt (FDR) in November 1932. Once in office, Roosevelt initiated and supported an explosion of legislation aimed at economic recovery and regulation and the creation of a social safety net based on a variety of government support programs—in short, FDR's New Deal.[4]

The 1930s represented a sea change for U.S. workers. With the passage of the National Labor Relations Act in 1935, the country's wage earners finally received federal legal protections to form unions of their choice and have those unions represent them in collective bargaining. American workers during the historic decade of the thirties also participated in the building of unions in the nation's mass-production industries, under the aegis of the newly formed Congress of Industrial Organizations (CIO). Led by John L. Lewis of the United Mine Workers of America (UMWA), the CIO stood for the organizing of workers on an industrial not a craft basis—part of the tradition of unionism in coal mining dating back to the Workingmen's Benevolent Association. The CIO attracted workers by and large not touched by the mainstream American Federation of Labor: the semi- and unskilled, immigrants, African Americans, and women. Within a remarkably short time, such CIO affiliates as the United Automobile Workers, the United Steel Workers of America, the United Rubber Workers, the United Packinghouse Workers of America, and the International Longshoremen's and Warehousemen's Union established strong presences in their respective industries.[5]

The labor history of the 1930s is the story of the CIO, the building of large-scale national labor organizations in the mass-production industries, and the shifting of labor conflict from the shop floor and the streets to institutional settings (normally, hotel rooms where corporate and union bureaucrats with government mediators negotiated contracts). A different kind of labor history unfolded in the Pennsylvania anthracite region during the 1930s. In response to the now deepening collapse of the anthracite coal market, mineworkers engaged in unique protest and survival activities during the Great Depression—operating primarily at the community level, both within and outside their union, and both within and outside the industrial relations regime established by the New Deal. Their efforts included dramatic campaigns for the equalization of work, a violent insurgency within the UMWA, and coal bootlegging. Behind these varied strategies stood common core beliefs and values that asserted the necessity of collective action rather than individual responses.

The alternative strategies adopted by the anthracite miners emerged within three distinct areas of the Pennsylvania anthracite region and are respectively illustrated in case studies of the economic circumstances, management practices, and labor responses at three leading anthracite firms, the Lehigh Coal & Navigation Company (LC&N), the Glen Alden Coal Company, and the Philadelphia & Reading Coal & Iron Company (P&RC&I). This history begins in the 1920s with the responses of these mine operators to the initial descent of the anthracite trade.

◆

Equalization of work campaigns erupted across the anthracite region with the onset of the Great Depression, but they had their greatest impact in the Panther Valley, home to LC&N at the eastern end of the southern field. In the 1920s, Jessy B. Warriner, the general manager of LC&N, filed reports from his office in Lansford with the president of the company, his cousin and fellow mining engineer, Samuel D. Warriner, who resided in Philadelphia. The accounts offer an inside view on the company's operations and management's concerns in the unfolding business crisis.

Although strikes in the first half of the 1920s led to wide annual fluctuations in operating hours and output, a sense of optimism emerges from J. B. Warriner's annual reports. Operations were suspended briefly in September 1923 until a settlement brokered by Pennsylvania governor Gifford Pinchot brought the miners back to work; still, the year as a whole was a good one for LC&N. Commercial output exceeded 4 million tons, making it the company's best year since 1918. Profits of $3.5 million exceeded levels during the war years. The tone of the report reflected the good bottom line: market demand was "very strong," and operations "nearly continuous." Productivity "was at a new high," and unit costs "were quite low." Future prospects were strong, and a "large program of development and improvement work was carried on."[6] At each of the firm's large collieries, working time approached 2,200 hours for the year. In addition, fully 15 percent of total output came from the reprocessing of earlier waste dumped on culm banks across the Panther Valley. Annual turnover was significant, as the company lost more than 2,500 employees to resignation, discharge, or suspension. Still, labor was plentiful and the firm hired more than 2,800 workers, and the overall workforce grew by 273 over the course of the year. The conclusion of the 1923 report was guardedly optimistic. J. B. Warriner informed LC&N managers in Philadelphia, "Granted reasonable freedom, however, from explosions, fires, and similar catastrophes, and a market demand that will keep the collieries in steady operation, results for the next few years should be at least fairly satisfactory."[7]

As the decade progressed, the company made notable strides in productivity and safety but could not control market demand for anthracite. In his 1927 annual report, J. B. Warriner noted that the year "was marked by unfavorable market conditions," a judgment that he repeated the following year as well. In both years, the company suspended operations and laid off mineworkers for significant periods of time. The 1928 report noted that closures averaged sixty-four days over all of the company's collieries, more than 20 percent of possible operating days, costing the company some 728,000 tons of anthracite coal. These suspensions raised operating costs despite management's best efforts. Warriner's frustration was apparent in the conclusion to his 1927 report: "The collieries are in a condition to operate at a very low cost under normal conditions, but it is difficult to maintain efficiency when operation is intermittent." In other words, for two years before the onset of the Depression, unfavorable market conditions adversely affected LC&N operations, cost mineworkers 20 percent of potential earnings, and significantly cut into the company's profits. The Depression, of course, only added to the company's and the industry's woes and led Panther Valley miners to propose drastic solutions to the problems they faced.[8]

The general manager's argument here was that inadequate demand contributed to rising costs and declining profits—his and his fellow managers' primary concern. But even in his accounts to the company president, the consequences of the suspensions on mineworkers are evident. J. B. Warriner wrote about the laid-off workers in his 1928 report: "These men were fitted into existing vacancies at other collieries, but the greater number were without employment until work was resumed at No. 5 and No. 6 operations in November. Many of the No. 4 employes are still idle, altho some have found employment in other industries."[9]

Warriner's generalizations become more real in light of the impact on specific individuals of the July 1928 suspension of operations at the Lansford Colliery.[10] John Dibero, a thirty-nine-year-old Italian-born track helper working in the No. 6 mine, lost four months of work in 1928. The layoff must have been a real hardship for an unskilled worker whose earnings averaged only $25 a week and had to support a family that included two teenagers. Fortunately, this suspension was only temporary for Dibero, who resumed employment in November and was promoted a year later with a substantial increase in wages to work as a mucker, removing rock debris with the drilling of new gangways; Dibero subsequently worked steadily for LC&N until the company closed underground operations in April 1954.

For many at the company, the closing of the Lansford Colliery proved more disruptive. Edward Kissner had begun work as a ten-year-old picking slate in a breaker at LC&N in 1887. He was fifty-one years old and worked as a pumpman in the No. 5 mine when operations were suspended in July 1928. Perhaps his age made him less employable than Dibero, but his layoff lasted until March 1934.[11] When he was rehired, Kissner found work in the breaker once again, picking slate with teenagers, probably earning a wage considerably below what he had made before the suspension. The work was steady, though, and Kissner worked in the Lansford breaker until he was discharged at age seventy-five in 1952. His personnel record noted the reason, "no light work." Clearly Kissner worked at a low-wage job until he could work no more. Although he had worked almost sixty years for LC&N, there is no notation of a company pension, though he would have been eligible for a UMWA pension of $100 a month.

While Dibero and Kissner remained at LC&N after the suspension of work, others used the occasion to search for steadier or better-paying work. William James Allen was a mucker earning seventy-five cents an hour when the Lansford Colliery closed down in July 1928. An Englishman, Allen had been working almost seven years at the company, but he never returned to LC&N. He may have been among those whom Warriner mentioned as finding work in other industries during the company's idle periods. In any event, the company had quite a substantial waiting list of idled workers in late 1928 and with the onset of the Depression a year later, the number of unemployed men in the Panther Valley grew still larger. Increasingly, it became clear to the miners of the Panther Valley that they would need collective rather than individual solutions to their problems. And as the miners looked for more systemic solutions to their ills, they found themselves on a collision course with Warriner and LC&N, the Panther Valley's dominant employer.

The general manager's annual reports provide an unusual view of the mounting conflict during the Depression decade. The key issue that divided workers and management concerned how the company would deal with declining demand for coal. Would management have a free hand to apply strictly economic criteria in shutting down a particular mine, or would it be compelled to "spread the work" in an egalitarian fashion that would share the burden evenly among all the company's workers? Management insisted on exercising the right to close higher-cost operations. The company maintained elaborate cost accounts for each mine, and sought to shift work to the lower-cost mines as demand weakened. Furthermore, the company tried to keep stripping operations going, and even opened new ones, while laying off underground miners. Both of these strategies were aimed at containing costs in the face of declining output and maintaining profits on the sale of less coal in a deflationary period.

Throughout the company's internal reports, the logic of management's efforts is clear. In 1929, for instance, the general manager recorded a notable victory in this struggle with miners: "An important development in the labor situation was the establishment, thru the general strike of July, of the Company's right to suspend operations at single collieries as economic conditions require, instead of at the entire group as has been the previous practice." Warriner was referring here to a companywide strike in July 1929 through which LC&N workers attempted to limit the firm's ability to shift work among its collieries in periods of lower demand for coal. A year later Warriner anticipated weaker demand and proposed a solution: "a drastic revision of working time must be made, allocating the tonnage to be produced to the more profitable collieries." The same principle concerned management in the rehiring of laid-off workers. As Warriner noted in 1929, "concern for the welfare of the employe[e]s could not be allowed to outweigh the consideration of the Company's interest and operating efficiency."[12]

As the general economic downturn of the Depression began to take its toll, the problems for LC&N became more severe. During 1930, collieries in the Panther Valley lost 119 days on average due to weak demand for anthracite; the next year lost time amounted to fully 160 days. The company's collieries were operating at less than half capacity even before the depths of the Depression. To the miners, though, company practices must have seemed contradictory. More than 800 workers were laid off with the 1931 closing of the Rahn Colliery, viewed by management as more expensive to operate than the company's other mines. At the same time, however, the company set a "new record" for output per hour at the Nesquehoning colliery because of increased stripping operations. The next year, the Lansford Colliery set a record for

output per hour, due largely to high production at the nearby Summit Hill stripping. While underground miners worked less than half of the time, workers on stripping operations enjoyed "steady working time."[13] Between 1930 and 1933, stripping tonnage tripled while overall output declined from 3.2 to 2.8 million tons. In 1933, a year of intense mineworker protest, stripping accounted for fully a third of company output, up from roughly 10 percent in the late 1920s. Organized mineworkers' judgment that the company was shifting operations from underground to strip mining was right on the mark.[14]

Miners refused to accept what they viewed as the inequities of this distribution and demanded the equalization of work across the company's collieries. At numerous local meetings across the Panther Valley in December 1932, miners' groups called for an equalization of working time. Periodically, brief strikes closed down stripping operations, as miners demanded reopening of underground mines to give work to the unemployed. These sporadic efforts were not successful, however, as the Lehigh Navigation Coal Company (LNC)—the wholly owned subsidiary of LC&N, created in 1930 to operate the company's coal division—refused to reopen its closed Rahn Colliery without major concessions by the union to cut labor costs.[15]

With the inauguration of FDR in March 1933 and the passage of the National Industrial Recovery Act (NIRA) three months later, the landscape of labor struggle in the anthracite region changed decisively. The NIRA called for the writing of industrywide Codes of Fair Competition aimed at mitigating competitive forces resulting from overproduction. Code provisions typically limited hours of work, regulated working conditions, and established floors under wages—all measures aimed at stabilizing employment. The legislation sought to shift labor conflict from the picket line into administrative settings. Delays, however, in the code-writing process for anthracite led to protests and the founding of a mass organization, the Panther Valley Equalization Committee, which demanded equal sharing of working time first in all LNC mines, but also across the region generally.

Mass marches on August 17 and 18, 1933, led to the closing of all the company's mines. Local reports compared the protests to events three decades earlier that had accompanied the rise of the UMWA, noting, "The parade of the unemployed today was one of the greatest crowds seen in a demonstration in Lansford since the strife-torn days of 1902." On the first day, 4,000 protesters marched from Tamaqua to Coaldale, where they were met by another 3,000 men and women. Mounted state troopers patrolled the march route, as the groups converged and marched to Lansford, where the crowd swelled to more than 10,000. Demonstrators marched past LNC headquarters, where a delegation of leaders presented the company superintendent with their demand for equalization of work across the firm's operations. The delegation included both local union officials and community supporters—James Gildea, the outspoken editor of the *Coaldale Observer*; a Catholic priest, Daniel Dailey; the Baptist minister, John Pounder; and a local merchant, David Sherman. The company answered that it could not comply with the request, whereupon the assembled demonstrators voted to call a general strike across the valley's mines.[16]

After the meeting some 3,000 protesters marched to nearby Nesquehoning to bring out the remaining miners there. The next day a car caravan joined by marchers proceeded to outly-ing LNC mines in West Hazleton and the patch town of Kaska William, where they convinced miners to lay down their tools. With all of the LNC mines shut down, the demonstrators proposed to carry the protest across the region, into the southern field in Schuylkill County and into the northern field around Wilkes-Barre and Scranton.[17]

These mass demonstrations catalyzed federal and state officials, who applied pressure on the Panther Valley Equalization Committee to call off the demonstrations and permit the anthracite code-writing to proceed. Governor Pinchot, who had actively intervened in earlier anthracite strikes, met with representatives of the protesters and the federal agency established by the NIRA, the National Recovery Administration (NRA). Responding to Pinchot's appeal, a meeting of 8,000 LNC mineworkers in Coaldale voted to suspend further demonstrations until an NRA anthracite code had been adopted. Lehigh Navigation Coal agreed to suspend several of its stripping operations as long as the associated underground mines were closed. With these assurances, miners called off the general strike and mining operations across the Panther Valley resumed. In calling off the campaign, however, demonstrators articulated their intention to win equalization of working time through

Equalization demonstration in Lansford, turning onto Walnut St. from Patterson St., August 17, 1933. *Valley Gazette*, no. 37 (July 1975): 29.

Equalization protesters with signs, Lansford, August 17, 1933. James Gildea in inset. *Valley Gazette*, no. 37 (July 1975): 29.

the code-making process. As protest leader Gildea declared, "We want the anthracite code to give us the things we are seeking. If it does not, the tramp of marching feet goes on! We will not take anything but employment for our answer." In response to the demonstrations and pressure from the governor, the Philadelphia & Reading Coal & Iron Company soon reopened three collieries, while LNC announced a week later that it was reopening the No. 10 mine, which had been closed for seven months. The demand for equalization was making a difference.[18]

The August demonstrations infused new life into the code-writing process. On the day after Panther Valley protesters agreed to call off their demonstrations and return to work, anthracite operators held their first closed-door meeting to consider a Code of Fair Competition.[19] Adoption of an anthracite code, however, was delayed pending the outcome of negotiations for a code for bituminous coal. When that code was adopted in mid-September, the way was clear for progress. Anthracite coal operators, organized through the Anthracite Institute, submitted a proposed industry code to the NRA in late

October. The NRA, in turn, published this proposed Code of Fair Competition and set a public hearing for November 17, 1933, in Washington, D.C.[20]

The anthracite operators' proposed code represented management's strategies to deal with the downward spiral of prices and production that gripped the anthracite industry in the first years of the Depression. Nothing in the code, however, reflected the previous summer's crisis. There was no evidence that mine operators had any intention of addressing the widespread demand for equalization of employment in the region's mines.

The proposed NRA code differed from those adopted in many other industries because about 90 percent of the mining workforce in the anthracite region was already covered by a collective bargaining contract.[21] Article IV, Section 1 of the code acknowledged "the rights or obligations of any of the parties created by or arising out of the collective bargaining contract . . . between operators and districts 1, 7, and 9, United Mine Workers of America." For workers not covered by the union agreement—a tiny proportion in anthracite—the code proposed a weekly maximum of forty-eight hours and minimum wages ranging from $15 to $12 weekly depending on the size of the community. Succeeding articles required the regular submission by code signers of reports and statistics to the Anthracite Institute and established a variety of pricing and sales policies intended to abolish "unfair" trading practices. Basically these provisions required open, published prices and sales policies and the uniform adherence by operators to these stated prices and policies in all sales. The aim here was to abolish various secret rebates or discounts in the sale of anthracite coal, and, by publishing prices and abolishing discrimination among customers, to do away with unfair competition among the sellers of anthracite. Since no prices were set by the code, price competition remained possible, but all such competition would remain public and above board. Administration of the code was to be carried out by a committee of operators, with each operator producing 2.5 million or more tons of anthracite to have one representative on the committee. Smaller companies were to be represented as a group, with one representative for each 2.5 million tons of anthracite they produced. A provision for a weighting of votes by tonnage gave a further advantage to the region's largest producers, not surprising given the highly concentrated nature of the industry and the power of the industry's leading corporate actors.[22]

In terms of the treatment of labor there was little new in the proposed code, nothing that deviated significantly from provisions of successive collective bargaining agreements with the UMWA, and nothing that began to respond to the demand for an equalization of work. The novel aspects of the code were confined to the sections that outlawed unfair trade practices. Operators could not agree entirely on these measures, however, and two major producers, the Philadelphia & Reading Coal & Iron and the Lehigh Valley Coal companies, refused to sign the proposed code. They objected to enforcement features of the code that required operators found in violation of the code to revise their pricing and sales policies to come into compliance. In addition, some of the smaller operators felt the code favored the larger producers and sought amendments to protect their interests.[23]

These differences were minor, though, in comparison to the storm of counterproposals that community and labor groups sent to the NRA. While management sought to restrict various forms of unfair competition, workers and their allies in mining communities were most concerned about unemployment. With a third of mineworkers unemployed and the remainder working only 50–60 percent of full-time, the NRA was flooded with letters, resolutions, and

Equalization delegation to Lehigh Navigation Coal, August 1933. James Gildea, Coaldale editor, is second from left; Dave Sherman, Tamaqua merchant, is third from right. *Valley Gazette*, no. 6 (Dec. 1972): 1.

petitions calling for efforts to distribute the work evenly among all miners.

James Gildea served as the chairman of the Panther Valley Equalization Committee. The son of a blacklisted miner, Gildea became the spokesman for miners in the area and in November 1934 was elected to Congress to represent the district that included the Panther Valley.[24] There he proved a strong supporter of FDR's New Deal and an advocate of a thirty-hour work week to counter unemployment. Re-elected to a second term in 1936, he became a respected community leader whose influence persisted for more than two decades. Throughout the period he proved to be a thorn in the side of managers at LNC who had to contend with his arguments and his leadership. Writing to the NRA he outlined workers' demands regarding the anthracite code: "Equalization of Working Time is our main battle." *Equalization*, as miners used the term, called for a spreading of available work by a given company equally among all of its mines, and by extension to all of its miners. The Equalization Committee opposed management efforts to concentrate operations at lower-cost mines or stripping operations, both of which left thousands of miners out of work.[25]

To spread out available work as much as possible, the Equalization Committee called for a maximum working week of four 8-hour shifts, well below the 48-hour figure proposed in the code submitted by anthracite operators. Furthermore, all strippings and coal preparation were to be limited to the same 32-hour week, and no new stripping operations were to be opened as long as there were unemployed miners.[26]

The Panther Valley Equalization Committee, the major umbrella organization in the area, was joined by numerous local organizations petitioning for the equalization of work. The Lithuanian Catholic Action Convention appealed to President Roosevelt to "compel the coal companies to equalize the time in distribution of work, so that all miners may have some work." The Tamaqua NRA Committee telegraphed General Hugh Johnson, head of the NRA: "Equalization of working time in the anthracite industry is absolutely necessary and is all that is needed to restore prosperity in Tamaqua and all of eastern part of Pennsylvania."[27] Equalization clearly enjoyed enormous community support.

Officers of the UMWA had not been early advocates of equalization of work, but by the time of the NRA hearings on the anthracite code, the animated pressure of union members had brought their leaders along. Thomas Kennedy, International Secretary-Treasurer of the union and a native of the anthracite region, testified on behalf of the union before the NRA. He called for a five-day, 30-hour week with overtime prohibited. In addition, he argued that the code should require "any operator with two or more collieries" to "equitably distribute the available working time among the various collieries of the company." Kennedy proposed the reopening of closed mines, the restriction of stripping operations, and a curtailment of the hiring of new workers until all currently unemployed mineworkers were back at work. He called for such actions not just to assist mineworkers but to aid the anthracite region as a whole, including "merchants, professional men, and all other classes directly or indirectly dependent upon the anthracite industry."[28]

Kennedy was not engaging in flights of rhetorical fancy when he claimed that equalization would help middle-class merchants and professionals as well as mineworkers. Equalization appealed across the class spectrum in the Panther Valley, as evident in a letter from David Sherman, proprietor of Sherman's Army & Navy Store in Tamaqua, to President Roosevelt in August 1933. The major point of his letter was that the equal spreading of work throughout the valley would prove a far more effective way of supporting families than the present overstretched combination of federal and private relief. He offered an initial calculation of likely earnings with equalization: "The minimum wage paid to the laboring class in the mines is $4.62 per eight hours, or $41.58 per month if 'EQUALIZATION OF TIME' were in effect." Comparing this likely income to the figure of $14 per month in relief for "an ordinary family consisting of husband, wife, and two children," he concluded "that with 'EQUALIZATION OF TIME', Relief would be unnecessary." That Sherman did not speak for himself alone is reinforced by the fact that his letter was endorsed at the bottom by Herman Fenstermacher, President of the

Tamaqua Merchants Association.[29] Moreover, Sherman went beyond words in supporting the equalization campaign. As noted earlier, he participated in the mass march in August 1933 on the Lansford headquarters of LNC and was a member of the delegation that met that afternoon with company officials to communicate the march's demands. Sherman's support of equalization was indicative of the broad appeal that the campaign had during the Depression crisis.

Advocates considered equalization a necessary response to unemployment. Given the erosion of the anthracite market during the 1920s, they did not view the Depression as the sole problem. They also argued that specific policies of anthracite management exacerbated the problems faced by the region's workers. Thomas Kennedy made the point in his testimony that some firms had operated strip mines "without regard to maximum hours, or, to the working time of the collieries with which they are connected. In many cases, they have been operated 24 hours per day with three 8-hour shifts. Obviously such practices reduce the operating time of the collieries and should not be permitted."[30]

In its testimony before the NRA, the Panther Valley Equalization Committee offered specific examples from the operations of the Lehigh Navigation Coal Company of the substitution of strip-mined coal for underground coal. While the company closed its Rahn Colliery in the spring of 1931, putting 860 men out of work, it expanded operations at the Lansford Colliery by more than 100 percent. The new coal produced at Lansford came from strip-mining operations in nearby Summit Hill. In other words, mechanized surface mining was replacing more labor-intensive underground mining. In the crisis of a worldwide depression, machinery was displacing men, and relief rolls swelled accordingly.[31]

John Pounder, a Baptist minister and spokesman for the Equalization Committee, called into question the corporate logic that placed profits before people. "The well-being and welfare of human beings," he wrote, "is far more important than profit, and should be given the right of way in the present emergency. The so-called 'high' cost or lesser efficient collieries must be allowed to stay in business if the purpose of the N.R.A. to get people back to work

and to increase the purchasing power of the people is to be accomplished." The broader significance of the conflict seemed evident to Pounder: "Is not the employment of the big army of idle mine-workers of vastly greater importance to the nation than profit to the few?"

To accomplish this goal Pounder advocated a variety of measures—not all of them consistent with one another—that struck at the heart of a capitalist economy. In calling for all closed mines to be reopened and work spread evenly, he asserted "the freedom of choice on the part of coal companies to operate only their low cost collieries, mines, and strippings [must] be denied." At the same time he advocated that the NRA take over "abandoned" mines "during the emergency" and lease them to operators who would bring them into service. Finally, he called on the federal government to set prices for anthracite coal that would be high enough to permit all the mines in the region to operate at a profit.[32] Between curbs on the managerial prerogatives of mine operators, forced leasing of closed mines, and government price-fixing, Pounder's plan challenged management authority on multiple fronts. Clearly, the public good was to take precedence over the private economic interests of the anthracite operators.

Anthracite management mounted a fierce counterattack, arguing basically that any measures that tended to increase mining costs would disadvantage anthracite sales in competition with cheaper alternative fuels and thus in the end lead to declining demand for anthracite and decreased employment in the region. J. B. Warriner, now president of LNC, replied at length to Pounder's equalization proposal. He attacked the figures used by Pounder in making his case for the economic impracticality of equalization and noted various efforts the company had made to reopen facilities or expand production only to be turned down by union locals dissatisfied with the terms offered. He devoted particular attention to his company's increased reliance on stripping operations, acknowledging, "stripping coal has in a minor degree replaced mined coal, both in this locality and in other parts of the anthracite region." He admitted as well that the lower costs of stripping were one of the factors considered by management, but that in the depressed times he argued such savings were needed to compensate for high labor costs and fixed charges. He noted in his argument the positive aspects of this development: "The utilization of strippings . . . is not only a measure of conservation of natural resources, but it is a phase of mechanization of industry which works in the long run to the benefit of consumers." He concluded, with a clear sense of his audience, NRA officials: "It is not our understanding that the federal government favors the restriction of labor-saving devices, as such, in any industry."[33]

Grassroots activists supplemented their appeals to Washington by lobbying local NRA officials and organizing local and regional demonstrations. At a regional NRA meeting held in Coaldale on October 9, 1933, Gildea proposed that government purchases of coal to be distributed to families on relief be limited to firms adopting the equalization of work. The Panther Valley Equalization Committee held meetings across the region to build support and even urged local NRA representatives to lobby UMWA President John L. Lewis to assure union support for the demand. Finally, caravans of Panther Valley miners drove to other anthracite communities to extend the campaign for the equalization of work.[34]

Miners in the Panther Valley, though focused on achieving an equalization provision in the NRA code, also challenged anthracite companies directly in pursuit of their goal. At a mass meeting of miners on November 12, 1933, those in attendance "deposed" the UMWA official presiding, "put in their own chairman, and voted to call out the collieries." The key demand was to require that the company divide work equally among its collieries.[35] A resolution called on LNC "to work no colliery more than two weeks while others are idle," and then to give work to the idled facilities. The group resolved, "that if the request is not granted, the unemployed union will take steps to [en]force it." In early December, Panther Valley miners made good their threat by setting up picket lines at LNC mines in Greenwood and Tamaqua to protest the company's failure to equalize work across its operations. Despite pressure from UMWA local officers to resume work, the men stood by the resolution they had passed a month earlier.

Local union officers opposed the strike as a violation of the UMWA contract with the operators, and the national union threatened Panther Valley locals with the revocation of their charters.[36]

Leaders of the UMWA appear to have opposed the local actions, but they did not come down hard against the locals as they did against protests in other areas of the anthracite region. LNC president, J. B. Warriner, reported that UMWA leaders "are very much disturbed over the affair, and want to take any steps they can to break it up."[37] But supporters of equalization also reached Lewis on their own, and their arguments appear to have given Lewis second thoughts. Lewis and the Reverend Pounder had an interesting exchange as Lewis claimed that the strike was "doing substantial harm to the cause . . . of equalization."[38] Pounder's letter in response revealed a savvy ally for the mineworkers. He indicated that he realized that Lewis's hands were tied and that he could not enthusiastically support the local campaign for equalization. "[Y]ou are under contract with the operators," he wrote, "and . . . are bound to keep the terms thereof." While he did not go so far as to state it explicitly, his letter was a clear indication that he and Gildea, as nonmembers of the UMWA, could promote equalization for mineworkers in ways that Lewis could not.[39]

That Lewis and Pounder would be exchanging views was indicative of the way the equalization struggle had become a community issue as well as a union one. At a mass gathering in Coaldale in December 1933, after two weeks of picketing, the miners adjourned their union meeting and reconvened as a meeting of the Equalization Committee to permit Gildea, rather than their local officers, to chair the proceedings. Gildea (and not local union officers) subsequently met with LNC officials and reached a formal agreement that the company would reopen Nesquehoning and Coaldale operations if the striking men would return to work in Greenwood and Tamaqua. With this concession, the men returned to work on December 20 and also elected local representatives to a regionwide committee proposed by Gildea to promote the cause of equalization within the UMWA. A week later, citing colder weather and improving markets, LNC announced the reopening of the

Lansford Colliery, thereby complying with the mineworkers' demand for an equalization of working time across its facilities.[40]

The events of November–December 1933 were a turning point for the equalization movement. United miners had succeeded in forcing LNC to accede to their definition of equalization of work. In correspondence in January 1934 with Hugh Johnson, head of the NRA, Gildea claimed as much:

In the Panther Valley we have equalization. The five mines of the Lehigh Coal and Navigation Company here have equalized their working time. . . . At the end of each half month we publish an equalization table and having demonstrated the idea to be practical we feel qualified to offer our plan of action to you and to the United Mine Workers for region wide adoption. Our men here will never again permit a 7-month lockout at certain collieries while so called low-cost operations work full time or almost full time. They have agreed amongst themselves when called upon to work at the end of each two weeks to simply drop their tools if the working schedule is contrary to the adopted equalization table.[41]

Gildea offered a succinct summary of developments in the Panther Valley. LNC had first accepted equalization as the price for the termination of the August general strike. Still, it took a mass meeting of unemployed miners in November to establish a set of standards for the long-term accomplishment of equalization and then a two-week strike to force the company to accept those principles. Finally, by urging the miners to establish a valleywide equalization committee within the UMWA, Gildea succeeded in bringing the UMWA into the movement and in institutionalizing the monitoring and compliance mechanisms needed to realize equalization.

The success of the equalization movement had resulted from almost continuous agitation on the part of rank-and-file miners in the Panther Valley, both those working at the LNC collieries and the unemployed. Collective bargaining had been important, but not the traditional sort controlled by the UMWA. As mine managers pointed out repeatedly, all the equalization strikes were in violation of the 1930 contract

that the UMWA had signed with LC&N. But still there were repeated, brief strikes in the winter and spring of 1934. At their mass meetings, miners responded much more to the arguments of Gildea than to those of their union's district or national officers. And Gildea's intervention in November and December had been responsible for the real compromise that got miners back to work. Finally, as strikes resumed, the UMWA sent a three-person commission to the Panther Valley to work out an agreement that would get the men back into the mines. Panther Valley miners demanded the curtailing of stripping operations that were depriving underground miners of work, and union officials had little success reminding the men of their responsibilities under the contract. It was the continuous pressure exerted by this sustained insurgency that led LC&N managers to make concessions and, in effect, accede to the demand for the equalization of work.[42]

Local agitation that followed the establishment of the NRA, more than collective bargaining or governmental code-making, led to the implementation of equalization of work in the Panther Valley. In the end, the NRA code-making effort proved a failure. Operators and the UMWA could never agree on labor provisions, and no code was adopted before a May 1935 Supreme Court decision found the NRA's code-making authority unconstitutional.[43] Still, in the process of lobbying the NRA for an equalization of work provision, Gildea and other activists in the Panther Valley Equalization Committee created a social movement with far more clout in the region than the UMWA. They involved fraternal organizations, religious groups, local businessmen, as well as employed and unemployed miners in a mass movement. Equalization became as much a community issue as a union or class issue; this development isolated LNC and made it far more difficult for the company to hold out.

The success of the strikes in November and December 1933 and the establishment of an equalization committee among all UMWA locals in the Panther Valley made possible the institutionalization of the equalization of work across the Lehigh Navigation Coal Company. The power of the equalization committee permitted a degree of flexibility in the implementation of its terms. Rather than enforcing an absolute equality of work-sharing among mines over a two-week or one-month period, the union equalization committee called for all the company's mines to work "on a basis of equality as far as humanly possible."[44] This language might seem vague on the surface, but it is apparent that the UMWA equalization committee assumed responsibilities that had traditionally been the exclusive province of management. Consider a newspaper account of the committee's deliberations:

At a meeting of the equalization committee of the United Mine Workers of the Panther Creek Valley . . . yesterday afternoon it was decided that the No. 8, No. 10 and Nesquehoning collieries having worked the same number of days, the working schedule for the second half of October should call for the operation of Lansford and Tamaqua collieries. As long as these two collieries operate the company may take a choice of any of the other three should the demand call for operation of more than the two plants fixed by motion.[45]

The assertion of authority evident in the description of this meeting is remarkable. The equalization committee "fixed by motion" the fact that the Lansford and Tamaqua collieries were to be operated during the next two weeks and the other collieries only if there were sufficient demand.

That the miners of the Panther Valley were able to impose their demands on LNC is confirmed by a Washington, D.C., account of developments. As Jett Lauck, a long-time consultant employed for the UMWA, was working on material related to the NRA's deliberations on an anthracite code, he met with a Labor Department official in late February 1934 and recorded in his diary what he learned:

Panther Valley had about equalized work through alternately working collieries. He said there were about 10,000 men in this district who were in complete control of the situation having put up a large bell in the mountains and if there was any deviation from the policy which they had imposed upon the

operators, the bell was rung and they mobilized and picketed the mines where the trouble took place and continued in a state of insurrection until matters were settled.[46]

Parts of this account seem apocryphal. The bell is a wonderfully concrete detail, but no other account offers any confirmation of that mechanism to "rally the troops." But the judgment that the mineworkers "were in complete control of the situation" is right on the mark. And to mine management, it must have indeed seemed that their workers were continually in a "state of insurrection."[47]

Against their better judgment, LNC management accepted the role reversal evident here and cooperated in good faith with the union on this issue. In the company's mining report for 1934, J. B. Warriner acknowledged: "The plan for equalization of working time, adopted by the men under outside leadership in 1933, was again enforced rigidly at the expense of some lost time at several collieries for short periods." Later in the same report, he admitted that the firm had had to accommodate to the workers' demands: "Tacit acceptance, to prevent serious interruption of work, has strengthened the belief of the men that a plan is permanently established, and it will be upset only thru a serious strike."[48]

Warriner's report makes clear that the mineworkers in their equalization struggle were following "outside leadership," rather than union officers in this campaign. This fact was a constant refrain in Warriner's reports. In commenting about the "unusual turbulence" in labor relations, he wrote: "This condition gave rise to a movement, organized outside the ranks of the United Mine Workers and headed by persons with no responsibility, demanding equal distribution of working time at the Panther Valley Collieries." Pounder and Gildea, of course had "no responsibility" to live by the contracts that the UMWA had signed with LNC, and that was precisely what permitted them to challenge the company's policies. What particularly irked management was the union's acquiescence in these unorthodox practices. "One of the disturbing features of this movement," wrote Warriner, "was the apparent weakness of the Union leaders to hold the men to lawful and orderly procedure under trying circumstances."[49] Yet it was the community nature of the equalization movement that made it so formidable and forced LNC to acquiesce in the men's demands.

Not only did LNC management acknowledge company acceptance of equalization of work, but contemporary statistics demonstrate that subsequent practice resulted in the desired outcome. Gildea addressed the District 9 UMWA convention in October of 1934, two weeks before the one-year anniversary of the first imposition of equalization of work by a mass meeting of unemployed miners. Summarizing the campaign's success, he reported, "Today the Panther Creek Valley collieries, all five of them are working, all five of them have worked their 149 days. . . . that is an accomplished fact." Shortly after the district meeting, the UMWA equalization committee reported in greater detail, slightly modifying Gildea's earlier claims. Four of the company's collieries had operated either 156 or 157 days in the past year, with only the Greenwood colliery, which had suffered a damaging fire, deviating significantly from the others. The equalization committee expressed its "appreciation of the efforts put forth by the Lehigh Navigation Coal Co. to furnish equal time to all employees." They "decided to function for another year," but noted that "it is expected the time will be equalized on the basis of each month."[50] From the first equalization proposal in November 1933 to its modification a year later, miners adopted the policy unilaterally and through their solidarity imposed the principle upon LNC management. They managed to place broad human needs in the Panther Valley above corporate profits. They had succeeded in curbing "the freedom of choice on the part of coal companies to operate only their low cost collieries, mines, and strippings," as Pounder had proposed at the anthracite code hearings a year earlier. They had demonstrated in the face of opposition from mine operators and their own union that miners did not accept the logic of corporate capital and that in a crisis the united power of the valley's workers and residents could prevail.[51]

Community support, rather than the power of the UMWA, made the workers' success pos-

sible. That the miners turned to editor Gildea rather than their union leaders frustrated management at LNC. Their strikes ended inconclusively, but the steady pressure that they generated and the overwhelming community support they enjoyed led Warriner to realize that if the company was to produce coal it would have to give up a measure of managerial authority and meet the demand for equalization of work.

Equalization dramatically changed the lives of miners' families in the Panther Valley. In 1931–32, when the LNC unilaterally set its collieries' work schedules, the "high-cost" Tamaqua operations worked only 73 days over the course of the year; by contrast, the Lansford Colliery had 185 starts over the year. Both sets of miners worked under UMWA contracts, but workers at Lansford earned more than 2½ times what their fellow workers at Tamaqua brought home. The next year saw some improvement, but still Coaldale miners worked 75 percent more days than did those at Tamaqua. Equalization of work, enforced with the threat of general valleywide work stoppages, ended these disparities. In the 1935–36 year, for instance, days worked ranged between 136 and 143 days at the company's five collieries; in 1936–37, the comparable range was between 113 and 119 days.[52] Times were not good for miners in the Panther Valley, but at least all shared the burdens of short time equally.

Equalization was a success in the Panther Valley, and its influence soon extended beyond the LNC. First, the campaign brought to the fore Coaldale editor Gildea, who successfully campaigned as a Democrat for the local congressional seat for Schuylkill and Northumberland counties. In a traditionally Republican district, Gildea served for two terms and brought miners' support of FDR into the halls of Congress. Beyond the Panther Valley and its legislative district, equalization also became the official policy of the UMWA throughout the anthracite region. The regionwide UMWA contract with the anthracite operators that went into effect in May 1936 included a provision for the "Equalization of Working Time." It provided for a division of "working time equitably" among the operating collieries of a given firm, "subject to a tolerance of not more than 20% from the high to the low to compensate for economic conditions and market demand."[53] The provision was not as stringent as that enforced in the Panther Valley, but it extended the basic principle of equalization to anthracite firms across the region.

Even as equalization was extended beyond the Panther Valley, it remained hotly contested within the operations of LNC. Conflict continued because working miners often interpreted the restrictions imposed by equalization differently than did either UMWA officials or LNC managers. For instance, on September 1, 1936, LNC, "acting on the recommendation of [UMWA] District President [Hugh] Brown," kept its Nesquehoning colliery in operation "in spite of the fact that some of the other collieries have had less time worked." Nesquehoning miners refused to work. While LNC President J. B. Warriner and UMWA leader Brown had agreed on "a change in our equalization plan," the workers did not. The outcome of this particular struggle is unclear, but Warriner was prepared to reopen the Coaldale Colliery, the one "with the least working time to date," if that was what it would take to bring the men back to work at Nesquehoning. Moreover, a company memo, written in July 1937, demonstrates that the company kept pressing against its miners' insistence on "the present rigid equalization schedule." Both of these conflicts confirm that equalization resulted from the determination of Panther Valley miners to enforce equalization, even in the face of the persistent opposition of mine managers and the equivocations of UMWA leaders.[54]

These conflicts reveal the radicalism of the equalization movement—both in its demand for a sharing of working time and as an expression of working people's solidarity. Reverend Pounder expressed the fundamental logic behind the demand when he argued before the NRA code hearings that coal companies must be denied the "freedom of choice" to manage their operations solely with a view to their profitability. The mineworkers themselves, though less focused on the morality of alternative solutions, showed their radicalism by their actions. When offered the possibility of reopening the Rahn Colliery in March 1932, provided they accepted concessions that would have required that they work at

wages below those provided in the union contract, they refused. Even more to the point, mineworkers at all the collieries operated by LNC were united in the demand for equalization of working time, even those employed at the more efficient Lansford and Coaldale collieries. These latter workers had benefited from company policy in the 1931–33 period that disproportionately concentrated work at their facilities. Lansford miners, for instance, worked 315 days overall in that two-year period, almost double the figure for miners at Tamaqua.[55] Still, Lansford miners demanded equalization of work, so that available work would be shared equitably among the company's employees. They also united in a demand for a shortening of the hours of labor so that currently unemployed miners might be called back to work as well. Evident in the Panther Valley during the Depression period was a clear call for policies that provided for the collective well-being of mineworkers, rather than individual solutions. Group interest rather than self-interest motivated the demand for an equalization of work. The policy was radical in its demand for the curbing of the prerogatives of management and in its placing of collective well-being ahead of individual self-interest. The equalization of work campaigns had no analogues in the national labor story of the Great Depression: the building of unions in the mass-production industries. The CIO-led mobilizations of the era certainly rested on community support, but the way that the protests in the Panther Valley skirted the union orbit was as distinct as its radical character.

The call for equalization of work succeeded in the Panther Valley due to specific conditions in that part of the anthracite region. The LNC linked five communities of anthracite mineworkers within a twelve-mile stretch. Although each of these communities had its own individual character, residents of Nesquehoning, Lansford, Summit Hill, Coaldale, and Tamaqua had a shared associational life, family ties, and history of labor grievances and struggle. They worked for one firm, with various and extended facilities, but they could easily see a total employment situation where a balancing of operations would permit the equitable spreading of work.

◆

Equalization gained support in other locales of the anthracite region, but without successful implementation as in the Panther Valley. In other sections the failure of equalization led to the emergence of alternative strategies to cope with mine closings and massive unemployment. In the northern field, in the anthracite mines of the Scranton and Wilkes-Barre areas, an insurgency developed first within UMWA locals and later as a movement to create a "dual," or competing, union, the United Anthracite Miners of Pennsylvania. Like equalization of work, the dual-union campaign emerged to address the deteriorating circumstances of anthracite miners during the Great Depression in the face of the UMWA's seeming incapacity. While mineworkers in the Panther Valley region had significant success with their equalization strategy, in the northern field dual unionism proved a failure with tragic consequences.

The northern field of the anthracite region differed in important respects from the Panther Valley at the eastern end of region's southern field. With coal deposits stretching the length of the Wyoming and Lackawanna valleys, the northern field was much larger than the Panther Valley and output correspondingly greater. Anthracite production in the northern field amounted to more than forty-four million tons in 1927, a figure more than ten times that for the Panther Valley. Moreover, a number of major anthracite producers operated collieries in the northern field—Glen Alden, Hudson Coal, Lehigh Valley Coal, Pittston, and Susquehanna being the largest—making for distinct economic empires within this larger, more diverse region.[56] The combination of geographical dispersion and competition among major firms meant that the northern field was less homogeneous and less unified than the compact Panther Valley with its single producer. Equalization faced a correspondingly more difficult battle in the northern field.

The first responses to the economic crisis in the northern field were similar to those in the Panther Valley and also entailed a call for the equalization of employment across the field's mines. The insurgency arose among miners employed by the Glen Alden Coal Company, with

operations principally in the Wyoming Valley around the city of Wilkes-Barre. The Glen Alden Coal Company was the product of twentieth-century antitrust actions that required the Delaware, Lackawanna & Western Railroad (DL&W) to divest itself of its coal lands and operations.[57] In September 1921 the newly founded Glen Alden corporation took title to the company's coal lands and other properties, in exchange for a $60 million forty-year mortgage. While Glen Alden and the DL&W were technically independent of one another, railroad stockholders were given an early opportunity to purchase stocks in the new corporation and Glen Alden sold its coal to the DL&W Coal Company, which had a long-term agreement to ship its coal on the DL&W railroad.[58] The economic fortunes of coal company and railroad were inextricably intertwined.

Production at Glen Alden mines, as at all other major anthracite firms, declined modestly in the 1920s and plummeted with the onset of the Depression. Output declined from more than eleven million tons annually in 1930 to less than seven million tons in 1932. Glen Alden closed the Pyne, Taylor, Hyde Park, and Pettebone collieries in 1931 and the Storrs, Hollenback, Stanton, Sugar Notch, Diamond, and Archbald collieries the next year. Fully a third of Glen Alden mineworkers employed in 1930 were out of work two years later.[59]

The closings demonstrated the impotence of the UMWA under its contract with the anthracite operators. In July 1930, UMWA President John L. Lewis negotiated a five-and-a-half-year extension of the union's contract. That contract sustained basic wage rates among mineworkers and achieved a partial dues check-off for the national union, but Lewis did not press forcefully for issues that had been raised by local activists from the region. The debate over the ratification of the contract was divisive, especially concerning a series of provisions that dramatically weakened local grievance committees and required the union to promote efficiency and discipline members and local leaders who supported strikes violating the agreement.[60] As the Depression deepened, Lewis took a conciliatory approach toward anthracite bargaining, "seeking to gain security for his mine workers through a policy of partnership and conciliation with the operators."[61] But this approach abandoned anthracite miners in the face of management strategies of concentrating production in lower-cost collieries. If Lewis and the union would not defend anthracite miners, the miners would have to defend themselves.

Following the ratification of the agreement in 1930, there was an outbreak of strikes in District 1, the northern anthracite field, protesting the mine closings and calling for the equalization of work across all operations at Glen Alden. Thomas Maloney, a local officer at the closed Stanton Colliery, led this effort along with the former district president, Rinaldo Capellini. Working through the Glen Alden General Grievance Committee, Maloney and his supporters called for equalization of work across that company's collieries as an alternative to inadequate public relief for the unemployed. In response to the failure of District officials or the UMWA to support these actions, Maloney ran for district president in opposition to incumbent and Lewis supporter, John Boylan. District officials made a heavy-handed effort to keep Maloney off the ballot on technical grounds, and it took a local judge's court order to permit his candidacy. Boylan won reelection handily, but charges of fraud circulated widely. As a Nanticoke miner wrote, "Through their control of election machinery and usually through control of officers of the locals . . . district officers were able to chalk up a generous lead for themselves . . . over Tom Maloney."[62] Maloney attempted to challenge the fairness of the election at the following district convention, but by refusing to accept the credentials of Maloney supporters as delegates to the convention, Boylan easily secured the certification of his reelection. By quashing the insurgency rather than addressing its causes, Boylan widened the breach between the union and its members that would eventually drive disgruntled miners to found an alternative organization.[63] One loyal UMWA member from Plains captured well the roots of what he called a "wide spread rebellion" in a letter to Lewis: "I am a sincere supporter of the District officers; yet, it seems to me that either they don't realize the danger of the situation, or are negligent in the performance of their duties. They never appear in the field to discuss the many complaints directly with the men and

make some earnest efforts to settle the various disputes."[64] Nor, one might add, did Lewis or other national officers ever urge district officers to take their members' concerns seriously.

Disgruntled UMWA members from more than thirty locals in District 1 met in Wilkes-Barre in February 1932 and called for a district convention to demand equalization of work. With district officials refusing to convene a meeting, Maloney called for a strike to press for equalization of work from the Glen Alden Company. The strike lasted less than three weeks, doomed from the outset by the combined opposition of the company and UMWA district and national leaders. Violence between striking and nonstriking miners provided a foretaste of the toll that the fratricidal division among UMWA members in District 1 would take. The union only compounded the problem by expelling Maloney and thirty other leaders for their part in the unauthorized strike.[65] The UMWA took a much more hostile and punitive stance toward District 1 insurgents than would be the case with the uprising in the Panther Valley the next year, and the consequences of the breach in the northern field were disastrous for the union and miners alike.

The expulsion of the leaders of the insurgency left the Lewis-loyalist officers of District 1 firmly in charge, and Boylan ran unopposed for district president in 1933. The annual district convention was free of the bitter struggles of recent years because Boylan and his allies had effectively excluded any challengers from the proceedings. With no possibility of reforming policy from within the UMWA, Maloney and his insurgent supporters called a rump convention that met in Wilkes-Barre in August with representatives from some fifty-five locals in attendance. There the dissident miners formed the United Anthracite Miners of Pennsylvania (UAMP). But the new group was entirely reactive, calling strikes successively against the Penn Anthracite Coal Company, the Delaware & Hudson Coal Company, and the Glen Alden Coal Company. The goal this time was no longer equalization of work but the rehiring of UAMP members fired by the coal companies tied with Lewis and the accommodationist UMWA. Finally, in November, Maloney called off the strikes in exchange for a federal investigation and arbitration of what had by now become essentially a struggle between the powerful, entrenched UMWA and the weak challenger, the UAMP. The investigation dragged on but ultimately, in October 1934, the arbitrator ruled that the UMWA contract took precedence over the rights of miners affiliated with the UAMP. Following one more strike in February 1935, for which Maloney and other UAMP leaders were jailed for contempt of court, and further internecine violence among miners, the UAMP effectively died in late 1935.[66]

This struggle over dual unionism in District 1 tore the UMWA and mineworkers in the northern anthracite field apart in the 1932–1935 period. While the equalization campaign in the closely knit Panther Valley brought together unemployed and working miners and united mineworkers and their middle-class allies, such solidarity did not emerge in the larger and more diverse northern field. An unbridgeable divide between unemployed and working miners proved the heart of the problem and contributed to widespread violence. Locals at idled collieries were the main supporters of Maloney and the UAMP. Unemployed miners picketed working mines in an effort to close them down and to equalize work across all collieries. Newspapers reported battles between picketing miners and those trying to get to work. In early November 1934 at the Susquehanna collieries in Nanticoke, 400 employed miners fought 600 picketers with fists, rocks, and bricks.[67] Beyond this hand-to-hand combat, there were also late-night porch bombings as striking miners attacked the homes of working miners. Dynamite also damaged the home of the UMWA district president and destroyed the automobile of a judge who had taken a hard line against striking dual unionists. Finally, in April 1936—some months after the UAMP gave up the fight—Thomas Maloney, his son, and a local school official died opening package bombs sent through the mails. Doubts remain as to the guilt of the person convicted and executed for the murders; still the mail bombings reveal how badly the dual-union campaign tore apart the UMWA in the northern anthracite field.[68]

With the demise of the UAMP, Lewis engineered the peaceful departure of District 1's president Boylan, who had lost the confidence of union members during the struggle, and the union readmitted former members who had participated in the dual-union movement. Still, district mineworkers no doubt learned the lesson that it did no good to challenge union authority.[69] The internecine battle within the UMWA in the northern field flowed from the failure of the early equalization efforts at Glen Alden. In contrast to the Panther Valley, mineworkers in the Wilkes-Barre area were unable to bridge the divides between collieries to implement proposals to spread work equitably.

◆

Equalization of work and dual unionism were alternative solutions to the crisis of the Depression in the Panther Valley and the northern anthracite field; in the western half of the middle and southern fields—west and southwest of the Panther Valley—coal bootlegging became the dominant response to colliery closings. In communities across Schuylkill and Northumberland counties, unemployed miners occupied idle coal lands, dug shafts and drifts into mountainsides, blasted and removed coal, established primitive breakers, and sold the roughly sized and cleaned coal to independent truckers who peddled anthracite in Philadelphia and New York and elsewhere in the northeast. By the mid-to-late 1930s, "bootleg" coal amounted to a significant share of all anthracite, exceeding 10 percent of the region's output in some years. A shift of focus to the operations of the Philadelphia & Reading Coal & Iron Company, in the southern anthracite field, illuminates this third Depression-era strategy among anthracite miners.[70]

The Philadelphia & Reading Coal & Iron Company (P&RC&I) formally began operations in 1923, when the Philadelphia and Reading Railroad complied with a 1920 Supreme Court ruling requiring that company to separate its coal producing and transporting activities. The coal operations of the Philadelphia & Reading Railroad had begun in the early 1870s, when state legislation permitted the railroad to expand into coal mining for the first time. The company's mining grew rapidly in the last third of the nineteenth century, and it soon became the region's largest producer. The Supreme Court's dissolution order had minimal effect on the company's operations. The dissolution was a fictive separation because the initial stockholders of the two firms were identical and the interlocking boards of directors made it impossible to separate out the "interests" of the railroad and the coal and iron company. Virtually all the coal produced by the company continued to be transported by the associated railroad, and high freight rates (double comparable rates for bituminous coal, for instance) continued to ensure that the railroad firm made the lion's share of the formal profits.[71]

The decade of the 1920s was a difficult period for P&RC&I, as it was for most of the major anthracite-producing firms. Annual coal output declined significantly over the course of the decade, from 11.5 million tons produced by the company and its lessees in 1923 to 9.6 million tons in 1928, a year before the onset of the Depression. Still, the company remained by far the largest producer in the region, owning more than 160,000 acres of coal lands with forty-four active mines and thirty-two operating breakers in 1927.[72] Like the Lehigh Coal & Navigation Company, P&RC&I had the capacity at the end of the 1920s to produce far more coal than the market could absorb. This overcapacity meant that the company's many mines and breakers were operating at far below their optimal efficiency and thus the company's costs of production were considerably higher than might otherwise have been the case and profit margins correspondingly lower. Aware of these difficulties, company management fixed on a dramatic course of action intended to improve the bottom line. In March 1929 the company proposed to its stockholders taking on $30.8 million of bonded debt to erect two mammoth central breakers, at Locust Summit and St. Nicholas, and to centralize virtually all coal preparation at these two modern, efficient plants. The plan also called for the modernization of some of the company's existing breakers and the closing of most of the company's antiquated wooden structures.[73]

Stockholders approved the plan, and the company began an ambitious modernization program just months before the onset of the Depression. The timing, of course, could not have been worse. The company brought the Locust Summit and St. Nicholas breakers on line in April 1930 and July 1933, respectively, but overall coal production by the company and its

lessees in 1933 sank to 6.7 million tons. With variation from year to year, between 1933 and 1940 annual production averaged just over 7 million tons, a figure totally inadequate to permit the company to pay the interest on its bonded debt and still make any profit. In the face of the continuing weakness in the anthracite market, the company closed underground mines and breakers and increasingly turned to stripping operations in a futile effort to reduce labor and overhead costs and turn a profit. Between 1935 and 1938 the company's losses—virtually all of which came from its coal operations—averaged more than $6 million a year. In February 1937 the company filed for bankruptcy protection, and its operations came under the supervision of the courts. The company adopted a plan that involved selling off the vast majority of its coal lands and cutting back operations still further. By 1940, the company had scaled operations back and had only seven active underground mines and eleven strippings still in operation.[74]

In one community after another across Schuylkill and Northumberland counties, P&RC&I closed collieries and left miners and their families without any means of support. In Mahanoy City, a Schuylkill County community of 15,000, six of seven collieries closed between 1930 and 1932, throwing more than 3,400 miners out of work. A 1934 study revealed that the business of local merchants had declined by more than 70 percent, and a reasonable estimate placed unemployment among adult males in the community at 75 percent. In Shamokin, twenty-five miles to the west, four of five collieries of the P&RC&I were closed in early 1934, leaving more than 5,000 mineworkers unemployed. Finally, in Lykens, at the extreme southwestern end of the southern field, the only colliery in town closed down; by October 1933 almost half the town's population was on relief. Schuylkill and Northumberland counties were truly disaster areas during the depths of the Depression, and, not coincidentally, they were the areas with the highest costs of production in both the deep mining and screening of coal. In comparison, the economies in the middle and northern fields of the anthracite region—with unemployment hovering around 25 to 30 percent—offered distinctly better prospects.[75]

The devastating colliery closings undercut the prospects for equalization of work that proved so effective a strategy for miners in the Panther Valley. Equalization rallies took place in Mahanoy City, and Gildea, leader of the Panther Valley Equalization Committee, even spoke there in August 1933 as Panther Valley miners marched to neighboring communities to build support. And when the Panther Valley miners called off their strike at the urging of Governor Pinchot, the governor met with a delegation that included representatives from Schuylkill County towns as well as the Panther Valley.[76] But while Panther Valley miners were able to force LNC effectively to spread the work among all the company's miners, workers at the P&RC&I had no such success. The Philadelphia & Reading's recent major investment in its two mammoth central breakers committed the company to closing virtually all of its small, decentralized breakers and their accompanying underground mines. Having drastically cut and centralized their operations, they were not about to accommodate the call for equalization, which would have meant that all collieries would have been in operation perhaps fifty days a year. Given this reality, unemployed miners from the P&RC&I and the Lehigh Valley Coal companies in Schuylkill and Northumberland counties adopted a different strategy: They invaded idle coal lands and began mining on their own accounts.

Coal bootlegging, as it was known (or *independent mining*, to its supporters), became a mainstay of local communities where the legitimate collieries had been shut down. The illegal gathering of coal from company properties by women and children for direct family consumption had long been customary in mining in Europe and the United States and during anthracite strikes in 1902 and 1925–26, striking anthracite mineworkers did illegally mine and sell coal to cover lost income.[77] But there was little precedent for the extent of bootlegging that developed during the Great Depression. The Pennsylvania Anthracite Coal Industry Commission conducted a census of bootleg mines and breakers in May and June 1937, providing the best surviving evidence on the scope and character of these operations. The census, conducted in summer months when operations would have been decidedly scaled back from winter peaks, provided estimates of 7,000 bootleg miners, another 2,000 picking over culm and refuse banks, 1,300 working in bootleg coal breakers, and 2,700 engaged in trucking bootleg coal—a total of 13,000 men supporting themselves and their families through the production of some 2.4 million tons of anthracite coal annually.[78] Investigators estimated 2,000–2,500 separate operations, noting that, "Most of the bootleg holes were concentrated in and around abandoned collieries in which the bootleg miners involved had formerly worked."[79] About two-thirds of bootleg miners operated within Schuylkill County, while another fifth lived and worked in Northumberland County. These counties were home to the P&RC&I, whose lands accounted for almost 62 percent of bootleg holes.[80]

The bootleg holes were small, often family affairs, typically employing three to five miners, working and dividing the proceeds of the sale of the coal cooperatively.[81] Geology played a role: Outcroppings and coal veins close to the surface still existed in the isolated hills of Schuylkill and Northumberland counties, allowing for often clandestine family-run operations.[82] Equipment was minimal. Typically, a hand windlass and a metal bucket were employed to lower miners into their holes and to bring the coal up to the surface. As operations went deeper and became more substantial, bootleggers would often place an old car or truck up on blocks, attach a rope over a wheel, and use the car engine to power an improved lift. Still, in most instances, they tapped into quickly exhausted veins of coal; the bootleg holes operated for less than eight months on average.[83]

Following production, bootleggers might sell their output directly to truckers who then sold the coal in nearby urban centers such as Philadelphia, New York, and Baltimore; in 1936, an estimated 1,000 trucks carrying bootleg coal rolled into New York City each week.[84] As likely, though, the miners would sell the coal to small-scale bootleg breakers that developed in the area, where the coal would be cleaned and sized, and would then fetch a better price from truckers heading out of the region. The third element in the overall production and distribution were the truckers who would transport and market the coal, usually for $1–$3 less per ton than the

Accident at a bootleg mine,
Wiggans Patch, near
Shenandoah, 1936. RG-13,
Pennsylvania Historical and
Museum Commission and
the State Archives,
Harrisburg.

price charged by the major anthracite railroad distributors (with savings of about 20 percent for consumers).[85]

Each bootleg miner typically worked for himself and used the proceeds to help support his family during a period when mining jobs were scarce in this part of the anthracite region. When asked by the numerous journalists and magazine writers who covered the story of bootlegging to justify their seeming illegal and radical activity, the men involved spoke of family survival and their right to access coal through their hard work. Mike McCloskey, a miner, thus reacted when questioned about the illegality of bootlegging:

Illegal? Are we supposed to starve to death just 'cause the collieries are closed? These hills are full of coal and there's millions of people who wants it. We're miners, without jobs, and our bellies are empty. We don't know and we don't care who's supposed to own the land. God put that coal there—not the Philadelphia and Reading Coal Company. . . . I don't want nobody's charity. . . . All I want is a chance to work so's our families can live.[86]

Another miner interviewed, a "young Hungarian" who was using his car to hoist coal, echoed the sentiments:

Stealing! No, I don't think this is stealing. If I go into a grocery store and take a loaf of bread some other man made just because it's easier than to make some—that's stealing. But, I'm working for this coal.[87]

Speaking directly of the P&RC&I and his family, a third bootlegger justified his activity similarly:

What did this coal company do for this coal? Bought this land years ago for a song. They've got enough coal here to last 200 years. They may get around to this coal I'm working a hundred years from now. But we'll all be dead then. They'll never use it. But we've got families that are hungry now. Should I leave here so as some guy can enjoy it a hundred or two hundred years from now—some guy that won't pay a penny for it, that won't put in a lick of work? Should I let my family starve now for the sake of some guy whose grandfather isn't born yet?[88]

Bootleggers cast their justifications in personal terms, but bootlegging was very much a collective endeavor. The bootleggers acted in concert, helping each other in mining and sharing jerry-rigged breakers. More important, bootleggers rallied to protect each other from periodic efforts at law enforcement. In 1934, a group of company men sent to close a bootleg operation, with a force of Coal and Iron Police in tow, found themselves surrounded by 500 angry miners. Women in the community had sounded the alarm and stood fast with the men. Powerless in the face of the unified resistance, the company hands and the police withdrew. A few months later in Goodsprings, another contingent of company workers, intent on demolishing a bootleg breaker and accompanied by forty state police, faced a crowd of 3,500 indignant miners and community supporters. The ringing of a tocsin on a truck had summoned bootleggers from the surrounding hills. The company agents succeeded in partially dismantling the breaker, but with a battle imminent they hurried back to Pottsville, hounded all the way by a car and truck caravan of aroused and taunting citizens.[89]

Bootleggers engaged in more violent actions. In December 1934, the Stevens Coal Company, in an effort to close bootleg operations, decided to strip-mine a tract of land near Shamokin where 1,700 illegal miners eked out a living. In response, the bootleggers repeatedly tried to block the installation of a huge steam shovel. They eventually blew up the shovel with sixteen sticks of dynamite, and Stevens dropped the plan to strip-mine.[90] In another incident at Gilberton, bootleggers dynamited the car of a Coal and Iron policeman after he had blown up a number of bootleg operations.[91] The attempt to close bootleg mines generally proved futile. Agents of the P&RC&I, for example, blew up 1,200 holes in 1934 during a period when investigators estimated that at least 4,000 new ones had been established on company property.[92]

Collective, violent actions of bootleggers continued throughout the Great Depression. In December 1940, William Greenough attempted to open a strip-mine operation on holdings where 700 bootleggers had mined coal for years. The bootleggers threatened to destroy two steam shovels that had been placed on the property. Greenough sought a court injunction to prevent the destruction of property, but to no avail. An organization of bootleggers formed in 1934, the Independent Miners' and Truckers' Association, then intervened with a proposal for Greenough to reach lease agreements with the bootleggers. Greenough balked at first, but on January 5, 1941, he signed an accord with the Association to lease tracts on a royalty basis, legalizing, in effect, bootlegging.[93]

The Independent Miners' and Truckers' Association provides evidence of the formal side to the collective character of bootlegging, a counterpart to the spontaneous, local resistances. The Association first displayed its muscle in 1935 when the anthracite railroad companies lobbied for the passage of a state law prohibiting the conveying of coal by trucks beyond twenty-five-mile distances from mine sites. In response, the Association led a demonstration of more than 10,000 independent truckers, miners, and community supporters at the state capital in Harrisburg. The demonstrators successfully prevailed on state legislators, and the legislation was defeated.[94]

The bootleggers enjoyed widespread community support. While managers and owners of the coal companies decried "stolen coal" and asserted that coal bootlegging undercut the moral fiber of coal communities, most residents of Schuylkill and Northumberland counties thought otherwise.[95] When local sheriffs or the coal police arrested coal bootleggers, local juries invariably refused to return convictions or justices of the peace let offenders off with token fines. John T. Flynn, writing for *Collier's,* described the following typical court scenes in bootlegging areas. Addressing a group of bootleggers convicted of trespassing, one judge, resenting the arrest of the men, declared, "Well, I reckon I'll have to fine you. Have you got ten dollars?" With "no" for the answer, the judge continued, "Well, have you got five dollars?" Blank stares from the men, brought, "How about a dollar? Haven't even got a dollar. That's too bad. Anyhow, I guess I'll have to fine you a dollar and you can bring it in when you get it." At another trial, when a group of convicted bootleggers could not pay the fines, they were marched off to jail, fed a meal, and then let out the back door by the sheriff.[96] Judges generally gave the police a hard time. When Coal and Iron Police were unable at one trial to produce maps and deeds to prove that arrested miners were actually trespassing on company property, the judge immediately dismissed the case and released the men.[97] With unsympathetic judges and juries, local police stopped bringing cases to courts.[98] Finally, local merchants, lawyers, doctors, and clergy who relied on the business and affiliation of their working-class neighbors also aided the independent miners and truckers and contributed to a climate that accepted coal bootlegging and isolated the formerly powerful managers and owners of the legitimate coal companies.

The bootleggers had support in the governor's mansion as well. Gifford Pinchot held off sending state police into the region to do the operators' bidding. One contemporary account explained his seeming failure to uphold law and order: "He said it wasn't customary to send in the state police unless the local authorities certified they were unable to maintain order. Back of his attitude there was considerable sympathy for the miners."[99] Things did not improve for the operators with Pinchot's successor. In 1934, George H. Earle III, a former Republican but a strong supporter of FDR, became the first Democratic governor in Pennsylvania in forty years and brought the "Little New Deal" to the state.[100] With his running mate, Thomas Kennedy, an anthracite-region native and long-time officer of the UMWA, Earle solidified the emergent alliance between the Democratic Party and organized labor and energized the party's base. Unable to enact his New Deal program in the first two years of his administration, due to a substantial Republican majority in the State Senate, Earle campaigned hard for FDR and the Democratic state ticket in November 1936 and was rewarded with a sweeping Democratic victory in Pennsylvania. For the first time in more than sixty years, the Democrats controlled both houses of the state legislature, and Earle was poised to press his legislative agenda.[101]

Following the election success, Earle called a series of public meetings as part of a tour he made of bootleg coalfields in December 1936; he would follow up his personal inspection with the appointment of the Pennsylvania Anthracite Coal Industry Commission to document the practice of bootlegging and make recommendations for government action. The presence at these meetings of local Catholic priests and Protestant ministers was particularly striking, as they rallied to support the bootleg miners in their congregations and communities. Father Bius Chesna, of a Lithuanian Catholic parish in Mahanoy City, sounded a common theme: "If there is no legitimate work for them, I think the men should be permitted to operate bootleg mines." J. H. McMillian, minister of a Welsh Congregational Church reinforced Chesna's testimony. "I am interested in this question from a moral standpoint," he began. "I believe that human life is more important than the legal or illegal mining of coal. . . . I don't think the operators should be allowed to work on their cheapest properties when there are mines idle."[102] Finally, E. O. Butkofsky, pastor of St. John's Reformed Church in Shamokin, offered something of a parable that summarized the dominant clerical perspective on coal bootlegging: "While there are legal questions involved with regard to property rights, no human being will condemn another when a man who is out of work and hungry will walk along a road, see an apple on a tree, and take that apple. We call it stealing, but morally the man is justified."[103] Respected religious leaders supported the efforts of unemployed miners to help themselves and their families. Their justifications for coal bootlegging were strikingly similar to the arguments for the equalization of work offered in the nearby Panther Valley.

So similar in fact, that Gildea, leader of the equalization movement, advocated on behalf of coal bootleggers in the pages of his newspaper, the *Coaldale Observer*. He reiterated the broader arguments offered at the governor's hearings:

Stealing of coal is not thievery in the accepted sense of the word, when it is done in broad daylight at the expense of intense physical exertion. Holes that idle men have spent months in sinking are not to be blown shut at a wave of official hands in some far distant office, especially when these same officials have already robbed men of their jobs, stolen the necessities of life from distressed women and deprived growing children of the opportunity to live richer and more fuller lives.[104]

In addition to emphasizing bootleggers' hard labor and the distress of their families, Gildea pointed to the absentee status of anthracite operators who made decisions to dynamite bootleg holes from a "distant office," presumably P&RC&I company headquarters in Philadelphia.

Gildea and the clergymen testifying in support of coal bootleggers were expressing commonly held sentiments. Journalists and others who came to the bootleg region found many adherents to these ideas. The journalist Louis Adamic interviewed Father Weaver, a Catholic priest from Mount Carmel, who spoke about coal bootleggers: "Some of them . . . are my parishioners; honest, upstanding working people. I'm proud to be their priest. It is absolutely untrue that this bootleg coal situation is having a bad effect on the bootleggers' character. . . . Coal bootlegging has no bad moral effect on the people. It keeps them from starving and turning into criminals." Weaver concluded, "They risk their lives every minute they work in those holes, and deserve everyone's respect and admiration. They have mine."[105] In a second inter-

view, with a writer employed by the Works Progress Administration (WPA), an area minister quoted Leviticus when asked about the Eighth Commandment and coal bootlegging: "And when ye reap the harvest of your land, thou shalt not make clean riddance of the corners of thy field when thou reapest, neither shalt thou gather any gleaning of thy harvest; thou shalt leave them unto the poor." The minister explicated this passage, saying, "Are not the miners the poor, gleaning a corner of the coal fields? I did not read that to justify the miners, necessarily, but to show that there is a spirit as well as a letter in the commandments of God."[106] In the struggle for public opinion, the support of Catholic clergy and Protestant ministers went far in neutralizing the charge of "stolen coal" that came from mine operators.

Bootleggers took the sermons of their religious leaders on faith and often constructed a religious justification for their actions. "God put the coal in these hills," said one miner. "He put it there for men to use. The companies fell down on their job. They say they can't mine the coal at a profit. They close the mine. But God put this coal here for use. We're using it. Sure, we'd rather mine it for the company. Better pay, better hours, safer work. But, if they won't mine it, we will."[107] And so religious and secular arguments complemented each other in defense of coal bootlegging.

Leaders of various local voluntary organizations similarly defended organized miners and attacked the P&RC&I management. In January 1934 officers of the United Polish Societies of Northumberland County wrote to President Roosevelt to protest "the abandonment of local collieries" by the P&RC&I. This ethnic fraternal organization minced no words in urging the president to find some means "to rid [the region of] the industrial despots now meting out a great calamity to a suffering humanity under its tutalege [*sic*]."[108] Less than two weeks later, the secretary of the Shamokin Association of Insurance Agents wrote to Secretary of Labor Frances Perkins, calling for an investigation of local conditions and urging action to alleviate the distress that was bound to follow the recent decision by P&RC&I to abandon five mines in the Shamokin area.[109] Just as in the equalization campaign in the nearby Panther Valley, residents in the bootlegging areas in the southern anthracite field united behind the miners in their communities.

The fact is that coal bootlegging became the major business in communities in the southern and western middle fields abandoned by the legitimate coal operators. Many neighbors supported coal bootlegging because it contributed to their economic survival. Farmers in the area sometimes set up a bootleg breaker in a barn and made a little money on the side. One farmer advertised at his gate, "FRESH EGGS AND COAL FOR SALE."[110] Former bootleg miners, investing one or two hundred dollars, might set up their own cleaning plant. A small operation on a raised platform, powered by an old car engine and using rollers to crush the coal and shaker-screens to separate it into different sizes, could provide a modest living for its operator and a laborer.[111] In addition, some area residents supported themselves trucking bootleg coal to surrounding cities. And finally, area merchants benefited from the money that the mining, processing, and sale of bootleg coal placed in circulation in their communities. Typically, it seemed that only coal operators and managers had anything bad to say about "stolen coal."

Morris Ernst, one of the commissioners appointed by Governor Earle to study coal bootlegging and make proposals to the state to deal with the problem, captured well the broader public sentiment in his report to the governor: "bootlegging is condoned by an increasing number, including the clergy, court juries, and other sections of responsible opinion [as] there has been a subtle process of differentiation between primary property and social property, induced by the raw necessities of making a living and keeping one's family out of the poorhouse or off the relief rolls."[112] The Pennsylvania Anthracite Coal Industry Commission issued its final report in 1938. The report is less noteworthy for its recommendations—none of which were acted on, including one for a state takeover of the mines—than for the documentation of the nature and scope of bootlegging and, as echoed in the words of Ernst, the reinforcement of a deep critique of private property and corporate power widely shared through the southern coalfields of the anthracite region.[113]

A coordinated, institutional response to bootlegging finally emerged in late 1940. Bootleg

coal production had grown to an estimated five million tons in 1939, and there were predictions that it would reach six million in 1940, more than 10 percent of the total of anthracite mined and processed. The rising figures greatly worried UMWA officials, who had remained largely silent on the matter and now saw significant numbers of union members working independently and not on contract. Equally troubled at the time was Ralph E. Taggart, president of P&RC&I. On December 9, 1940, Taggart met with UMWA officers to develop an answer to bootlegging. The meeting produced recommendations for the Commonwealth of Pennsylvania to hire unemployed miners in reforestation work and to close illegal operations, and for the coal companies to make concerted efforts to rehire men. The parties also asked the Republican governor, Arthur H. James, to get involved.[114]

Governor James immediately utilized a quasi-public agency that he had seen established through state legislation in 1939, the Anthracite Emergency Committee (often referred to as the Allocation Committee). The Committee—comprised respectively of three union, industry, and public representatives, all appointed by the governor—was empowered to set overall weekly production levels of anthracite based on market forecasts. Companies voluntarily agreed to produce an assigned quota of coal based proportionally on their prior sales, an arrangement that greatly favored major producers. Within weeks of the legislation, seventy-two coal producers, who accounted for 95 percent of all production, had signed on and agreed to abide by the decisions of the Emergency Committee.[115] The Committee represented the latest means to control anthracite output and boost prices, the Commonwealth of Pennsylvania now in effect supporting old cartel practices (something the unsuccessful anthracite code had not been able to accomplish during the NRA period).

The Anthracite Emergency Committee assumed the responsibility to formulate an answer to bootlegging. In early February 1941, the Committee announced a plan approved by the industry, the UMWA, and the governor. Coal operators agreed to reemploy as many certified miners who had been involved in bootlegging as possible, to lease land to remaining small-scale operators, and to buy their output (and the current store of bootleg coal) for processing and sale. The Committee, in return, would raise quotas for firms that successfully absorbed bootleg production.[116] The implementation of the plan generated some discord as UMWA miners attacked diehard bootleggers, whose organization had not been party to the agreement. However, by late July 1941, the Anthracite Emergency Committee reported that 60 percent of the coal that had been mined on a bootleg basis was now handled by established companies.[117]

The scheme developed by the Committee undercut coal bootlegging, but other factors helped as well. Established firms began to develop ways to lease coal lands to independent miners, provided that they sold their coal to the companies for processing. Increased production of major firms during World War II and the enlistment of men reduced the numbers of bootleggers apart from the institutional arrangement implemented by the Committee. Still, bootleg production persisted. The organization of those involved in coal bootlegging expanded in succeeding years, becoming known as the Anthracite Tri-County Independent Miners, Breakermen, and Truckers Association. The group maintained a presence in Schuylkill and Northumberland counties during the war, representing small operators in their lease arrangements, and kept up a lobbying presence in Harrisburg.[118] However, after 1941, bootlegging declined, became isolated, and was never again the mass movement it had been during the Great Depression.[119]

◆

Equalization, dual unionism, and coal bootlegging were three distinct strategies that emerged among anthracite mineworkers during the Depression decade as they sought to counter the mine closings and growing unemployment that came with the economic crisis. These strategies provide a new dimension to understanding labor struggles during the Great Depression. Developments within the anthracite region take labor struggles out beyond factory floor and union hall into the broader community. They reveal, as well, ordinary workers, rather than union leaders, as the key agents responding to economic crisis.

Responses to crisis in the anthracite region, however, were not monolithic, and it is important to consider the roots of the divergent strategies that emerged. The question boils down to this: Why did workers' responses divide along geographical lines within anthracite? The answer begins with intraregional differences in geology and economics. Firms in the northern field, the Panther Valley, and the Schuylkill Basin faced different circumstances and thus adopted different strategies to cope with overproduction. In the Panther Valley, the LNC sought to solve the problem of overproduction by shifting work from higher- to lower-cost operations, but still kept the vast majority of its collieries open. This strategy permitted miners to argue for equalizing work across all the company's mines. Two related factors then contributed to the success of the equalization movement—the fact that workers in the Panther Valley had to deal with only one employer and the fact that the Panther Valley was a relatively small geographical area with a limited number of towns that had strong connections historically. When Lansford and Coaldale miners marched to Nesquehoning in August 1933 to close down LNC's mines there, the mineworkers in Nesquehoning knew the protesters and joined their cause. The locals in the Panther Valley had for some time constituted a subdistrict within District 7 of the UMWA and had a history of working together to represent their interests. Finally, the union hierarchy in the larger UMWA district did not oppose the independent actions of the Panther Valley miners. Rather than declaring the local actions illegal and enforcing union discipline, district officials slowly came around, a bit reluctantly, but still they ultimately defended their members' actions, something that did not happen in District 1 in anthracite's northern field.

In the northern field, dissident miners attempted to launch an equalization campaign but met head-on the opposition of both district officials and employed miners. In the northern field, the first strikes deteriorated into battles between working and unemployed miners. No doubt the much larger size of the northern field and the lack of a history of cooperative relations among miners from different collieries undermined solidarity as the equalization movement took its first unsteady steps in the Wyoming and

Lackawanna valleys. Instead of facing a neutral union hierarchy, Thomas Maloney and his allies faced an actively hostile group of district officials, who succeeded in isolating the dissidents from the rest of the district's rank and file. Thus what began as a movement for equalization of work at the Glen Alden Coal Company degenerated into a factional fight that the District 1 insurgents were bound to lose.

Finally, the economic circumstances of anthracite mining in the Schuylkill basin played a crucial role in steering local miners from their first efforts to secure equalization of work toward coal bootlegging that became that area's dominant response to economic crisis. The Philadelphia & Reading Coal & Iron Company faced a much more profound crisis than did LNC; moreover, the setting in Schuylkill and Northumberland counties differed dramatically from what prevailed in the Panther Valley. In the summer of 1933 when Panther Valley miners marched on the headquarters of LNC, only one of the company's six major mines—the Rahn Colliery—had been closed down. While management strongly opposed the equalization of work across company operations, miners and their community allies could make a plausible case for the spreading of work across all the collieries in the Panther Valley. To the south and west, in the domain of the P&RC&I, the situation in mining communities was far more catastrophic. In Mahanoy City, the company had closed six of seven collieries; in Shamokin four of five collieries stood closed; in Lykens, the community's lone mine had been shut down. Miners simply could not make a case for equalization of work when such a large share of operations stood idle. The P&RC&I problem was particularly acute because of the company's disastrous centralization program, which had led to the construction of mammoth central breakers at St. Nicholas and Locust Summit. The heavy interest charges that the firm incurred for these projects made it next to impossible for the company to respond to pleas of unemployed miners and their neighbors to reopen small mines and spread the work. Central breakers and increasing strip-mining operations could not accommodate the call for the equalization of work.

So miners idled by management policies at P&RC&I took the law into their own hands,

opened up veins common in the area, and became coal bootleggers, in the language of the day. And like their fellow mineworkers in the Panther Valley, they had a measure of success. Their tactics resonated with neighbors, including middle-class owners of small businesses and local clergy, and coal bootlegging gained community backing. The movement brought income into their communities and helped keep families off relief. While mine officials viewed coal bootleggers as insurgents outside the law, community sentiment was quite different. Neighbors emphasized the strong work ethic and the striving for economic independence that characterized coal bootleggers rather than the formal illegality of their actions. Thus there was little conflict in the Schuylkill area such as that which erupted in the northern field between working and unemployed miners. Virtually everyone at the western end of the southern field was unemployed, and virtually everyone went into business for themselves as coal bootleggers. In response, coal company managers expressed outrage over "stolen coal," but theirs were lonely voices that generated scant sympathy for the monopolistic coal companies they represented.

The differing circumstances in these three areas of the anthracite region explain the distinct paths of protest that emerged during the Depression. Different as these three approaches were, they were based on fundamentally compatible values and beliefs. The miners acted, as one local union put it in a resolution that it sent to John L. Lewis, with "the purpose of securing an equitable distribution of the fruits of labor."[120] Coal bootlegging and dual unionism emerged only after months of unsuccessful efforts to reopen mines or to convince coal companies to equalize production across all collieries. Miners wanted work, not relief, and they urged their employers and their union to find a way to distribute what limited work was available in an equitable fashion. In doing so, working miners questioned the purely economic calculus of their employers and the business unionism and autocratic rule of the UMWA. With the equalization of work and coal bootlegging, insurgent miners presented alternatives to the prerogatives of capital that helped them cope with the rigors of the Depression. They could not address the long-term causes of eco-

nomic decline in Pennsylvania's anthracite region, but they could help miners piece together
a living that helped their families survive hard
times until the renewed demand for coal that accompanied war preparedness campaigns and
U.S. direct involvement in World War II provided a boost to the local economy. Given the
failure of New Deal and state efforts to revive
the anthracite industry, even the partial success
of miners' struggles is quite remarkable.

The lesson that emerges from their stories is
that they succeeded by turning their individual
and class needs into community struggles. Thus,
when Panther Valley mineworkers took the
cause of equalization beyond the limits of the
UMWA and it became a community campaign
joined by local clergy and professionals, only
then did the movement have any chance of success. So, too, coal bootlegging worked only because the miners could depend on overwhelming community support to shield them from
strict enforcement of the coal companies' property rights. Clergy acting from time-honored
traditions of social justice and storekeepers desperately in need of miners' purchasing power
were powerful allies in support of coal bootleggers. Conversely, when District 1 insurgents
neglected to mobilize community support for
their campaign against mine management, they
were quickly outmatched and soundly defeated,
with devastating, even deadly, consequences. By

reaching out to the broader community, by iden-
tifying their specific interests with those of their
families, friends, and neighbors, mineworkers
became an almost irresistible force.

The high point of labor mobilization and
protest in U.S. history occurred during the
Great Depression of the 1930s. The organization of millions of unskilled and semiskilled
workers in the nation's mass-production industries was the signature achievement of the period. Like mineworkers in the Pennsylvania anthracite region, workers in the steel, automobile,
rubber, electrical, and other major industries
drew vital support from sympathetic community
members, pro-labor civic leaders and politicians,
and New Deal legislation. In contrast, the celebrated labor campaigns of the era featured national labor figures and the building of bureaucratic unions; in anthracite, union officials faced
an energized rank and file intent more on shaping concrete responses to the Depression crisis
than on formal union victories. And while the
CIO achievement of union recognition and contracts in the mass-production industries seriously challenged management prerogatives, the
equalization of work movement and coal bootlegging not only disputed capital's right to manage but also the very sanctity of private property.
In the course of their struggles in the 1930s,
Pennsylvania anthracite mineworkers forged
their own New Deal.

CHAPTER 4

Reprieve and Final Collapse, 1940–1970
Capital and Labor Respond

THE MINERS' NEW DEAL RE-
vealed working-class solidarity in the face
of economic crisis and an anthracite working class that could speak to and for the broader
communities in the region. The mineworkers'
struggles represented a response to the national
Depression as much as to anthracite's decline,
and drew strength and legitimacy from the recovery and reform measures of the Roosevelt
administration. As the nation and region moved
beyond the Depression, the crisis in anthracite
persisted in the postwar period and new initiatives would be needed to stabilize the region.
The collective grassroots responses that characterized the Depression decade gave way to the
institutional responses of organized capital and
labor. Unfortunately, the anthracite operators
and the UMWA lacked the commitment to the
industry and region that might have enabled
them to meet the new challenges of the postwar
period. Instead, their actions contributed to further decline.

The Depression did not end at a singular moment. Yet, with the outbreak of war in Europe in
August of 1939 and an increasing emphasis on
war preparedness by the federal government,
employment in the anthracite region steadily
improved. Adoption of an Allocation Plan by the
Anthracite Emergency Committee simultaneously reduced cut-throat competition among
producers and undercut coal bootlegging. Conflict over the equalization of work and coal boot-

legging diminished as unemployment shrank. In
1941 production expanded significantly as anthracite output exceeded 56 million tons and
miners worked on average more than 200 days.
Production then moved steadily upward, peak-
ing at 64 million tons in 1944, when mines operated on average 292 days. Anthracite firms made
these gains in output in the face of a tight labor
market caused by the departure of tens of thousands of men into the military. The anthracite
labor force contracted steadily throughout the
Depression and World War II. The region's
mines had employed 109,000 workers in 1934,
but that number declined to 88,000 in 1941 and
to 73,000 in 1945. Yet while operating fewer
mines and employing fewer mineworkers, anthracite companies achieved tonnage increases,
by using their facilities and labor power more
steadily across the year. Between 1935 and 1937,
for instance, mines had operated 190 days a year
on average; between 1943 and 1945, mines remained open 277 days a year.[1] Mineworkers'
earnings and company profits both improved
dramatically during World War II.

While anthracite mining continued as the region's dominant industry, averaging roughly
80,000 employees over the course of World War
II, its trajectory was clearly downward as employment slipped steadily, even with the wartime expansion of demand for coal. At the same
time, manufacturing increased in importance
and in the 1940s displaced coal mining as the

85

dominant economic sector in the region. In 1939 industrial employment stood at about 40,000 in the leading anthracite counties, Lackawanna, Luzerne, and Schuylkill; by 1947, that figure had grown to 77,000.[2] About a third of manufacturing employees—primarily women—found work in the growing apparel sector of the regional economy.[3]

Despite the transition to a wartime economy, the anthracite region was not well situated to take advantage of growing defense-related demand in the early years of World War II, due to the dominance of coal mining in the regional economy. One contemporary study described the growth of anthracite operations in 1942 as "disappointing," and noted as well "critical material shortages" in the region's textile industry. In November 1942 industrial activity in the state of Pennsylvania reached a level more than 80 percent above that of the late 1930s; yet for Scranton and Wilkes-Barre corresponding increases were only about 20 and 40 percent respectively.[4] A 1942 report of the Anthracite Coal Commission commented on declining job prospects in the regional silk industry decimated by the cutoff of silk imports. With stronger job prospects elsewhere, many young people in the anthracite region were leaving. The report noted: "From the Scranton region alone an estimated 5,000 workers have left during the past year for employment outside of Pennsylvania. In the cities of Bridgeport, Philadelphia, Baltimore, and Washington, where surveys were made in October 1941, a total of 6,500 workers were found who had left the anthracite counties within the year."[5] The Commission called for the locating of defense plants in the region to stem the flow of out-migration and provide employment and income to families.

But the migration did not diminish, and the younger generation in the anthracite region, whether single or married, left in great numbers. Northern New Jersey, with strong defense-related employment, was one common destination. Kathryn Sudol, for example, the child of Italian immigrants, grew up in the anthracite town of Mahanoy City. The prospect of piecework in a local shirt factory did not appeal, so at the age of twenty-one in 1942, Kathryn moved to Paterson, New Jersey, where she worked as a

secretary for the Army Chemical Warfare division. She sent money home from each paycheck, small contributions to help her parents and younger siblings. Better job prospects, higher pay, and the independence of living away from home as a single woman in her early twenties all appealed to this miner's daughter.[6]

Stanley and Selina Woodring also migrated from the anthracite region to New Jersey during World War II. They came from the Wilkes-Barre area, and Stanley had worked around the mines for seven years when he first came to New Jersey to work in the shipyards at Kearny. After he had been employed for a year, his wife and children joined him. Near the end of the war they qualified for subsidized housing in a federal government-sponsored development in Winfield. Selina recalled years later that this was the first house she had lived in with indoor plumbing and central heating—quite an improvement over the outhouses and coal stoves of her childhood home. Selina had worked in a Wilkes-Barre silk mill before she moved, but she found engine testing at Lawrence Engineering a better job opportunity.[7] Defense employment was the draw for anthracite region out-migrants, for without such work, economic prospects in the region were modest, and the war-induced growth and prosperity evident in many urban areas largely bypassed the region.

As wage increases failed to keep pace with the skyrocketing cost of living, labor conflict again erupted in anthracite, reflecting workers' continued dissatisfaction with their union. A three-week strike in January 1943 saw some 15,000 to 25,000 anthracite mineworkers protesting a fifty-cent-a-month dues increase that had been passed by a United Mine Workers of America (UMWA) national convention the previous fall.[8] The issue of an increased levy on members by the International union (the official term for the UMWA, which included two Canadian districts) had been festering for more than a year.[9] Concentrated in the Wilkes-Barre operations of the Glen Alden Coal Company, the strike seriously cut into production and led the War Labor Board to hold hearings and lobby intensively with leaders of UMWA locals.[10] The protest was in fact a wildcat strike, opposed by district and national UMWA officials. It drew support regionally from less than a third of anthracite miners, and after intense governmental and UMWA pressure striking miners returned to work after three weeks. John L. Lewis and local leaders convinced striking miners to call off the strike to permit Lewis and UMWA negotiators to press for substantial wage increases when industrywide contracts expired on May 31. On June 1, with negotiations stalled, 83,000 anthracite mineworkers joined more than 400,000 in bituminous in the first nationwide shutdown of the coal industry since the strike of 1922. With confusion over the oversight of coal production by the War Labor Board, a threatened total government takeover of the mines, the intervention of Secretary of Interior Harold Ickes to dictate an accord, and several announced and then aborted truces, the ranks of striking miners nationally swelled and ebbed through the month of June. By the end of the first week of July, with negotiations renewed through Ickes's office, the final diehard miners returned to work.[11] In characteristic fashion, Lewis then moved first to negotiate an accord for bituminous, a contract renewal approved by the War Labor Board in November. With this precedent set, anthracite operators and the UWMA reached a settlement that raised anthracite mineworkers' wages by slightly more than a dollar a day.[12]

Indicative of the union's role in anthracite, the January 1943 wildcat strike was aimed at the UMWA rather than operators. Miners also demanded a wage increase of two dollars a day to address the rising cost of living in the anthracite region, but it was their anger at the recent dues increase that led them to take action. The dues increase, in the view of one local president, was "the last straw."[13] Anthracite miners felt neglected and exploited by union officials, their dues going to fill the UMWA's national coffers while miners and their families struggled with rampant inflation and an uneven regional economy. John L. Lewis had not visited the region in years, and anthracite miners felt slighted.[14] After strenuous UMWA efforts extinguished this fire, contract negotiations still dragged on—to the frustration of anthracite miners. Anthracite mineworkers resented the way their wage talks were invariably postponed each year until the much more economically important negotiations for bituminous were completed. An-

thracite's 75,000 workers held minor national significance in comparison to more than 400,000 in bituminous. A local newspaper editorial expressed the common feeling that anthracite was generally "regarded as the tail of the kite," with little attention focused on the region's distinct conditions and needs.[15]

Still, the World War II period provided a measure of respite to the industry and mineworkers ravaged by a decade of Depression. Two years, 1943 and 1944, were particularly prosperous. Working six days a week, anthracite mines reached production levels that had not been achieved since the late 1920s. The Lehigh Coal & Navigation Company, through its subsidiary Lehigh Navigation Coal (LNC), produced almost 4.7 million tons of coal in 1944, more than 90 percent above its output for 1939. LNC enjoyed profits of more than $1.25 million, while consolidated profits for the parent company as a whole—primarily railroad and coal operations—reached almost $3.4 million. The dividend of $1 per share was the highest the parent company had paid since 1931. Workers, too, enjoyed unaccustomed prosperity, typically earning time-and-a-half overtime pay for their sixth day of work during the week.[16]

Yet victory in Europe and Japan and conversion to a peacetime economy revealed the slender basis for the anthracite region's wartime prosperity. Income in the region shot up in the last years of the war, but full employment depended on the exodus of residents into the armed services or into war industries beyond the region. What would happen when wartime demand for coal eased, defense industries laid off workers, the armed forces demobilized, and veterans and recent out-migrants returned? The Panther Valley faced problems indicative of difficulties throughout the region. By the end of 1944, some 1,200 employees of LNC were serving in the armed forces, their former jobs assured to them upon their return, while somewhat more than 4,600 continued to work for the company.[17] The wartime experience of anthracite was no indicator of postwar prosperity.

Developments at LNC provide a useful starting point to survey the postwar period for the anthracite industry and its mineworkers. As demobilized veterans returned, LNC management struggled to maintain the company's profitability in changing economic circumstances. The company made extensive investments to increase strip-mining operations and develop new uses for anthracite wastes. Employment rose with the return of valley veterans who had worked previously for the company, but overall company output declined. In 1948 LNC had a workforce of 6,200 and an output of 4.3 million tons, down modestly from the wartime peak of 4.7 million tons in 1944. With declining demand for anthracite, the company closed its Nesquehoning breaker and processed coal from Nesquehoning mines at its more modern breaker in Lansford. Despite such cost-cutting efforts, profits for the coal company declined sharply, from a peak of more than a $1,250,000 in 1944 to a scant $19,000 in 1948.[18] Company policy in this period entailed a concerted effort to maintain output even in the face of slumping demand. Management undertook a series of steps intended to cut costs and thereby maintain a margin of profit even in the face of weakening prices. The increasing reliance on strip mining was a major element of the cost-cutting effort as were investments to improve coal cleaning in the breakers and to recover larger amounts of fine coals that were increasingly in demand. Finally, LNC management tried also to reduce labor costs in 1947 by ordering wage deductions for mineworkers who left their posts before what was considered proper quitting time, a move that led to three years of intermittent sit-down strikes before management relented.[19]

The one area of their business that management could not control was the overall demand for anthracite coal. As one former officer of LNC put it, "the bottom fell out of the market for anthracite" in the winter of 1948–49 and never recovered. An unusually warm winter and increasing competition in the home heating market from alternative fuels led to a cut in anthracite sales in 1949 of almost 25 percent. By 1953, sales had fallen by another third, and at 27 million tons, regional sales were almost exactly half of the 1948 level.[20] These successive declines in sales led LNC and other producers to cut back on operations; once again the mines were operating only two and three days a week.[21] Miners' paychecks sagged accordingly, and young people saw no future for themselves in the region.

In the face of declining demand for coal, LNC closed its mines at the end of April 1954 and opened negotiations with the UMWA and its local unions on a plan to continue scaled-down operations. Until that date three breakers had been operating, with a combination of strippings and underground mines supplying coal to Tamaqua, Coaldale, and Lansford breakers. The company proposed to cease all strip mining, close two of the three breakers, and transport all coal to the Coaldale breaker for processing, leaving 1,000 Panther Valley mineworkers unemployed. In addition, the plan called for the payment of contract wages proportional to miners' actual output instead of "consideration wages" that had been common. Companies paid consideration wages to contract miners, normally paid by tonnage, in those circumstances when the work assigned did not permit miners to cut enough coal to meet their customary earnings. When firms had to pay consideration wages, their wage costs skyrocketed and they had difficulty operating in the black. Over time, LNC had struggled with UMWA locals over this issue, with management feeling that the practice gave miners inadequate incentives to increase their productivity.[22]

Negotiations ebbed and flowed in May and into June, and ultimately five of the six locals and the International union accepted the company plan. Dissatisfied mineworkers in the Tamaqua Local 1571, however, picketed collieries across the valley when the company attempted to implement the plan and resume operations. Members of the Tamaqua local balked at the proposed settlement, arguing that it violated their UMWA contract. They called for further negotiations and argued that equalization was the proper response rather than the plan proposed by LNC's president Julian Parton. Union mineworkers refused to cross the picket lines set up by the Tamaqua local members, even in the face of an order to return to work from UMWA International president John L. Lewis, and in late June LNC management announced the closing of all underground operations indefinitely. The mines never reopened under LNC auspices.[23]

While fellow mineworkers would not cross the Tamaqua local's picket lines, Panther Valley mineworkers did not receive the support of oth-

End of the shift, No. 6 mine, LC&N, 1953. Photograph by George Harvan.

ers across the anthracite region as they had during the equalization campaign of the 1930s. "You won't find too much sympathy here for the Panther Valley men," said one Pottsville miner. "They're strike happy." A former contract miner had similar comments: "in the Panther Valley they carry this 'all for one' business to extremes." The spirit that had animated the earlier equalization struggles had survived among the Tamaqua stalwarts in May and June of 1954, but they were now a distinct minority in the Panther Valley and were isolated from the mainstream sentiment among anthracite mineworkers more generally at this date. Times had changed even as some miners kept the old faith.[24]

On one level, the closing of the Panther Valley mines can be read as a tragic misunderstanding, and that is the interpretation that former president of LNC, Julian Parton, offers in his memoir, *Death of a Great Company*. From Parton's perspective, militant miners misjudged the seriousness of the company's condition and management's resolve to wrest concessions if operations were to resume. James Gildea, local newspaper editor and former head of the Equalization Movement during the 1930s, receives pointed criticism for his leadership in the crisis. Parton suggests that if democracy had prevailed, with less grandstanding, majority sentiment

among the miners would have prevailed and miners would have returned to work, and they and the company would have been better off.[25]

At the same time, though, Parton provides a crucial account of a second story unfolding simultaneously, one concerning financial machinations of the New York investment firm of Model, Roland and Stone that were also influential in determining the outcome of this conflict. Finding that the value of Lehigh Coal & Navigation Company's assets (fully $26 a share) far exceeded its share price of $8.50, the firm and its principals began to purchase large blocks of company stock in 1951.[26] These investors anticipated breaking up the company and selling its component parts for far more than they paid for its shares. The proposed sale of the company's Lehigh and Susquehanna Railroad to the Central Railroad of New Jersey alone would have provided stockholders with a payout of $13 a share. At the same time, another large block of shares came into the hands of C. Millard Dodson when the parent company, LC&N, purchased the Weston Dodson Company, an independent anthracite mining firm, and Dodson joined the company's Board of Managers. At the end of 1953, Leo Model, now representing owners of 20 percent of LC&N stock, also joined the Board of Managers. The new majority on the

Two miners drilling in the Mammoth Vein, Coaldale No. 8, ca. 1950. Photograph by George Harvan.

Board requested the resignation of the parent company's president and steered LC&N toward its breakup, seeing more profit in the sale of its assets, particularly its railroads, than in its continued operation as a coal mining and transportation company. Certainly labor conflict and declining markets for anthracite contributed to the company's demise, but even had the mining operations been profitable, it is very likely that Model and his associates would have followed their course in dismembering Lehigh Coal & Navigation and selling its parts.[27]

With mining discontinued and the prospect fading that LNC would ever reopen the Panther Valley mines, the company arranged to lease its mines and breakers. Two stripping contractors, James and Frank Fauzio, teamed up with the former LNC president and vice president, W. Julian Parton and Joseph Crane, to form the Panther Valley Coal Company and lease the Lansford Colliery and the Nesquehoning mine from the coal company. Mining resumed, though on much reduced terms, in October 1954. A second operator, the Coaldale Mining Company, leased the Coaldale mines and breaker. By the end of 1955, with leases on 70 percent of its coal operations, LC&N began to turn a small profit.[28]

The leasing arrangements lasted only five and half years, though, and employment continued to decline. There had been some 4,000 miners

employed sporadically by LNC in the Panther Valley in its last year of operations, but only 2,300 returned to work when the Panther Valley Mining and Coaldale Mining companies leased LNC mines. The new mines signed contracts with the UMWA and hiring was based on seniority at the earlier LNC operation.[29] The two new companies merged in 1955, with mining at the Panther Valley Coal Company closing down at the end of 1957, while the Coaldale Mining Company remained in business through February 1960. At their peak in 1957, the two new companies produced between them only about 1.1 million tons of coal, less than half of the 2.5 million tons that LNC had produced in 1952. Short work weeks and suspended production were common in the period. In its last full year of operation, the Coaldale Mining Company produced less than 600,000 tons of coal and its workforce had shrunk to 1,200 men.[30]

The Coaldale Mining Company, despite a substantial loan of more than $5 million from the UMWA, simply could not make a steady profit in the shrinking market for anthracite.[31] Declining demand for anthracite doomed the effort to provide continuing employment in underground mining in the Panther Valley. In 1962 the Fauzio brothers leased the LNC coal lands, and four years later they purchased them outright from the company. Organized as the Greenwood

Motorman at No. 6 mine,
LC&N, 1952. Photograph by
George Harvan.

Mining Company, they undertook strip mining and built a modern breaker at Greenwood, near the western end of the valley. Throughout the 1960s the output of the Greenwood Mining Company averaged only 640,000 tons annually, and its workforce numbered a few hundred.[32]

With mining lands providing modest leasing income for a decade after the closing of their underground operations, the Board of Managers of LC&N was free to search for other ways to turn a profit. Diversification and dismantling were the competing strategies, and by the early 1960s the latter approach had won out. C. Millard Dodson, who had joined LC&N from his own anthracite mining and selling company, favored the route of economic diversification. As president of the company, he mapped out a strategy that included purchasing a fleet of ocean-going tankers and buying a bituminous coal company and coal lands in West Virginia. Although these two ventures never proved particularly profitable, earnings from land sales and resort development in the Poconos and from two railroads provided modest returns for the company's stockholders.[33]

Still, the New York City investment banker, Leo Model, chafed at what seemed to a financier the slim pickings of industrial operations. By 1961 he had successfully arranged the sale of the Lehigh and New England Railroad. The next year the firm reorganized, shifting all its active assets to a new subsidiary, the L.N.C. Com-

pany, while keeping only the Lehigh and Susquehanna Railroad, under lease to the Central Railroad of New Jersey (CRRNJ), with the original parent firm, LC&N. The new subsidiary proceeded to sell off its remaining properties and to make substantial payoffs to its stockholders. LC&N paid out dividends to its shareholders based on its railroad rental income, but with the bankruptcy of the CRRNJ in 1967, that cash flow was suspended. For a brief period the firm tried to capitalize on losses that could be carried forward for tax purposes by investing in candy manufacturing, while renewed railroad rentals provided the firm with some small reason to continue in operation. In 1978, however, LC&N sold its remaining track and railroad stock to ConRail and in 1986 the firm dissolved, selling its last candy-making operations to Tootsie Rolls.[34] Lehigh Coal & Navigation, having mined its last anthracite in April 1954, zigzagged through three decades of fitful strategies that bled its resources without ever providing the windfall profits that New York investors had counted on making when they swooped down on the company to make a killing.

◆

While the conflicting imperatives of disgruntled mineworkers and outside stock speculators played out in the Panther Valley and the board rooms of LC&N, similar scenarios emerged in

Inclined planes, No. 6
breaker, Lansford, 1950.
Photograph by George
Harvan.

the operations of the Glen Alden Coal Company, the leading anthracite firm in the northern
anthracite field. The product of the segregation
of the coal lands and mining operations of the
Delaware, Lackawanna & Western Railroad
(DL&W) from its transportation activities, the
Glen Alden Coal Company accounted for about
15 percent of all anthracite production in the
1920s and 1930s. Glen Alden rapidly became a
major player among anthracite firms over the
next forty years, though its output declined
along with that of the industry as a whole.
Throughout the period Glen Alden stood as the
first or second largest producer of anthracite
coal, holding onto its share of overall output.[35]
Its president, W. W. Inglis, played a leading role
among anthracite operators in this period, and
he had a solid, cooperative relationship with the
president of the UMWA, John L. Lewis.[36]

Good working relations with fellow operators
and with the UMWA, however, did not spare the
Glen Alden Coal Company the difficulties of ad-
justing to declining demand for anthracite coal.
When Glen Alden purchased the DL&W Railroad's coal lands and collieries, it acquired mining
operations that at their peak had produced 11.5
million tons of coal annually; by 1950 total output
was less than half that figure.[37] Over the decades
of decline, Glen Alden's managers made difficult
decisions. To reduce their costs of production and
maintain profits in the face of slackening demand,

they repeatedly closed mines, laid off mineworkers, and consolidated operations at fewer locations. In 1917, DL&W had operated twenty-
three large collieries, each producing on average,
half a million tons of coal annually. By 1931, its
successor, Glen Alden, operated twenty major
collieries each producing on average 400,000 tons
a year. By 1937 the company had only fourteen
working collieries and had cut its overall workforce from more than 29,000 to just under
18,000. During World War II, increased demand
enabled the company to keep sixteen collieries in
operation, but its workforce declined by more
than 30 percent to just over 12,000.[38]

Glen Alden's response to continuing slippage
in demand for anthracite in the postwar period
was further consolidation and efforts to reduce
labor costs and overhead expenses. In 1954 the
company sold off its retail sales outlets—Household Fuel Corporation and Burns Brothers—
and tried to focus on its mining operations. At
the same time the firm sold its coal lands, accompanying buildings, and equipment in the
Scranton area to concentrate its operations entirely in the Wyoming Valley.[39]

The company streamlined in anticipation of
efforts to diversify the company's product line;
the firm's strategy became apparent in the ultimate renaming of the company in 1967 as the
Glen Alden Corporation. After the firm acquired a Texas air conditioner manufacturer, the

Skeleton of Coaldale No. 8 breaker; sold for scrap, 1965. Photograph by George Harvan.

Mathes Company, and a manufacturer of specialty and fire trucks, Ward LaFrance Truck Corporation, it dropped "Coal" from its corporate name. This strategy involved a conscious choice to balance risks and cease to be dependent solely on the recovery of the anthracite home heating market. In addition, the move permitted the firm to employ the losses in its coal operations to counterbalance for tax purposes potential profits in new lines of business. These moves, though, suggested that the firm's commitment to production of anthracite coal and operation in the anthracite region was beginning to wane. By 1958 the company derived more than 40 percent of its sales revenues from air conditioner and truck sales. Furthermore, annual reports revealed continuing sales of surface lands and coal reserves, moves that improved the company's short-term bottom line while they positioned the company for the eventual abandonment of mining operations.[40]

Glen Alden managers and stockholders initially exerted a measure of control over these strategies, but eventually other actors came into the picture with far greater resources. The first acquisitions seem to have been well-considered moves into more promising sectors of the economy, which also permitted the company to take advantage of its cumulative losses in mining operations. In 1959 the company merged with List Industries Corporation, owner of the RKO chain of movie theaters. While the first two acquisitions had been the conscious decisions of Glen Alden management, the List merger resulted from an effort on the part of the principals at List Industries to acquire sizable blocks of Glen Alden common stock. In what was essentially an unsolicited takeover bid, List Industries acquired ownership of 38 percent of Glen Alden stock. As one contemporary commentator noted in 1958, "Now it is List that calls the tune."[41]

And management's tune changed abruptly. With the ascension of Albert A. List as the company's president and chairman of the board, the firm entered a period of feverish activity on the financial front. The company purchased the coal lands and operations of the Hudson Coal Company and sold off its Mathes and Ward LaFrance divisions as well as textile finishing operations that had been part of the List empire. Financial transactions rather than production became the central focus of management in the early 1960s as the company bought and sold operations in rapid succession. Annual reports reveal the increasing importance of profits from the purchase and resale of businesses rather than from the operations themselves. Indicative of this new strategy, management reported that the company held increasing shares of its assets in cash and bank certificates of deposit—more than $25 million in January 1964—a development that placed the company in an advantageous position to respond rapidly on prospective acquisitions or other investment opportunities.[42] As in the case when the New York City investment firm of Model, Roland and Stone gained a large stake in LC&N, the merger with List Industries steered Glen Alden in new directions.

The final two steps in the transformation of Glen Alden came in rapid succession. On January 1, 1966, Glen Alden divested itself of its remaining coal lands and operations by sale to the Blue Coal Corporation, a new entry into the anthracite mining business. The next year the company merged with three firms, the BVD Company, Philip Carey, and Stanley Warner, national manufacturers of men's clothing, building products, and women's undergarments and family products, respectively. From a struggling anthracite coal company seeking to diversify its product line, the Glen Alden Corporation emerged in 1967 as a giant conglomerate with $537 million in annual sales and no coal operation. With the addition of the liquor giant Schenley Industries the next year, annual revenues soared to almost $1.25 billion. Only the "Glen Alden" in the company's name provided a link to the former coal giant that had struggled to survive a decade earlier.[43]

Between the end of World War II and Glen Alden's sale of its coal lands and operations January 1, 1966, output and employment moved steadily downward. Between the closing of collieries to accommodate declining demand for anthracite and the company's increasing reliance on mechanization, the size of the mining workforce shrank precipitously in these years. At the end of the war, the company employed just over 12,000 workers and operated sixteen breakers with an annual output amounting to almost 7 million tons of coal in 1945. Although the first three postwar years saw increased output and growing employment with the return of veterans to the region, by 1950 output had declined below the 1945 level.

When Francis O. Case took over as company president in early 1953 he inherited an operation that had drifted along. Output had declined to six million tons the previous year, though employment remained at the 12,000 level. Case committed the company to extensive investment in expensive mechanical mining equipment that offered the possibility of dramatic improvements in productivity. As a manager, W. W. Everett, explained to Case in May 1954, investment in continuous mining machines, universal cutters, and loading machines required multishift operations "to obtain the economies which are necessary for us to stay in business." The tradeoff between cutting costs and maintaining employment was striking. Everett noted the need to "close down places which are tied to bad practices" and that meant cutting back the workforce. In the end he called for a compromise that would "permit the company to operate at a profit and still furnish reasonable employment to a number of men who, otherwise, would have no earnings at all." This compromise, however, proved difficult to sustain as annual output for the company declined to four million tons precisely in the period that use of continuous mining machinery provided an opportunity to further reduce the size of the workforce.[44]

Beginning with the efforts to mechanize coal cutting at the face, total employment at Glen Alden shrank rapidly. An internal company memo in January 1954 noted less than 7,000 employees and predicted that "further reduction [in employment] can be achieved without diminishing production." By May 1956 the company's workforce stood at 4,800. And as coal began to take a backseat in the plans of the Glen Alden Corporation, and product diversification became the management byword, that number declined still further. By the time that Glen Alden sold its coal lands and operations in January 1966, its anthracite workforce had dwindled to only 2,000 employees.[45]

With the sale of Glen Alden's coal lands to Blue Coal Corporation in 1966, coal mining in the Wyoming Valley sputtered along. The new company went through numerous ownership changes in the next quarter century, with the Central States Pension Fund of the Teamsters' Union playing a major role in financing one of the purchases. As the company's debts to the

Anthracite Health and Welfare Fund, to Luzerne and Lackawanna counties, and to local municipalities mushroomed, the company went into bankruptcy in 1976, from which it emerged only in 1992 with its sale to a prospective developer, the Earth Conservancy. After Glen Alden's exit from anthracite in 1966, coal mining in the Wyoming Valley came to an end, for all practical purposes.[46]

◆

The Schuylkill coal basin provides a third case study of the closing of mine operations in the anthracite region, one that echoes the examples of LC&N and the Glen Alden Coal Company but with notable differences—including the place of bankruptcy proceedings and state agencies and the efforts of an independent coal company to survive in the face of Philadelphia & Reading Coal & Iron's determination to liquidate coal operations and diversify holdings. This story pitted a long-time leaseholder, the St. Clair Coal Company, against its lessor, the P&RC&I, the anthracite region's leading owner of coal lands. The conflict between leaseholder and landowner adds a new element to the role of capital in regional economic decline, but the central dynamics evident in the LC&N and Glen Alden cases emerged in the southern anthracite field as well.

The St. Clair Coal Company traced its origins to 1853, when Enoch McGinness sank a deep-shaft mine in the town of St. Clair that linked up with an existing slope mine. Poor construction and various mishaps limited operations and financial returns for McGinness, who was forced to sell the operation in 1855 to recoup his losses. Succeeding owners fared no better, and in 1864 they leased the mine to a group of Boston investors who organized the St. Clair Coal Company. In 1871, the Philadelphia and Reading Coal and Iron Company (PRCI) purchased the St. Clair tract and took over the operation of the mine. The purchase was part of an aggressive campaign by Franklin B. Gowen, president of the Philadelphia and Reading Railroad, to acquire extensive coal lands in the southern coal basin and thereby exercise a measure of control over anthracite production and prices. The PRCI managed mine operations in St. Clair directly at first and then through lease. In 1895, a reestablished St. Clair Coal Company signed a long-term lease with the PRCI.[47]

The St. Clair Coal Company maintained mine operations for more than six decades as the Philadelphia and Reading Railroad, and its mining subsidiary underwent successive reorganizations. J. P. Morgan rescued the railroad in 1896, creating a holding company that included the railroad and the PRCI.[48] A court-ordered separation of transport and mining activities in 1923 then transformed the PRCI ostensibly into an independent entity (with a slightly altered name, the Philadelphia & Reading Coal & Iron Company [P&RC&I]).[49] The P&RC&I faced deteriorating circumstances with the postwar decline of the anthracite coal market and the Great Depression. With $4 million in losses in 1936 and unable to meet debt obligations to its bond and mortgage holders, the company filed for bankruptcy in February 1937.[50]

The P&RC&I remained under court protection for seven years as its fiscal solvency was restored. The bankruptcy proceedings began a two-decade process of the selling off of assets that would eventually lead to the demise of the St. Clair Coal Company. The court-administered reorganization of the P&RC&I overwhelmingly favored the interests of the major creditors of the company (and other influential businesses that had claims on the firm). Petitions for redress by individual residents of the anthracite region, anthracite communities, and the UMWA received scant attention and no action. The bankruptcy filing also protected the company from demands on the part of UMWA locals at P&RC&I collieries to abide by the equalization of work provision contained in the union's regional contract.[51]

Judge Oliver Dickinson presided over the P&RC&I's bankruptcy case and in a controversial move he left the company's management in place without appointing an independent trustee with authority to judge claims, dispose of property, and dispense awards.[52] The company's executives maintained control of operations and decisions to sell assets, though court permission was required for their actions. They could even recommend their own salary increases, which received regular approval; Ralph Taggart, president of the P&RC&I earned $60,000 in 1938, more than sixty times the average income of employees of the firm.[53] Various court challenges to

Dickinson's handling of the case and objections from the Securities and Exchange Commission led to the appointment of a Special Examiner in 1940, Nicholas Roosevelt, nephew of Franklin Roosevelt. But Roosevelt's authority permitted him only to investigate charges and recommend a plan of reorganization, not to administer the company. He did prepare a three-hundred-page report for the court, which contained exactly one paragraph on the impact that the liquidation of operations to meet the claims of creditors had on mineworkers in the employ of the firm.[54]

Paying off debts owed principal bond and mortgage holders became the single objective of Judge Dickinson and his court officers.[55] The major creditors, banks in New York City and Philadelphia, organized themselves into "committees" to collectively petition for first consideration. Any aggrieved party could approach the court to pursue a claim, but powerful groups armed with top legal assistance went to the head of the line.[56] The P&RC&I paid off its debts largely by selling land holdings, its iron operation, and repair shops and by cutting expenses though the closing of facilities deemed unprofitable by company executives.[57]

Judge Dickinson heard various complaints about the management of the P&RC&I, but the court's decisions on awards were based purely on restoring the credit rating of the company. Charges of mismanagement—including poor decision making on equipment purchases and the sizes of coal produced and marketed, the unwarranted high salaries of top managers, the misuse of unused lands, and unwise leasing arrangements—were dismissed either as unfounded or irrelevant to the situation.[58] One claimant argued that financial interests still controlled both the mining and transport of coal and that coal prices remained uncompetitive. The petition quoted findings from a study of the Bureau of Economic Research that there was a "continuation, in modified form, of the same community of interest and control" that existed before court-ordered separations. The petition argued that forty men connected to the banking house of J. P. Morgan and the First National Bank of New York City formed a persisting interlocking directorate.[59] For the bankruptcy hearings, such charges were considered beside the point.

Individuals fared poorly in the proceedings. For example, William Yodin and Mary Ridely filed petitions seeking compensation for damages to their property caused by P&RC&I operations; their cases were never heard.[60] Frank Farsky, a retired miner, submitted a claim for $25,000 for injuries suffered inhaling "dust and deleterious substances" while in the employ of the company. The P&RC&I had opted out of the Pennsylvania workmen's compensation system, as allowable under the Pennsylvania Occupational Disease Act, but that left the company open to suits for negligence. Farsky could not draw benefits from the state's insurance fund and his sole resort was suing the P&RC&I. While under bankruptcy protection and court reorganization, suits could be brought only before the bankruptcy court. The court never ruled on Farsky's case.[61]

Communities, too, sought redress. In April 1939, eighteen school districts and townships in Northumberland County petitioned the bankruptcy court to compel the P&RC&I to pay taxes and utility charges owed them. The depression-ridden local public agencies had defaulted on their own loans, and community services had been severely curtailed; schoolteachers had not been paid for months. The total bill was $939,955. The P&RC&I challenged the figures presented, as well as the notion that its local tax obligations had to be met as the company liquidated its assets under court-administered reorganization. The court agreed with the company, recognizing that the local governments were left helpless—"Such is the painful situation here created"—but payments would have to await the restructuring of the P&RC&I's finances. The same town and school districts then sought redress through the U.S. Court of Appeals. In 1943, a circuit court ruled in their favor, but awarded the plaintiffs only $97,503, barely a tenth of their original claim.[62]

The UMWA also petitioned the bankruptcy court. In 1938, the union asked the court to stop the P&RC&I from closing mines in Bear Valley, Brookside, Gilberton, West Shenandoah, and Hammond. Union lawyers argued that the mines could be operated profitably; that vast investments would be lost, especially if the mines flooded; that the company would not be able readily to meet demand for coal with the return of good economic times; and that more than 3,000 miners and their families were affected and that they would become public charges to their already financially strapped communities. The court summarily dismissed the petition. The closings would result in "hardships more or less substantial," the court acknowledged, but restoring the fiscal solvency of the company required the closing of mines; more important, the court questioned whether the union had any standing in the bankruptcy proceedings because the UMWA was not a creditor of the P&RC&I.[63]

On March 14, 1944, the bankruptcy court declared the P&RC&I fiscally solvent and its reorganization complete.[64] The process included the issuing of new stock and bonds; holders of old equities could exchange them for new securities and cash. The existing management remained, again, in place. In the immediate post–World War II period, the fortunes of the P&RC&I continued to slide, sales dipping from $88 million in 1948 to less than $40 million by the mid-1950s.[65] A takeover of the firm's management by outside speculators, in similar fashion to the experiences of LC&N and the Glen Alden Coal Company, then transformed the business dramatically and led to the demise of the St. Clair Coal Company.

In 1952, the investment firm of Graham-Newman purchased 600,000 shares (close to 4 percent of the outstanding stock) of the P&R C&I. Two years later a group of investors from Baltimore bought another 11 percent of outstanding shares, and the new equity holders combined to oust the stodgy Philadelphia-based directorship of the company.[66] With the P&R C&I's stock undervalued, the new insiders could have profited purely by liquidating its $15.3 million net worth of assets. Instead, they moved to protect and boost the value of shares through a radical restructuring involving the phase-out of coal production, using losses on the coal business to provide a tax umbrella for new investments, raising funds with improved credit ratings, and the building of a diversified holding company. In 1955, the new directors approved the closing of the Locust Gap and Potts mines, the Locust Summit central breaker, and two stripping operations (with the loss of 1,700 jobs).[67] In characteristic fashion, they changed

the name of the firm in 1956 to the Philadelphia and Reading Corporation (PRC) with no trace of coal in its title—a change that anticipated by eleven years the parallel development for the Glen Alden company. Management, at the same time, placed the remaining coal operations under a new subsidiary, Reading Anthracite, which would be sold in 1961.[68]

While liquidating its coal business, PRC went on a buying spree, initially acquiring Union Underwear and the Acme Boot Manufacturing Company and then other clothing firms: Blue Ridge Manufacturers, Imperial Shirt, Marlboro Shirt, Boys Tone Shirt, and its biggest apparel purchase, Fruit of the Loom.[69] The company also entered into optical, toy, and chemical manufacture.[70] By 1959, PRC-held companies had sales in excess of $135 million.[71] Other than Reading Anthracite, not a single firm in the new conglomerate maintained operations in the anthracite region.

The St. Clair Coal Company got caught in the vise of the selling and buying of companies and stock manipulations of a breed of investors historically not present in the anthracite region; financiers in the past had their grips on developments to be sure, but their dealings increased the capital capacities of the region. For more than a decade prior to its closing, St. Clair Coal had valiantly maintained a place in the anthracite industry in the face of the declining market for coal and the fiscal crises of its lessor. In 1957, the firm was the twelfth-largest producer of anthracite and had succeeded by efficiently mining and processing coal on lands owned and leased by the P&RC&I (which received tonnage royalties), by mining on its own properties, and by preparing for market coal mined by other companies.[72] St. Clair Coal had invested heavily in its breaker and in washing facilities to retrieve coal from culm debris.[73] More notably, the company had adopted in the 1940s a strategy to retail its coal directly to customers, transporting the coal on its own trucks rather than remaining beholden to rail lines (in its original leases, the company had been obligated to ship coal on the Reading Railroad, but after antitrust actions, it gained control of the marketing of the coal it produced).[74] St. Clair Coal's independence in sales became a source of conflict with Reading Anthracite.

The declining market for anthracite countered efforts at survival; in 1955, the company closed down its two deep-shaft mines, the first cutback of operations in the firm's sixty-year history.[75] St. Clair Coal also experienced the death of its long-term president, Harold M. Smyth, but his wife remained determined to sustain operations.[76] Then, on April 6, 1957, the board of directors of the company received notice that Reading Anthracite would terminate its lease on October 31 of that year. On June 20, Reading Anthracite gave further notice that St. Clair Coal would have to dismantle and remove its breaker and other machinery and facilities from properties owned by Reading Anthracite, granting a sixty-day extension beyond October 31 to complete the removal.[77]

The new owners of PRC, through Reading Anthracite, were determined to end a century of coal production in St. Clair. During the summer of 1957, the managers of St. Clair Coal offered to sell their breaker to Reading Anthracite or operate just the breaker under a lease agreement (processing coal produced by Reading Anthracite). The lessor showed no interest, and on September 17, 1957, the officers of St. Clair Coal publicly announced that having been denied an extension of their lease, the company's operations would close entirely on October 15. Community and UMWA officials offered to plead the case for renewal of the lease, but with no response from Reading Anthracite, St. Clair shut operations on October 15, ironically on the same day that the governor of Pennsylvania, George Leader, announced a campaign for a U.S. stamp honoring anthracite.[78]

Reading Anthracite remained silent on the decision. Clearly, in terminating the lease and ordering the dismantling of the breaker and other facilities, Reading Anthracite sought to eliminate a competitor and limit the production of coal for the market—in hopes, no doubt, of boosting prices. The chief accountant of St. Clair Coal, Russell Marr, who had tried to negotiate with Reading Anthracite during the summer of 1957, offered as much at a private meeting of community supporters of the company before the public announcement of the closing: "The only excuse or reason that we have been able to get from them, the Reading Company, is that they didn't want competition. That's the

only reason they have given us any time for kicking us out of here."[79]

The Allocation Plan implemented in 1940 to stabilize prices and profits probably contributed to Reading Anthracite's decision. According to the law establishing the plan, the quota assigned a lessee reverted to its lessor on termination of a lease agreement. When Reading Anthracite denied St. Clair Coal a renewal of its lease, the company automatically received authority to increase its own production by 10 percent.[80] In contrast, after October 31, 1957, the St. Clair Coal Company reduced its operations to a single office dedicated to selling its direct assets—land, buildings, and equipment—to pay off loans, taxes, and other claims against the firm.[81] A shell of a business by the 1960s, St. Clair Coal did not formally dissolve until May 1979, but the breaking of the lease effectively ended the company's mining operations.[82] Reading Anthracite, following this action, received authorization to produce and sell 10 percent more coal than had been permitted earlier. More important, it had rid the field of one more competitor.

The decisions of the financial interests behind the new Philadelphia and Reading Corporation and the managers of its subsidiary, Reading Anthracite, had a divisive effect on the last generation of anthracite mineworkers in Schuylkill County. There would be no mass equalization of work campaign this time around. For example, in June 1957, Reading Anthracite closed collieries in Oak Hill and Pine Knot—with the loss of more than 560 jobs.[83] Laid-off workers did not remain idle. They decided to picket and close down a large Reading Anthracite strip-mining operation in Wadesville to gain sympathy and support for the spreading of work. They met resistance on all fronts. Both St. Clair Coal Company managers and workers opposed the action because coal from the strip mine kept St. Clair's breaker operating full time. Workers at Wadesville similarly sought the UMWA's help in clamping down on the illegal picketing that was jeopardizing their jobs. Community leaders and members also voiced their disapproval. Ultimately, a court injunction put an end to the action.[84] Local newspapers reported resignation and sadness among St. Clair Coal Company employees, but no concrete actions, when they joined the ranks of displaced mineworkers. Minute books of the board of directors of the company do reveal, though, that on November 27, the company's boiler, wash, and power houses, recently sold to a bidder, were "set afire by workmen's torches," a fitting reminder that working people were being affected by the actions of Reading Anthracite.[85]

Lehigh Coal & Navigation, Glen Alden, and Philadelphia & Reading Coal & Iron provide examples of capital flight—of the loss to the region and its communities of investment potential, economic assets, and employment opportunities. The histories of these companies in the post–World War II era, of course, do not match the standard cases in today's newspapers of firms that close their doors, move to low-wage areas in the United States or abroad, and reestablish operations. Anthracite mining basically ended.

The role of finance capital is essential to this story. Outside financiers have always been critical in the history of the Pennsylvania anthracite region—in bankrolling transportation and mining development and attempting to stabilize markets and even labor relations. Venerable New York and Philadelphia investment firms profited enormously from the enterprise and toil of the area's residents, and they had a vested interest in the region's economic well-being. Yet investors who entered the scene after World War II had tangential concern for the region. They aimed at creating conglomerates and improving the share value of their diversified holding corporations through the buying and selling of companies. The collieries of the Pennsylvania anthracite region represented tax shelters for them and assets for the taking and selling, not productive facilities (even ones that could be adjusted to the realities of the declining market for anthracite). The human resources of the area did not enter into their calculations. Emblematically "Coal" disappeared from the corporate names of LC&N, Glen Alden, and P&RC&I as anthracite employment disappeared from northeastern Pennsylvania in the 1950s and 1960s.

◆

One might have expected resistance to corporate liquidations from the anthracite miners' union, the United Mine Workers of America.

The union, after all, represented mineworkers, not stockholders. In the postwar decades, the UMWA remained a major actor as the bituminous and anthracite coal industries accommodated to the changing energy marketplace. Unfortunately, the union's needs and policies contributed to the decline of anthracite in the post–World War II period.

There was little that officers of the UMWA could do when an individual anthracite firm closed down its mining operations or began the moves toward diversification that typically led to the eventual closing of mines. Union officials could see the larger pattern that was emerging from the actions of individual firms. Following the lead of the anthracite operators and their advocacy organizations, such as the Anthracite Institute, they lobbied for state and federal government policies that favored the industry. Most of these actions were defensive in nature, aimed at limiting the expansion of oil imports, opposing the construction of natural gas pipelines, or providing testimony before the Federal Power Commission against a petition by an anthracite-region firm to distribute natural gas in the area. By opposing the increased use of competing fuels, the Anthracite Institute and the UMWA sought to stave off anthracite's continuing decline.[86]

Prominent among the union's defensive actions in the postwar decades was its opposition to the construction of the St. Lawrence Seaway. Thomas Kennedy, secretary-treasurer of the International union, testified in opposition to the proposed seaway as early as July 1941, stressing the adverse effects the project would have on the anthracite region. An editorial in the *Anthracite Tri-District News,* a reprint of commentary from the *Shamokin News-Dispatch,* reinforced Kennedy's arguments in August 1945, pointing out that public tax money would "be used to stimulate competition against Anthracite." The *Tri-District News* played up opposition to the seaway in its pages, reporting on further Kennedy testimony in 1947 and 1954 and Anthracite Institute opposition in 1952. The Seaway, of course, was finally approved in 1954 as a joint U.S.–Canadian undertaking, opening up the Great Lakes to larger oceangoing vessels, including oil tankers. Traffic on the Seaway rose dramatically, and cargoes transported totaled a billion tons between 1959 and 1984 and another billion by 1996. And no doubt the Seaway did undercut anthracite sales in Canada over these years, contributing in part to anthracite's continuing decline.[87]

The UMWA also joined the Anthracite Institute, local chambers of commerce, and numerous other local organizations in November 1947 in opposing an application to the Federal Power Commission for the construction of three natural gas pipelines across Pennsylvania. Even more so than the St. Lawrence Seaway, natural gas competed directly with anthracite in the home and commercial heating markets in the state, as coverage in the *Tri-District News* noted regularly.[88] Like the Seaway opposition, union and Institute efforts here were ineffectual; federal officials would not permit local opposition from the anthracite region to stand in the way of a project that had national economic implications. By 1960 natural gas's share of Pennsylvania's energy use stood at 16.6 percent, while anthracite's contribution had declined to 10 percent. The gap between the two fuels' shares of energy consumption continued to widen in the last decades of the twentieth century.[89]

Despite the UMWA's involvement in efforts to promote anthracite and oppose competing fuels, anthracite represented only a minor component of John L. Lewis's concerns as World War II ended. Of 383,000 coal mineworkers in 1945, only 73,000—less than 20 percent—found work in Pennsylvania's anthracite region.[90] Measured in terms of tonnage, anthracite's share was even less, about 60 of 578 million tons produced in 1945, a tad more than 10 percent of overall output.[91] The priority of bituminous concerns was evident, reflected in the fact that the UMWA typically first negotiated a nationwide agreement with bituminous operators and then the anthracite region settlement took shape based on the earlier and more weighty labor contract.

In the last year of the war and in the immediate postwar period, the UMWA focused much of its energies in contract negotiations on the creation of a Welfare and Retirement Fund that would finance health care and pensions for miners and retirees. Lewis first raised the issue of such a fund during the 1945 negotiations with bituminous operators, but got nowhere with this

demand. In the spring of 1946, however, as negotiations stalled, Lewis called out 340,000 bituminous miners in support of his demand for an industry-financed health plan for mineworkers. The terrain of negotiations shifted dramatically when President Harry Truman, proclaiming a national emergency, seized the mines, and Lewis began bargaining with the Secretary of the Interior Julius Krug rather than bituminous operators. The Krug-Lewis accord, signed in late May 1946, established a jointly operated, union-government health plan, funded by a five-cent royalty per ton of coal, and provided for a medical survey of bituminous mining areas to see how best to implement such a plan.[92] With the accord, bituminous miners returned to work, but implementation of the health plan was undermined by a lack of commitment on the part of operators who had not been party to the settlement. With little to show in the way of implementation of the earlier accord, miners struck again in November, this time against the federal government that continued to operate the mines. Ignoring an injunction, Lewis and the union were held in contempt of court and received stiff fines. The union appealed the fines, but rather than exposing himself and the union to mounting liability, Lewis called off the strike and ordered the miners back to work. The miners' health plan remained a possibility but not a reality.[93]

While the UMWA's struggles with bituminous operators played out on a national stage, the union succeeded in establishing a measure of medical and old-age security for anthracite miners with much less fanfare. In the anthracite region, as in bituminous, mineworkers and their families eagerly anticipated the creation of a Health and Welfare Fund. As early as March 1944, officers of Local 1738 in Lansford wrote to Lewis, calling for payment by union operators of a fifteen-cents-per-ton royalty "to be used to create a fund so that miners in the future shall not be com[p]elled to suffer in poverty as in the past." Funds were to be used for the victims of industrial accidents, for miners suffering from occupational diseases, and for modest pensions. A year later, the Panther Valley General Mine Committee (which included representatives from Lansford) passed another resolution calling for a fund to relieve those suffering from miners' asthma.[94] The 1946 anthracite contract established the Anthracite Health and Welfare Fund, but locals continued to pepper Lewis and International officers with their concerns about the slowness of payments and with advice as to how best to proceed.[95] The Fund introduced benefits stepwise beginning with death benefits in 1946, disability payments in 1947, and pension support in 1948.[96]

Bituminous operators came around as well in negotiations that were concluded in July 1947. Over the next five years, Lewis succeeding in making the UMWA Welfare and Retirement Fund a reality; his efforts transformed the nature of health care delivery for bituminous mineworkers in the Appalachian and midwestern coalfields. Through a convoluted process Lewis succeeded first in getting operator support for the plan and then maneuvered adroitly to secure union control of the Fund. In a stroke of superb union generalship, Lewis secured company financing of the Fund through increasing royalty payments, beginning initially at five cents per ton of coal in 1947 and increasing to forty cents per ton by 1952. Rather than focusing on wage payments to mineworkers, the UMWA negotiated royalty payments pegged to the output of coal at unionized firms and built up a substantial trust fund to be used to improve health care and pensions for mineworkers and retirees. In return, Lewis and the union supported management efforts to mechanize the bituminous mining process and accepted the dramatic decline in the workforce that accompanied mechanization. The establishment of the fund, in turn, set an example that other major unions emulated in the postwar period as health insurance became an important element in negotiated wage-benefits packages.[97]

The result of these successes was a pathbreaking system for the delivery of health care to miners, retirees, and residents of the nation's bituminous coal regions. The Welfare and Retirement Fund established ten Area Medical Offices, scattered across the soft-coal producing areas of the country, with two in Pennsylvania, three in West Virginia, and others in Kentucky, Tennessee, Alabama, Missouri, and Colorado. The Fund pioneered the provision of rehabilitation services to severely disabled miners.[98] It established clinics, medical groups, and a chain of

state-of-the-art hospitals that dramatically improved the availability of medical care in coal-mining communities. The Fund instituted policies that kept administrative costs at a minimum and through various group and prepaid programs, avoided many of the problems associated with traditional health insurance and fee-for-service medical practice. One historian, calling the health plan a "noble experiment," expressed well the significance of the UMWA's achievement in the Fund's first decade. A minister in Hazard, Kentucky, summed up the feeling of Appalachian miners: "The health card was worshipped. . . . These hospitals mean everything."[99]

The union's second major accomplishment in the postwar period was the successful funding of a pension program for retired miners. Begun in 1948, and supported by the same tonnage royalties that funded the health plan, the pension program provided an annuity of $100 a month to miners who retired at or after the age of sixty and had worked twenty years or more in coal mining. Pension payments in the bituminous field consistently exceeded health care expenditures, averaging roughly $70 million annually in the 1950s and 1960s, compared to $50 million for medical payments. By 1970, with the retirement of older mineworkers, pension payments exceeded $100 million a year, threatening the solvency of the Fund.[100]

The pension effort revealed starkly two major shortcomings of the Welfare and Retirement Fund. First, by pegging Fund income to royalties on coal from union operators, the UMWA became vulnerable to reductions in coal output due to the vagaries of the business cycle and the growing substitution of alternative fuels. Compounding these uncertainties, as nonunion production of bituminous coal expanded in the 1960s and 1970s, the Fund received no royalties on that growing share of overall output. Second, the bargain that the UMWA struck with bituminous producers in 1950 acknowledged the union's stake in the economic health of the industry. Holding up his end of the bargain, John L. Lewis became, in effect, a junior partner of the largest coal producers, and in future negotiations eased his demands on the companies. Despite significant increases in the cost-of-living over the next two decades and significant wage

increases in bituminous contracts over the period, the UMWA accepted a steady forty-cent-
per-ton royalty between 1952 and 1971.[101] Total
bituminous output, on which royalties were
paid, increased in the period from 314,000 to
437,000 tons, but periodic declines in output
(1958–1963 being a particularly difficult stretch)
led to significant short-term declines in Fund income.[102] Since the UMWA did not subsidize the
Fund out of its own coffers, which themselves
were dwindling as the mining workforce shrank,
Fund managers had to cut medical benefits and
pension payments regularly to accommodate
these fluctuations.

The treatment of pension benefits for bituminous miners illustrates the problems that resulted from a pay-as-you-go system tied to coal
royalty payments. Pension benefits were at their
most generous in 1949 when Fund trustees provided a $100 monthly payment for miners who
retired at or after sixty years of age after working
in mining at least twenty years. In 1953 trustees
restricted eligibility by requiring miners to have
worked twenty of the last thirty years before
their retirement. The next year the Fund cut off
disabled miners and widows who had been receiving modest benefits. In 1960 the maximum
benefit was reduced to $75 monthly, part of a
continuing effort to limit expenditures to match
royalty payments into the Fund. In 1965, with
improving conditions in the industry, the
trustees raised the pension once again to $100 a
month, but also required that a miner work his
final year in a union mine to be eligible for a
pension.[103]

In addition to tightening pension eligibility
requirements, the trustees of the Welfare and
Retirement Fund significantly reduced medical
benefits provided to active mineworkers and retirees. In July 1960 the Fund ended health coverage for miners who had been unemployed for
a year or more, cutting off almost 60,000 miners,
or a fourth of those previously eligible for assistance. Two years later, the Fund cut off miners
whose employers had defaulted on their tonnage royalties. Finally, in 1963 and 1964 the
Fund sold its chain of hospitals, the Miners
Memorial Hospitals, as a further cost-cutting
measure. The United Presbyterian Church, long
active in mission work in Appalachia, helped negotiate the sale of the hospitals to a nonprofit

corporation, Appalachian Regional Hospitals.
The sale was supported by two federal loans totaling $8 million and additional subsidies from
the state of Kentucky. Between cutting off beneficiaries and reducing services, the Fund
brought income and expenditures into line, but
at the price of reneging on its commitments to
improve the quality of the lives of bituminous
miners and retirees.[104]

Medical and pension benefits for bituminous
miners suffered from the uneven output of soft
coal in the 1950s and 1960s, but these difficulties were nothing compared to those of anthracite miners. The key problem for the anthracite region stemmed from the fact that the
Anthracite Health and Welfare Fund was distinct from its bituminous counterpart, and each
was funded entirely by tonnage royalties from
their respective operators. And while bituminous tonnage subject to royalties increased by
39 percent between 1952 and 1971, anthracite
tonnage plummeted from more than 40 million
to less than 9 million tons, a decline of almost 80
percent.[105] These tonnage figures resulted in
dramatically different levels of royalty payments
to the respective funds. In the early 1950s bituminous royalty payments averaged about $130
million annually, compared to $11–12 million
annually for anthracite. By 1971, the disparity in
the resources available to the two funds had
mushroomed, with annual royalty income
amounting to $170 million for bituminous and
$5 million for anthracite.[106]

The segmentation of the two funds coupled
with anthracite's steep decline in the 1950s and
1960s meant that anthracite miners and retirees
saw their benefits slip away over the course of
two decades. Pension benefits declined from
$100 a month in 1949 to $30 in 1961; death
benefits for widows and family members of anthracite mineworkers were cut from $1,000 to
$500 and then suspended for four years' time
beginning in February 1958. The vanishing production took its toll. In 1961, tonnage royalties
from the production of 17,000 active anthracite
miners supported 16,000 retirees.[107] By 1974
barely 2,000 employed anthracite miners supported the remaining 15,000 retirees.[108] As anthracite production plummeted, tonnage royalties declined accordingly and the trustees of the
Health and Welfare Fund spent much of their

time devising ever-more-elaborate schemes to reduce the number of mineworkers eligible for UMWA pensions. For a period of time, if a retired mineworker returned to employment in the industry, he was permanently disqualified from future pension benefits. In the 1970s, if an anthracite operator stopped making payments to the Fund, some of the retirees of the company lost their pension benefits, based on their seniority. For more than four and a half years, between February 1958 and October 1962, the Fund stopped paying any death benefits. Arrears approached $4 million—intended benefits for the families of almost 8,000 deceased miners—when the Fund finally resumed making $500 payments to widows or other dependents in late 1962. The Fund and the union certainly did not look good in the eyes of members, retirees, and their families, when a local newspaper reported that the twelve widows of miners killed in the 1959 Knox Mine disaster had still not received death benefits three years after the catastrophe.[109] In a terrible zero-sum game, Fund trustees had to play off the pensions of one group of retirees against those of other retirees and against the wages of active mineworkers.[110] Even these strategies could not cover the shortfall, and in the 1980s a federal agency, the Pension Benefit Guaranty Corporation (PBGC), stepped in and provided more than $3 million of assistance to permit the Anthracite Health and Welfare Fund to meet its obligations to retirees.[111]

Given the limited funding base for anthracite, the Health and Welfare Fund could never undertake the kind of strong medical program that characterized its bituminous counterpart. Pension payments and death benefits together constituted almost 99 percent of benefits paid out by the Fund between 1952 and 1958, the period of the Fund's greatest success.[112] Medical benefits were limited to modest funding of black-lung clinics across the region that typically offered doctors' consultations for three or four hours per week and the support of research into the diagnosis and treatment of black lung by the Jefferson Medical College in Philadelphia.[113] Nothing undertaken in the anthracite region even remotely compared to the ambitious provision of health services offered to mineworkers and retirees

supported by the bituminous Welfare and Retirement Fund.

To its credit, the UMWA did subsidize the Anthracite Health and Welfare Fund as its trustees struggled with the industry's decline. Throughout the 1950s the union made loans totaling $5 million to assist the Fund. In September 1959 the union forgave these loans, indicating that it would not expect or require the Fund to repay the International.[114] Even while extending this short-term support, the union also explored the possibility of merging the anthracite and bituminous funds to provide a longer-term solution for anthracite miners and retirees. An exchange of memos in 1951 between John L. Lewis on behalf of the union and Harry M. Moses, of the Bituminous Coal Operators' Association (BCOA), documents one such effort, even as it reveals the difficulties in bringing off such a merger. From the operators' perspective, the move would have meant a diversion of bituminous tonnage revenues to support anthracite miners and retirees and hence a potential added cost of production.[115] The UMWA continued to raise this issue intermittently, with Lewis's successor as President, Thomas Kennedy, noting his commitment to the merger of the two funds at the UMWA's national convention in Cincinnati in October 1960.[116] However, once the two welfare funds had been established and funded through separate collective-bargaining efforts, insuperable barriers operated to keep them distinct. And given the UMWA's evident desire to forge a working alliance with the BCOA, the union would not seriously upset matters by pressing stubbornly on behalf of anthracite miners. Thus the establishment of separate welfare funds for the two coal regions left anthracite mineworkers without hope for decent medical and pension benefits, even as the UMWA struggled to secure such benefits for their bituminous brothers.

In viewing the sad fates of the bituminous Welfare and Retirement Fund and the Anthracite Health and Welfare Fund, it is tempting to explain the miners' misfortunes as the inexorable consequence of the decline of coal mining in the postwar period. With Fund income pegged to coal tonnage, what could the Funds' trustees do in the face of plummeting coal output? What choices were available other than cutting miners' benefits in the face of declining income? Yet the Funds' trustees and the International officers of the UMWA cannot be left off the hook so easily because the union's financial machinations contributed in no small part to the Funds' collapse. Lewis's stubborn determination led to the creation of the Funds, and for that he deserves credit; at the same time, with his vision for how to save the coal industry and maintain the union's and his own personal power, Lewis bears responsibility for the Funds' misfortunes.

The creation of the two Funds provided Lewis with a major cache of resources that he and the other trustees were legally obligated to use for the benefit of mineworkers and retirees. Once the two Funds were fully operating in the early 1950s, their combined income averaged about $140 million a year. Although the Funds operated on a pay-as-you-go basis, the trustees set benefit levels in a manner that permitted the Funds to create a reserve for future needs. At any time, the sums of money available for investment were substantial and those funds should have been invested in ways that provided a measure of security and income for the Funds' beneficiaries. But that was not how Lewis and those around him thought.

As Lewis launched the two Funds in the late 1940s, the UMWA purchased the National Bank of Washington. By mid-1950 the bank's deposits had grown from $22.6 to $83.7 million as the union shifted much of its financial resources into non-interest-bearing accounts at the bank. The Welfare and Retirement Fund also had a substantial reserve fund that might have been invested in ways that served the interests of its beneficiaries. Between 1955 and 1971, the reserves of the Fund declined below $100 million only once, peaking at $146 million in 1958 and $180 million in 1968.[117] Yet the Fund earned a return of less than 2.4 percent annually as interest or dividends on these reserves. The placement of a large share of the Fund's reserve in the UMWA-owned bank earning little or no interest gave that bank and the union a good deal of investment flexibility but did not meet the trustees' legal fiduciary responsibility. The trustees were not making a high priority of their responsibility to bituminous miners and retirees but were placing the union's and Lewis's interests first.[118]

By the laws governing trust funds, Lewis and his fellow trustees had a legal responsibility to use the income of the Welfare and Retirement Fund to provide health services and pensions to UMWA members and retirees. Instead, they took the tonnage royalties paid by the bituminous operators, deposited them in their own bank, accepted token returns on the Fund's reserves, and wearing their hats as bank officers made investments that brought substantial returns to the National Bank of Washington. Since the UMWA owned the bank, its coffers bulged. It became the wealthiest union in the nation at the same time that the Welfare and Retirement Fund sold off its chain of hospitals in Appalachia and drastically cut back on pension benefits for retirees.[119]

These actions, though, did not go unchallenged. In 1971 the District Court for Washington, D.C., ruled in favor of a class-action suit brought on behalf of a group of mineworkers and retirees. The Court found that the trustees for the UMWA Welfare and Retirement Fund had failed to meet their responsibilities to Fund beneficiaries, specifically because of the large sums of the Fund's reserve that they kept in non-interest-bearing accounts in the National Bank of Washington. Trustees—particularly John L. Lewis and Josephine Roche, but also Tony Boyle, during the period he had served as a Fund trustee—had placed the interests of the UMWA and its bank before those of the Fund's beneficiaries. The suit was successful, and liability was set at $11.5 million to be paid by the union and the bank to the Fund, as compensatory damages for the final five-year period between 1966 and 1971. There is no doubt that the actual losses suffered by mineworkers and retirees since the establishment of the Trust Fund were considerably greater than this figure, but the judge in the case, for complex reasons, limited the damages to a five-year period.[120]

The Anthracite Health and Welfare Fund followed similar practices that had been ruled a breach in trust by the District Court, but the resulting losses for Fund beneficiaries were far less than for the bituminous miners. A sense of the scale here emerges from an examination of the Fund's annual reports. In the period 1951–1960, for instance, the Anthracite Health and Welfare Fund maintained an average reserve of $628,000. Yet the Fund earned on average across the decade only $900 a year in interest on its reserves, for a return of 0.14 percent. If the Fund had earned only a modest 3 percent on its reserve in that decade, it would have increased its interest income twentyfold and over the decade have earned for its beneficiaries almost $180,000 in additional income.[121] Had the Fund invested its reserves, a common practice for pension funds, its returns for beneficiaries might have been still greater. But Lewis and his fellow trustees permitted the union-owned bank to enjoy interest-free use of these funds, seeking to shore up the finances of the UMWA rather than provide a financial cushion for mineworkers and retirees. Melvyn Dubofsky and Warren Van Tine, in their biography of Lewis, exonerate the UMWA president from a possible charge of using welfare funds for personal financial gain, but they remain critical of his actions: "Lewis was not seeking personal financial profits but rather trying to strengthen the union and, consequently, his own position of power. Unfortunately activities advantageous for the institution were not necessarily beneficial to the membership."[122]

Lewis used the resources placed in the National Bank of Washington to implement his vision of shoring up the coal industry and securing for himself a leading role as a modern-day captain of industry. The deposits provided Lewis "financial leverage," in the words of Dubofsky and Van Tine, as the UMWA worked through Cleveland industrialist Cyrus Eaton to purchase controlling interests in the West Kentucky Coal Company and the Nashville Coal Company. These nonunion companies promptly signed union contracts and formed the core of a far-flung operation that constituted by 1960 "the third largest bituminous producer in the country."[123] The union also invested more than $5 million in the Coaldale Mining Company in the anthracite region, to permit that firm to lease coal lands from LC&N and thus provide employment in the Panther Valley. These investments were followed by investments in coal-purchasing electric utilities as Lewis and the union played the role of capitalist financier to the hilt. Between its tonnage royalties, union dues, and its banking and investment activities, the UMWA had become, according to the *New York Times,* the "richest labor union in the land."

Yet bituminous miners displaced by mechanization and anthracite retirees (hard-pressed to get by on pensions of $30 a month) did not benefit from this union empire.[124]

The opportunities that the trustees of the benefit funds squandered strikingly reveal the failure of the UMWA in handling mineworkers' pensions. While other trade unions negotiated pension benefits for their members in the postwar period, few of those unions succeeded in controlling the investment and distribution of those funds. Given Lewis's success in establishing the bituminous and anthracite funds, the UMWA had a controlling voice in the investment of the funds built up through tonnage royalties and the payout of those funds to retirees. Not only did the trustees fail to meet their fiduciary responsibility to maximize earnings for current and future retirees, but they also failed to use investments for social purposes that might have benefited union members. For instance, the International Ladies' Garment Workers' Union "instructed the Chase Manhattan Bank to use its pension fund reserves to grant mortgages to working-class applicants." What a contrast to Lewis's strategy of placing benefit funds' reserves in non-interest-bearing accounts with the UMWA-controlled bank.[125]

While the UMWA was building up its banking and bituminous coal interests and neglecting its fiduciary responsibility to the beneficiaries of its health and pension funds, it also opposed efforts in another area vital to the interests of mineworkers—the mounting campaign for expanding black-lung compensation benefits. Union concern for its alliance with mine operators, the foundation of its postwar labor peace with the BCOA, led it to drag its heels even after the establishment of the UMWA Welfare and Retirement Fund. While the Fund invested in improved health care facilities for bituminous miners, its director, Josephine Roche, prohibited political activity on the part of her staff with regard to workmen's compensation coverage, occupational disease prevention programs, or safety legislation related to black lung. Acceptance of the advancing mechanization of mining, regardless of its impact on the health and safety of miners, was the union's mantra, and the price of its collaboration with organized mine operators.[126]

The movement that ultimately succeeded in winning the passage of unprecedented federal legislation providing benefits for victims of black lung began in Pennsylvania and West Virginia as grassroots campaigns for broadening state workmen's compensation programs. In 1965 Pennsylvania amended its Occupational Disease Act to compensate anthracite and bituminous miners who had contracted black lung. UMWA president, Tony Boyle, concerned to protect bituminous operators with whom the UMWA had become cozy, opposed the extension of the legislation to bituminous miners.[127] However, the legislation did pass, and the movement spread to neighboring West Virginia. There, bituminous miners, drawing support from community organizers from Volunteers in Service to America (VISTA) and medical professionals, established the Black Lung Association in January 1969. Meetings in mining communities spread across the state, and protesters converged on the state capital of Charleston in late January to demand action by the state legislature. Legislative hearings provided experts an opportunity to argue for reform of state compensation laws to permit benefits for a wide variety of respiratory diseases among miners and to accept a presumption that the long-term exposure to coal dust while working underground provided an adequate basis for mineworkers' eligibility for compensation. In mid-February, mineworkers embraced the movement and launched wildcat strikes that brought 40,000 out of the mines within two weeks, virtually closing down all the mines in the state. However, these actions received precious little support from the UMWA. Boyle treated the Black Lung Association as if it were a dual union, and the editor of the *UMWA Journal* excused the union's earlier silence on the compensation issue by claiming, "we didn't know the disease existed." Even without the backing of the UMWA, the movement grew. By March 11, 1969, West Virginia had met the miners' demands and made black lung a compensable disease under the state's system of workmen's compensation. Kentucky followed in February 1970, meaning that three major coal-producing states—Pennsylvania, West Virginia, and Kentucky—were now providing a measure of relief for the victims of black lung.[128]

Federal action followed swiftly on the heels of the passage of legislation in West Virginia. In response to the pressure of West Virginia congressman Ken Hechler and Pennsylvania's Daniel J. Flood, and over the unified opposition of coal operators and the UMWA, Congress considered the Federal Coal Mine Health and Safety Act. Typical of the UMWA's stance on coal mine safety legislation, Boyle alleged that only "self-proclaimed do-gooders," "outside agitators," and "overnight experts" supported the legislation. With the pressure of the miners' social movement and the threat of continuous disruption of supplies of coal, Congress passed and President Nixon signed the Act in December 1969 that set a permissible limit on coal dust at three milligrams per cubic meter of air and required that the limit be reduced to two milligrams within three years. Mine operators would have to adopt improved ventilation or use water with mechanical coal-cutting machinery to reduce airborne dust and protect miners' lungs. Moreover, the Act included a provision, coauthored by Flood, that established a black-lung benefits program to provide federal assistance to miners who suffered from black lung. The law provided for the first time consistent, national treatment of black-lung victims.[129] As had been the case earlier in West Virginia, threats of work stoppages in the mines figured significantly in the president's final decision to sign the new law. Protest among miners, rather than testimony or lobbying by UMWA officials, led to this signal victory. As the leading historian of the struggle for black-lung compensation, Alan Derickson, has written:

An extraordinary militant worker-centered movement made [the federal law] a reality. However indispensable they were to the cause of reform, leaders . . . rode a wave of protest whose power ultimately derived from the determination of rank-and-file miners and their families to shut off the nation's principal source of energy if their demands were not met. The unparalleled strike in West Virginia won state compensation and sent shock waves that proved decisive in winning federal legislation.[130]

Protests against the union's inadequate responses with regard to pensions and health and safety issues reflected a larger grassroots movement demanding greater democracy and accountability within the UMWA. These protests rejected the union's financial machinations aimed at becoming a major player in the bituminous coal industry and building up a mammoth union treasury. As the UMWA turned to its brand of business unionism, increasingly the power shifted from the districts and locals to the union's international headquarters in Washington, D.C. The costs of the centralization of power within the union included the undercutting of union democracy at the local and district level and the increasing alienation of working miners from the union's national leadership.

These developments, of course, were not new in the late 1960s. As early as the 1920s, Lewis had secured authority from the UMWA's national convention to suspend for cause regularly elected officers of the union's districts and appoint provisional officers to run the districts. Originally a temporary measure to reestablish control of the International union and correct abuses at the district level, the appointment of provisional leadership for districts increasingly became the norm, a means to ensure not only Lewis's control over district affairs but also his authority within the International union. The practice was strongly opposed by the districts—at the 1942 UMWA convention local districts submitted 108 resolutions for the restoration of local authority—but to no avail. As Dubofsky and Van Tine note: "by 1944 twenty-one of the UMW's thirty-one districts were run by officers named by him. On the International Executive Board, Lewis, Kennedy, and O'Leary transacted business with sixteen appointed members casting 287.5 roll-call votes against only ten elected members with a voting strength of 72.5."[131] And the practice of superseding district elections did not cease when Lewis retired; by 1969 "provisional" leadership prevailed in nineteen of the twenty-three districts that remained in the shrunken union.[132]

This national practice was evident in the anthracite region as well, where in the fall of 1941 the UMWA suspended the elected officers of District 7 and replaced them with provisional officers appointed by Lewis. The suspension came in response to a wildcat dues strike in the district that district officers did not adequately quash, in the view of International officers. In early Janu-

ary 1942, members of the International Executive Board convened in Hazleton and heard "evidence affecting the District 7 situation," at which time they formally ruled that the dues strike in the district was "illegal and in violation of the [anthracite wage] agreement and not in keeping with the provisions of the International Constitution of the United Mine Workers of America." Moreover, the International Executive Board censured the continued existence of an ad hoc protest committee, noting that all concerns had to be channeled through established union bodies, the local unions, the district, and the International union.[133] Despite the specific events that led Lewis to take over District 7, provisional leadership continued indefinitely. In 1954, more than twelve years after the initial imposition of "provisional" leaders, district President Mart Brennan and Secretary-Treasurer David J. Stevens remained in office. In fact, Lewis praised the two officers at a meeting of the International Executive Board in early May 1954, touting their having paid off the district's debt to the International to the tune of $600,000 over a twelve-year period.[134] In fact, District 7 never graduated from its "provisional" status. In 1969 the UMWA combined all three anthracite districts, whose memberships had shrunken dramatically with the industry's decline. The new combined district was designated a "provisional" district, and all anthracite mineworkers lost the right to elect district officers.[135]

Even in Districts 1 and 9, where (until 1969) officers were nominally elected by members of the districts, numerous complaints suggested that district officers were not responsive to members' concerns. By the post–World War II period the UMWA had evolved into a well-oiled machine operated by the International officers in Washington, D.C. Policy for anthracite was formally set by tri-district conventions held in the region, but district presidents and local leaders beholden to them faithfully controlled these affairs and saw to it that rank-and-file activism had little place in the proceedings. District presidents, in turn, knew where the power was in the UMWA and typically implemented the policies of the International officers in their districts, lest they, too, might be suspended and replaced. By controlling the seating of delegates, setting the agendas for the meetings, and imple-

menting parliamentary procedure at the meetings, district officers, in close consultation with
the International officers of the UMWA, ruled
anthracite region unionism with a strong, guiding hand.

During this period, Thomas Kennedy, former
president of District 7 and a native of the anthracite region, served as the UMWA's vice president. Lewis invariably relied on Kennedy to
communicate with district and local officers in
the anthracite region and passed issues along to
him for resolution. Kennedy, however, had been
an International officer since 1925, and while he
did spend some time in Hazleton and maintained a second home there, he lived and
worked in Washington, D.C., and his allegiances
were with Lewis and the International union.[136]

Business unionism emerged in the UMWA in
the postwar decades and displaced the more
rank-and-file-centered militant trade unionism
that had been the union's hallmark earlier. Business unionism of two distinct sorts flourished.
On the one hand, the business of unionism—the
collection of dues, the conducting of collective
bargaining, and the processing of workmen's
compensation cases and workplace disputes—
predominated in terms of union activities. In addition, with the purchase of a major Washington,
D.C., bank and with investments in coal companies and coal shipping, the UMWA became a
major player in the coal business. Local members and activists in the anthracite region were
well aware of these developments and grew pro-
foundly alienated from the UMWA as it evolved
in this new direction. As members challenged
the UMWA district and International hierarchy,
the union's officers became defensive and
worked increasingly to maintain their power and
privileges within the union. Power, which had
once been sought to achieve specific goals for
the union's membership, became a goal in its
own right, as union officers enjoyed a range of
privileges that separated them from the union's
rank and file. In this context, a widening gulf of
distrust and resentment grew up between the
union's leadership and its members, and a can-
cerous corruption emerged among the union's
leaders.

These practices prevailed in the union's highest ranks. During contract talks, Lewis increasingly resorted to secret negotiations with the

leading bituminous operators without the participation of other members of the negotiating
committee. Similarly, Lewis and the union's
secretary-treasurer attempted to hide the union's
investments in coal companies with which the
UMWA conducted collective bargaining.[137] And
in the same vein, the financial statements of the
UMWA Welfare and Retirement Fund and the
Anthracite Health and Welfare Fund concealed
what the funds' trustees did with the reserves
they maintained. All of these actions provided
maximum latitude and minimal accountability
on the part of the union's top officials as they
carried out their responsibilities.

The actions of Lewis and the International
officers established the framework within which
the union's district leaders operated; the aggran-
dizement of power and lack of accountability so
evident at the national level were equally apparent in the union's activities in the anthracite region. Examples of abuse of power by officers of
District 1 in the northern field of the anthracite
region provide ample evidence that these national trends within the UMWA in the 1950s
and 1960s were well represented at the district
level.

District 1 had long been a center of conflict
within the union, the heart of the earlier dual-
union movement organized by Thomas Maloney
and the United Anthracite Miners in the 1930s.
That conflict had its origins, at least in part, in
the alienation of the local membership from district staff. A report sent to Lewis by William
Sneed, an International staff member who spent
time visiting the three anthracite districts in November 1938, highlights the problem. District 1
played prominently in Sneed's remarks. Mixing
advice with observation, he wrote: "Have field
workers visit local unions and quit sitting in offices contin[u]ously telling of their exploits of
the past. Almost any day you go to the office, you
find Board Members there.... All of them
began telling stories of their accomplishments
since 1920." He commented that the anthracite
miners resented staff whose only experience was
in the bituminous regions. Sneed's criticism of
one union staffer revealed the chasm that separated paid staff and working miners: Sneed remarked that the staffer should "get used to
wearing sociable clothes, less diamonds among
the Membership." All in all, he felt the district's

offices were overstaffed with representatives who provided little or no service. As one might expect, "The Membership complains bitterly at having to pay them."[138]

In the depths of the Depression, the districts were generators of revenue for the UMWA. The 1936 anthracite contract had finally provided a dues checkoff system for the union. Mart Brennan, elected president of District 9, summarized in his annual report for 1938 the income that came to the union through required union membership. Over a two-year period, the district received $664,000 in members' dues, returning $136,000 to the locals and passing another $335,000 directly to the International. Still, even with these payments, the district had a budget of $173,000 for the period.[139] Given that miners were typically working two or three days a week in 1938, making perhaps $7–10 a week, they resented spending some $86,000 a year of their own money to support district staff and their "diamonds."[140]

What becomes abundantly clear in reading through union correspondence between the 1930s and 1960s is the growing distance between the International union in its Washington, D.C., headquarters and rank-and-file members in the anthracite region. The union operated on three levels—the International, the district, and the union local. If District 9's accounting is representative, more than half of anthracite members' monthly dues after 1936 went to the International. Another 26 percent funded the district office, while less than 21 percent remained with the union's locals. Union members elected their local officers and sometimes their district officers, though frequently districts were in "provisional" status, meaning that Lewis appointed their president and treasurer. Moreover, the International appointed district staff and organizers, and so maintained a loyal cadre of union officials at the district level. Finally, the International Executive Board made key decisions for the International union, but its membership was heavily weighted toward district presidents who owed their appointments to Lewis. The end result was a largely undemocratic union that siphoned dues away from the locals and ran the union with precious little input from the grass roots.

Occasionally, an independent observer tried to alert the union to its problems. In one such instance in the anthracite region, Stephen Moriak Jr., a Purdue University student majoring in labor management who grew up in a mining town, wrote to UMWA officials in Washington in 1950 after speaking with miners in the Wilkes-Barre area. His letter recalled his exchange:

I asked them about the union. I asked them if they were benefiting by it. They laughed. They said it was a joke. They said the company was running the union. They said it was corrupt. They said that union representatives were appointed and not elected as should be done. They said their grievances were laughed at and if a miner had too many of them he was fired. They said that they fared better when they had no union at all. They said that graft was on the rampage.

Moriak wrote as a pro-union observer, who took it upon himself to let union officials know what he had heard and to advise that the UMWA needed to give its "house a cleaning."[141]

Union officials could not hear what Moriak had to say. For his troubles, he received a lengthy response from John Owens, the union's secretary-treasurer. Owens replied that he must have been "misinformed," for all officers were elected by secret ballot, all expenditures were regularly audited, and the union had established elaborate grievance mechanisms to "adjudicate . . . disputes." In his reply, Owens recited the structures and practices that the UMWA constitution established without acknowledging that theory and practice might differ.[142]

Yet Owens must have been aware of various improprieties in the financial affairs of locals and districts. The same files that contain this correspondence have periodic reports of investigations of the financial malfeasance of union officials and the subsequent intervention of the International Executive Board to restore order and trust. Less than a year after Owens replied to Moriak's letter, an International representative reported back to the UMWA vice president on the results of his investigation of an anthracite local in Ashley. The officers in this case were suspended for a two-year period for their misuse of union funds but reinstated to full privileges in November 1953. The International Executive Board overruled the initial ruling of the district president even in the face of considerable protest from local members. In fact, Owens authored the letter that announced the Board's reinstatement of one of the former local officials.[143]

Owens was a paragon of judiciousness in the letter that restored the "rights and privileges" of the suspended Ashley local officers, but he had a self-righteous side as well. He wrote a sharp letter to District 1 president, August Lippi, in August 1953, about another case, a "disgraceful situation" concerning the misappropriation of funds by officers of another anthracite local. Particularly revealing was Owens's understanding of the general principle involved in the case. "We cannot tolerate," he wrote, "Local Unions spending money that rightfully belongs to the International Union."[144] What about money that rightfully belonged to the local's members or might have been reasonably expected to have been used for their benefit? Why didn't Owens in November 1953 demonstrate the same moral outrage for the misdeeds of the Ashley local officers and sustain the judgment of the district president that their suspension should continue? Apparently, upholding the prerogatives of the UMWA International Union was a considerably higher priority for International officers than defending the rights and privileges of the union's rank and file.

Still, anthracite union members and activists believed in the UMWA and periodically wrote to Lewis or other International officers to air their grievances. The district correspondence files of the UMWA contain many letters from local officials complaining about the inaction of district officers or representatives. "We have made repeated appeals to the District Office for leadership and have been ignored since the time [of] the colliery shutdown" wrote Stephen Sporay, of the West Mahanoy General Mine Board in March 1953. "Why can't we get a District representative to consult on the subject matter?" he asked Lewis. He concluded by asking for a conference with Lewis "to investigate District 9 for laxity in office."[145]

Local activists were particularly frustrated by their inability to change things. Whenever they attempted to protest policies or nominate candidates who might promote new policies, they found the weight of the union organization against them. Incumbent officials had various strategies for fending off members' efforts. One member who wrote anonymously, signing himself, "Fight for Right," described the way union officials operated:

They get in office by hook or by crook and the men have to like it. Sometimes they elect themselves into office by not posting notices that nominations are being held. Some locals have no meetings for the men, but meet at some barroom by themselves. There are no minutes, books, no financial records, nothing. Some locals are run by three men, only the President, Recording Secretary, and Financial Secretary. These same three men also serve as Committee Men. When they receive the check from dues, it is divided three ways and it is O.K. because there are no meetings and no reports. This is what we get put down our throats and if a man puts up a complaint or fuss to get officers in at the head of a local, he is put out of his job and there is no job for him at any mine then.[146]

Rank-and-file miners had enough experience with union "democracy" to be distinctly skeptical of the kinds of claims that Secretary-Treasurer Owens made in response to the well-meaning college student, Stephen Moriak Jr., who wrote to the International.

The drift toward autocratic rule of the UMWA in the anthracite districts was not widely known beyond union ranks. Usually, the exchanges found in the union's correspondence files remained private. On occasion, however, abuses of power reached a level that they could no longer be kept under wraps. Such was the case with the misconduct of District 1 President, August Lippi, and fellow district staff members, in relation to a major catastrophe in Port Griffith in the Wyoming Valley. An investigation following the Knox Mine disaster in January 1959 revealed a consistent pattern of financial malfeasance and double dealing by UMWA district and local officials. The events led to criminal indictments of district officials and rocked the union in the anthracite region for a decade.

The Knox Mine disaster took the lives of twelve miners in Port Griffith in the northern anthracite field on January 22, 1959. The Knox Coal Company generated its profits by extending its operations underneath the Susquehanna

River, mining dangerously and illegally close to the river bottom. The ice-carrying, swollen river broke through the roof of the mine and ten billion gallons of water flowed into the mine and into all the interconnecting mines of the Wyoming Valley. Seventy miners managed to escape the rising water underground, but twelve were swept away, and the resulting mine flooding proved the final blow that led to the end of all underground mining in this portion of the northern anthracite field.[147]

As in all mine disasters, human error played a role in the tragedy, but in this case subsequent investigation revealed the venality and systematic corruption of local mine officials, union officers, and UMWA district officers as root causes. A series of hearings and investigations revealed layers of responsibility. First, there was the misconduct of officials of the Knox Coal Company, lessees of the mine, who willfully disregarded state regulations that prohibited mining within thirty-five feet underneath a waterway. Next, an inspector for the Pennsylvania Coal Company,

the firm that leased the mine, had discovered the violations, but his superiors had not acted to halt the illegal mining. Moreover, the laxity of state mining inspectors permitted this illegal mining to continue until the State Department of Mines ordered the operation closed on January 13, 1959. Yet the mining still went on nine days later when the Susquehanna River broke through.[148]

The UMWA, too, bore a large share of responsibility for the disaster. Trial evidence showed that officers of the UMWA local at the Knox Mine and Lippi, the District 1 president, had taken bribes from company officials. Ultimately nine District officials were indicted, and seven convicted for criminal activity. These bribes had insulated the company from union claims regarding the safety of mining operations that might otherwise have served as a brake on its illegal mining. Lippi's involvement went even deeper than bribery; an investigation uncovered that he was one of the owners of the Knox Coal Company, in clear violation of the Taft-Hartley

Anxious wives of miners, Knox Mine disaster, Port Griffith, January 1959. Photograph by George Harvan.

Act. While never convicted for the ownership per se, Lippi was found guilty of corporate and personal tax evasion for failing to report and pay taxes on his income as an owner of the firm. He was also found guilty of "conspiring to defraud the Exeter National Bank of $39,000." After a string of unsuccessful appeals, Lippi entered federal prison in November 1965.[149]

The UMWA International union should have distanced itself from Lippi as the extent of his illegal dealings came to light, but such was not the case. In August 1960 Lippi successfully ran for another term as district president, and the UMWA refused to intercede. At this date Lippi had been convicted for bribe taking, though that case remained under appeal. In March 1965 Lippi was reelected to yet another term, and once again, allegedly because of outstanding appeals, the UMWA took no action. So much for the ethical standards of the International union that otherwise maintained the "provisional" status of three-fourths of its districts. Members in those districts did not have the right to elect their president or secretary-treasurer, but so long as District 1 members continued to reelect the trusted Lippi, the International Executive Board kept hands off. Going still further, the 1965 district convention raised Lippi's salary even while he was under the cloud of two convictions. Moreover, UMWA President Tony Boyle appointed Lippi's son as acting district secretary-treasurer with an annual salary of $16,000. Finally, while serving his prison term, Lippi angled to run again for district president after his release. Only in August 1968, with the U.S. Department of Labor insisting that Lippi's conviction made him ineligible to hold union office for five years, did the International Executive Board finally put the district under provisional status and appoint district officers.[150]

The chain of complicity in the Knox Mine disaster and for its consequences ran from the local union through the offices of District 1 and right up to the highest ranks of UMWA officialdom. And through it all, the Anthracite Health and Welfare Fund withheld death benefit payments to the widows of the victims of the disaster for more than four years. This sorry tale of a union's abuse of the trust and authority vested in it by its members demonstrates how a self-interested union bureaucracy hijacked the union

and its health and welfare fund from the membership. By the late 1960s, Lippi and other local and district officials were serving only themselves, and it took the deaths of twelve unsuspecting miners and the subsequent investigation to bring the corruption to light.

The venality and conflict of interest that characterized district officials in the Knox Mine disaster reached beyond the anthracite region to the union's International headquarters in Washington, D.C. There, the well-oiled bureaucratic machinery that had been perfected by Lewis to ensure his control of the union began to run less smoothly after his retirement in 1960. The trigger for the breakdown came when contenders emerged to challenge the leadership of Tony Boyle, who succeeded to the UMWA presidency with the death of Thomas Kennedy, Lewis's immediate successor. International official Joseph A. "Jock" Yablonski dared to run as a reform candidate against Boyle in 1969. After Yablonski announced his candidacy, Boyle attempted to fire him from his position in Washington, D.C., as director of Labor's Non-Partisan League. At a meeting with local unionists in Springfield, Illinois, Yablonski was attacked from behind and rendered unconscious. Then a well-organized and funded group of pro-Boyle followers heckled Yablonski supporters at a rally in Shenandoah in the anthracite region. Money flowed freely, and the Boyle machine pressured local leaders to secure nominations for their candidate. Journalist Brit Hume reported electioneering tactics in the anthracite region that were strongly reminiscent of the charges made by the anonymous anthracite miner, "Fight for Right" in the 1950s. In one community, nominations proceeded without any notice, and local officials refused to permit secret balloting when it became apparent that Yablonski had substantial support. In Shenandoah, a Boyle supporter engineered his nomination after Yablonski supporters had left the local meeting. Bribes, packing meetings, setting the clock forward to complete nominations before the scheduled start of the meeting—these were some of the tactics the Boyle camp employed to ensure their victory. In the end, Yablonski still secured nominations from 96 locals, a number dwarfed by Boyle's 1,056.[151]

While bribery and strong-arm tactics could

not keep Yablonski from securing the nominations necessary to mount a campaign for International president, they did ensure Boyle's victory at the polls in December 1969. Boyle won in loyal districts by overwhelming margins, and even in Yablonski's own district Boyle could count on strong margins in pensioner locals that had been kept on the union's books, even though they no longer had active members. The final vote count according to the UMWA tally was Boyle's 74,000 to Yablonski's 41,000.[152] Still, Yablonski, aided by Washington lawyer Joseph Rauh, challenged the outcome of the election, demanding further investigation of allegations of misuse of union funds and election fraud. The Department of Labor refused to impound the ballot boxes, but before it could complete its investigation of the charges, Yablonski, his wife, and daughter were slain in their rural home in Clarksville, Pennsylvania, in late December.[153]

The criminal investigation into the murders brought together local, state, and federal law enforcement and steadily revealed the accuracy of charges by Yablonski's sons that the murders were an assassination resulting from their father's direct challenge to Boyle. Ultimately, investigators found three Cleveland men responsible for the murders and determined that they had been hired by two UMWA district officials in Tennessee under orders from Boyle. The course of justice was slow but steady, and eventually nine individuals, including Boyle, were found guilty for the three murders. Prodded by the Yablonski murders and the ensuing criminal investigation, Department of Labor investigators began tracking Yablonski's original charges and found a systematic pattern of fraud in the recent union election, ruling that a new election would have to be held under federal government supervision. Further investigation revealed the misuse of union funds for campaign contributions and placed Boyle at the center of these actions. Boyle served two years in prison for embezzling union funds and channeling them to political candidates and then, after a second trial on the murder charges, spent the rest of his life in a Pennsylvania prison until being transferred to Wilkes-Barre General Hospital, where he died in May 1985.[154]

The UMWA's descent into corruption and murder occurred as residents of the anthracite region faced the final closing of the mines. Taken as a whole, the record of capital and labor in responding to the crisis in the post–World War II period was dismally inadequate. Leading anthracite firms acted helplessly with the changing patterns of fuel use in the Middle Atlantic and New England states. When the end neared, outside investors took over the anthracite firms, broke them up, and sold their component parts. With accumulated proceeds and tax advantages, the new corporate managers purchased attractive companies that had no connection with anthracite coal and whose operations were located elsewhere across the country. Anthracite mining died quietly, hardly noticed in the business press that covered the births of the new conglomerates. Moreover, in the turbulent period between the late 1950s and the early 1970s, the UMWA's tepid response to the decline—lobbying against the St. Lawrence Seaway, for example—was overwhelmed by the venality and criminality of its district and national officers. John L. Lewis, who retired in 1960 and died in 1969, remained largely unimplicated in the cascading wrongdoing, but there is no doubt that he was responsible for creating the union culture that bred the August Lippis and Tony Boyles of the union and for centralizing power at the top of the union hierarchy that permitted such figures to abuse the powers of their offices and disregard the interests of their members. The national union focused on buying up coal companies and building its Washington bank, while struggling anthracite miners and retirees in northeastern Pennsylvania made do with declining support from the union's near-bankrupt Health and Welfare Fund. Just as the failure of anthracite operators and the UMWA to bridge their differences in the 1920s had contributed to mining's initial retreat, so too their respective postwar failures propelled anthracite's final decline. In this context, former mineworkers looked to other allies to help breathe new life into their dying communities. If anthracite appeared to have no future, then perhaps economic redevelopment offered some measure of hope.

CHAPTER 5

Industrial Development Efforts
Community and Governmental Responses

THE DECLINE OF THE ANTHRA-
cite coal industry in northeastern Pennsylvania extended for eight decades after
coal mining's peak of production and employment in the region during World War I. As employment in anthracite declined from 175,000 in
1917 to 3,000 in 1970 and finally to less than
1,000 in 2000, losses of jobs and income ravaged
communities. The dominant industry in the region declined fitfully over an eighty-year period
and local chambers of commerce, governments,
and residents' groups proposed a variety of initiatives to turn things around. There were numerous efforts—led by campaigns in Scranton,
Wilkes-Barre, and Hazleton—to diversify the
regional economy and stanch the loss of jobs by
attracting new employers or by encouraging existing companies to expand operations. These
efforts quickly became community projects as
residents of the anthracite region responded to
proposals from local elites to raise funds to finance recruitment campaigns. While campaigns
depended on community support, they were not
entirely free of conflict, and sometimes area residents felt imposed on by campaign organizers
and civic leaders. Nonetheless, the campaigns to
retain and attract businesses bore fruit, and they
add an important dimension to an understanding of regional economic decline.[1]

Interest in attracting new industry predated
the initial decline in anthracite production following World War I. This history began in the

1880s as the region was entering a period of significant expansion. Area businessmen were already concerned about their extreme dependence on coal; slight downturns in coal
production affected the business of local merchants, who organized campaigns to lure new
firms—especially silk and, later, garment manufacturers—to their communities with a variety
of subsidies and tax abatements.[2] The first development initiatives sought to supplement the
earnings of male mineworkers by offering employment to women and children in mining
families—particularly in periods when demand
for coal was weak, due either to unseasonably
warm weather or the impact of business downturns.

Scranton emerged as the leading urban center in the anthracite region in the late nineteenth century, but enhancing its industrial
structure became a concerted endeavor as early
as the 1880s. Commercial leaders formed a
board of trade in that decade, which lobbied city
officials to offer ten-year tax abatements to new
companies that opened operations in the city. In
1913 and 1914, the board organized fundraising
drives that provided the new Scranton Industrial
Development Company more than $1 million to
assist in the establishment of new manufacturing enterprises. After the stock market crash in
October 1929, the board's successor, the chamber of commerce, set up a Credit Guarantee
Fund to support economic activity in the city.

114

Chamber programs culminated in a 1939 plan to encourage young residents to find work locally and remain in Scranton.[3] Redevelopment efforts that drew great attention in the postwar period built on decades of earlier activity.

After Pearl Harbor, the entry of the United States into World War II led Scranton-area businessmen to intensify their efforts to attract new business. In the early 1940s, the chamber of commerce prepared a recruitment pamphlet, "The Scranton Area and the Advantages It Offers as an Industrial Site," that made the city's best case, emphasizing the decline of anthracite output and employment during the 1930s and pointing out that Scranton offered more than 35,000 employable persons within a fifteen-mile radius. A second pamphlet set forth the chief theme of the chamber's activities: "Boost Scranton, Make Jobs," its title proclaimed. Whereas the first pamphlet spoke to managers of firms considering relocation, the second appealed more broadly to Scranton-area residents to convince them of the need to raise funds and "sell" Scranton as an industrial location. The second pamphlet may have also served as a training manual for individuals destined to go out among their neighbors and make a sales pitch for the local fundraising effort.[4]

The high point of Scranton's efforts to recruit industry came during World War II as the chamber of commerce and local elites successfully lobbied the federal government to build a major defense plant in the area. To coordinate their multiple efforts to recruit businesses, city leaders established the Scranton Plan. The Plan gained a national reputation that created considerable favorable publicity for the city and led other communities to follow Scranton's lead.[5] The Plan's first success came in June 1943, when the Murray Corporation broke ground for the construction of a 500,000-square-foot plant to manufacture wings for B-29 bombers. Chamber flyers exulted, "This is the 'Big One' which everyone has hoped for and worked for." Local leaders anticipated that the plant would employ 7,000 with an annual payroll of $10 million.[6] Ultimately the operation employed only 3,000 workers at its peak between its opening in February 1944 and the end of war, but those jobs made a significant contribution to Scranton's wartime prosperity.[7]

After World War II, with many signs pointing to the permanent decline of anthracite coal, there were renewed efforts to recruit new businesses to diversify the city's economy. In the initial campaign at the war's end, the Greater Scranton Chamber of Commerce approached the federal government, seeking to purchase the Murray Corporation's plant, with a view to leasing it back to the company, now reconverted to the peacetime production of bathtubs, kitchen sinks, and washing machines. The project promised to employ some 4,000 workers with annual salaries and wages reaching $8 million.[8] As wartime exigencies receded, the goal was to provide adequate employment and earnings for male heads of households, particularly returning servicemen. The headline on one page of a Scranton pamphlet was representative. "JOBS for Men . . . Jobs for Veterans" captured the common sentiment in the immediate postwar period.[9] As local business leaders negotiated with the federal government, they mounted a campaign that raised more than $1.7 million. In the end, they spent some $1.2 million to purchase the factory from the federal government, leaving the balance of the funds as seed capital to be administered by the Scranton Lackawanna Industrial Building Company (SLIBCO) for the construction of factory buildings for lease or sale to other companies coming into the area.[10]

The conversion of the wartime plant became the first of a string of successes in Scranton. The Murray Corporation was a major employer in the city for three decades, though employment at the factory plummeted sharply from almost 3,500 in 1947 to 1,100 in 1953 and just over 500 in 1959. SLIBCO and the Scranton Plan coordinated their efforts and brought additional industrial employers to the city. By 1949 drives had raised more than $2.7 million from local businesses and more than 5,000 individual donors. SLIBCO developed a slick recruiting campaign and worked hard to sell industrial prospects on the advantages of locating in the Scranton area. Headlines in successive pamphlets touted the group's success: "31 New Plants Built, 28 Plants Expanded" was followed later by "39 New Plants Built, 55 Plants Expanded."[11] As a result, by 1949 SLIBCO owned eleven industrial buildings valued at more than $2 million. In addition, the Scranton Plan's Murray plant was appraised at $3.5 million. With the founding of the Lack-awanna Industrial Fund Enterprises (LIFE) in 1950, fundraising, the recruiting of firms, and the construction of industrial buildings increasingly proceeded within a framework that extended to all of Lackawanna County. LIFE held repeated fundraising campaigns across the 1950s, and by the decade's end the fund was constructing its sixteenth shell building for an incoming firm in nearby Carbondale. By 1960 these Scranton Chamber of Commerce efforts had led to the construction or conversion of thirty-one industrial buildings occupying more than three million square feet with a construction cost of almost $20 million.[12]

Scranton led in the establishment of industrial development agencies in the postwar period, and the methods that Scranton leaders employed set the framework in which other communities operated. Industrial development agencies were hybrid private–public entities. The organizations were established by chambers of commerce and private business leaders, but they raised funds through public appeals and spoke for the broader public in their communities. In the first years, these agencies typically built a factory for a prospective industrial employer and/or offered direct subsidies for relocation expenses. Over time, they moved away from that model and erected generic industrial shell buildings and leased or sold the buildings to relocating or expanding firms. By constructing shell buildings, the agencies accelerated the start-up process for new companies and provided flexibility for the buildings' use. Companies could finish the buildings to meet their individual needs. The leases typically provided subsidized low-interest mortgages, and the agencies offered connections with local commercial banks that supplied the main mortgage financing.

Industrial development agencies began with assistance to individual firms, expanded into the construction of generic shell buildings, and eventually moved to a model that included the preparation of industrial or office parks for relocating firms. The industrial park model permitted agencies to anticipate their clients' needs and simultaneously plan for a large number of industrial sites. Thus Scranton developed Keystone Industrial Park, Wilkes-Barre built Crest-

wood Industrial Park and followed up with Hanover Industrial Estates, and Hazleton had its Valmont and Humboldt Industrial Parks. The industrial agencies followed similar strategies in their work, seeking prospective employers, helping with site selection and financing, coordinating with architects and city planners, and assisting the families of key managers with moves.

Hazleton and Wilkes-Barre were SLIBCO's leading competitors for industrial recruits. The stories of these two cities offer instructive comparisons with Scranton. Hazleton is a considerably smaller mountaintop community located forty miles southwest of Scranton. Recruitment efforts in Hazleton also had a lengthy history dating back to the 1880s. Hazleton's "factory hill" was a visible sign of the community's success in achieving a measure of industrial diversity. By the mid-1920s, Hazleton area silk and shirt firms employed about 5,000 (mostly female) workers, roughly a fourth of mine employment in and around the city. In the interwar years, local elites stepped up efforts to retain industrial employers, keeping a local silk firm from moving south by guaranteeing a $50,000 mortgage. In addition, the city council in 1935 purchased part of the plant of the downsizing Duplan Silk Corporation to provide low-rent space for other industrial tenants. In contrast to other coal communities, Hazleton grew slightly across the Depression decade, and its population reached an all-time peak of 38,000 in 1940.[13]

As World War II ended, however, local business leaders realized that something had to be done to offer employment to returning veterans and young people entering the labor force. In 1947 leaders of the Hazleton Chamber of Commerce recruited Electric Auto-Lite, a midwestern manufacturer looking for a site for a northeastern branch plant with more than a thousand jobs. They raised some $650,000 through a grassroots fundraising campaign aimed at local residents and businesses. Of this sum, they used $500,000 as a direct subsidy toward the purchase of land and the building of a plant for Electric Auto-Lite and reserved the remaining $150,000 to provide the newly founded Hazleton Industrial Development Corporation with a revolving fund to recruit additional firms.[14] Hazleton's experience was strikingly similar to Scranton's campaign to raise funds to buy the

Murray Corporation bomber wing factory with remaining funds placed in the hands of SLIBCO for future use. The two declining mining communities followed the same model in their efforts to diversify their local economies as anthracite mining played out.

By all accounts Hazleton leaders and residents were disappointed with their experience with Electric Auto-Lite. The firm had projected employment of 1,000 in its new plant but scaled back its plans immediately and began operations with 250 workers, reaching a peak of only 329 in 1959.[15] Residents' optimistic hopes went unfulfilled, and many thought that the ultimate cost of the subsidy—$2,000 per job—was too much. Local leaders felt a need to recast their strategy in the face of the general disillusionment with the first recruitment. Rather than continue subsidizing individual firms, the Hazleton Chamber of Commerce took a new tack in 1956, raising $14,000 in small donations from more than 5,000 contributors to purchase land for a five hundred–acre industrial park followed by a much larger effort to create a substantial revolving fund to permit the construction of shell buildings to be leased or sold to prospective employers. Again the local chamber of commerce established an independent, nonprofit organization, the Community Area New Development Organization (CAN DO), to erect the shell buildings, recruit new firms, and arrange for the lease and/or sale of the buildings to new companies. Between 1956 and 1974 Hazleton (by CAN DO's count) added or expanded sixty industrial firms, employing more than 7,000 workers with annual earnings of almost $50 million. The improved employment situation contributed to a rebirth of the city, with the addition of a new public library and a YMCA–YWCA complex reviving the city's downtown. Additional community fundraising and volunteer work contributed to tree planting, neighborhood improvements, and the expansion of the local branch of Pennsylvania State University.[16] With the recruitment of new businesses, the creation of an industrial park, and urban improvements, Hazleton received a substantial facelift.

Industrial development efforts in Wilkes-Barre followed patterns similar to those in Scranton and Hazleton. The economies of all three communities were based on anthracite

coal, and they were close enough together so that elites knew what their counterparts were doing in the other communities. The Wilkes-Barre Board of Trade attempted to recruit businesses to complement coal mining in the 1880s and attracted lace and silk manufacturing in this period. The Matheson Motor Company moved to Wilkes-Barre in 1905 from Holyoke, Massachusetts, employing some five hundred workers producing automobiles. The Wilkes-Barre Wyoming Chamber of Commerce, the board of trade's successor, expanded on those efforts in September 1929 by setting up an Established Industries Committee and a New Industries Committee. The first committee aimed at increasing local employment by emphasizing to area residents and businessmen the importance of "buying and using Valley-manufactured products." The latter committee focused its efforts on investigating industrial prospects and serving as an honest broker between the community and these prospects by recruiting the best ones to the area.[17] At the same time the chamber hired Lockwood, Greene & Company, a New York City industrial engineering firm, to conduct a comprehensive economic survey of the Wyoming Valley for use in recruiting new businesses to the area.[18] The onset of the Depression underscored the importance of local efforts and led Chamber leaders to add fundraising to the voluntary efforts of these new committees. Contributions from local businesses established an Industrial Development Fund that channeled the chamber's efforts through the outbreak of World War II. After the war, successive efforts led to the establishment of a Veterans Administration hospital and the erection of several downtown buildings for industrial purposes. The creation of the Committee of One Hundred in 1951 and the Greater Wilkes-Barre Industrial Fund in 1953 signaled a new era in industrialization efforts. Successful fundraising permitted the purchase of a tract of six hundred acres suitable for the establishment of an industrial park.[19]

William O. Sword headed up the Committee of One Hundred's fundraising and recruitment efforts, which succeeded on both fronts. Using $250,000 of the $727,000 raised by the campaign for preparation of the first site in what would become Crestwood Industrial Park,

Sword and chamber associates convinced the Foster Wheeler Corporation, a maker of heavy industrial and military equipment, to move from Danville, New York. The effort mirrored the earlier successes of Scranton, which had attracted more than fifty new firms by this date, and anticipated the industrial-park successes for CAN DO in Hazleton that followed beginning in 1956. Ultimately, the three communities competed directly with one another for a limited number of new businesses, but with a sizable tract of land outside the city and a well-organized business and professional elite, Wilkes-Barre held its own in this intercity competition.[20]

Beginning in 1961 the Wilkes-Barre Wyoming Chamber of Commerce improved its competitive position by soliciting the support of organized labor and extending its fundraising to include the city's working and middle classes. Organized labor supplied contributors, but equally important, it joined a Labor-Management-Citizens Committee to support an atmosphere of labor-management cooperation intended to convince prospective employers that their operations in the Wyoming Valley would be free of the labor conflict that had characterized anthracite coal mining in earlier periods. By 1965 community fundraising had topped $5 million, and the value of industrial projects supported by the Chamber exceeded $40 million. Crestwood Industrial Park was the centerpiece for Wilkes-Barre's success; between 1953 and 1967 fourteen firms relocated their operations to the park. In 1965 the new firms attracted by the city's campaigns employed 13,000 workers with an annual payroll of more than $53 million.[21] The growth in manufacturing jobs did not quite match the loss of mining jobs in the postwar period but did cushion the impact of the precipitous decline of coal mining. And the combined growth of manufacturing, service, and government jobs outpaced job losses in the 1950s and 1960s, so that the rate of unemployment in Luzerne County (including Wilkes-Barre and Hazleton) declined from an average of almost 15 percent between 1954 and 1958 to less than 6 percent between 1964 and 1966.[22]

For a decade after World War II, the industrial development campaigns were strictly local undertakings. Initially local chambers of commerce raised funds on their own, prepared their recruitment materials, and contacted industrial prospects. They entered into negotiations with each prospective employer over the nature of the assistance that they could provide to land their catch. The local organizations might purchase land for a prospective plant, offer to build a plant, or help pay for relocation costs. Among communities in the anthracite region, Scranton, Hazleton, and Wilkes-Barre stood out as the best organized and most successful. Still their efforts were hampered by the limits of their fundraising in providing the seed capital for the subsidies they needed to offer. Often after landing one prospect, leaders had to halt their efforts, unwilling to go back to their local communities with yet another fundraising drive so close on the heels of a previous one. Moreover, their recruitment activities typically proceeded on a voluntary basis, and there were limits to how much the organizations might ask of their volunteers. Finally, the work could be frustrating, as communities competed with one another and with communities in other states for a limited number of firms poised to relocate. Underfunded and overly competitive, the industrial redevelopment effort taxed the capabilities of communities in the anthracite region in the postwar period.

Politicians in Harrisburg, led by Democratic governor George M. Leader, observed the economic difficulties many of their constituents faced back in their home districts and the limitations of the strategies they had developed to cope with anthracite's decline. Like George Earle twenty years earlier, Leader held grassroots meetings across the state to survey the conditions in distressed areas and to build up popular support for his program.[23] Leader proposed an approach previously employed by New England states to reinvigorate local economies, and in 1956 the Commonwealth established the Pennsylvania Industrial Development Authority (PIDA).[24] PIDA was conceived to assist firms in setting up or expanding their operations in economically depressed regions of the state. The authority offered low-interest second mortgages to eligible firms operating in areas of particularly high unemployment. State legislation specifically established the parameters of the funding package: Local banks provided 50 percent of development costs with commercial mortgage loans; PIDA itself provided 2-percent mortgage loans covering an additional 30 percent of project costs; finally, local industrial development authorities were required to provide the remaining 20 percent of mortgage financing. Over time the state increased PIDA's share to 40 percent and reduced the local authority's contribution to 10 percent. PIDA was limited to offering or denying loans to firms recommended by local industrial development corporations, thus encouraging communities to organize recruitment campaigns. By offering 100-percent financing to firms, including heavily subsidized second mortgage loans and substantial state assistance, PIDA vastly increased the resources available to assist firms moving into the state or expanding their operations. Between 1956 and 1978 PIDA in this way approved more than 1,300 loans totaling over $488,000,000.[25] Much of this assistance—more than 27 percent of all PIDA aid between 1956 and 1970—flowed into the three largest counties in the anthracite region, Lackawanna, Luzerne, and Schuylkill.[26]

In addition to the efforts of local industrial development authorities and the state of Pennsylvania, a third major actor came forward to promote industrial development in the anthracite region—the privately owned regional electric utility, Pennsylvania Power & Light Company (PP&L). From an early date PP&L executives saw the handwriting on the wall and realized that their major consumers of electric power (accounting for a third of demand in 1947), the anthracite companies, were in crisis. For the utility to enjoy continued prosperity, the region would have to attract additional industrial customers to make up for the lost demand as anthracite mining declined. In 1940 PP&L established an Industrial Development Department and kept close tabs on industrial prospects, offering a variety of services to attract companies into their service region. Moreover, PP&L staff worked closely with community leaders, supporting local bond drives, the founding of industrial development authorities, and community efforts to recruit new industry to the region.[27] Wanting to deliver kilowatts of electricity to these potential customers, PP&L proved to be an important ally for community-based recruitment efforts.

In a 1940 "Report on Industrial Development," PP&L first outlined the ways its Industrial Development Department would operate in its effort to "create and maintain prosperity within the territorial limits" the company served. It would seek to recruit new enterprises to its service region, maintain and expand industrial companies currently operating in its region, and, equally important, create "an Industrial Development consciousness on the part of areas, cities, and towns in Pennsylvania Power & Light Company territory." As part of meeting this last goal, staff made many local presentations and distributed pamphlets and flyers. In 1950, for instance, the department "gave talks before civic groups in 43 different communities" and distributed 1,000 copies of a Federal Reserve Bank publication, "Operation Boot Strap," promoting area programs aimed at local economic development. Starting as early as it did, PP&L played a real leadership role in the industrial development field in central and northeastern Pennsylvania.[28]

In line with the philosophy articulated in their 1940 report, PP&L officials played major roles in industrial development in all three of the communities that led the efforts to diversify the economy of the anthracite region. PP&L executive John Davidson began working for regional industrial development in Hazleton in 1940. Beginning in 1947 he served as president of the Hazleton Industrial Development Corporation and headed up the fundraising campaign that permitted Hazleton to land the Electric Auto-Lite Company. The company transferred Davidson to Scranton in 1957, where he led both company and community industrial development efforts for the next three decades. In 1959 he directed a fundraising campaign for Lackawanna Industrial Fund Enterprises that raised almost $1.7 million. He served for a period as president of the chamber of commerce and then as vice chairman of LIFE. In these capacities he promoted development of the Keystone Industrial Park in nearby Dunmore. When he finally retired in 1988, he had been involved in industrial development work for almost fifty years.[29]

PP&L made similar contributions in Wilkes-Barre, where the company's vice president, Frank Meuller, joined with Percy Brown, long-time owner of a downtown grocery store, to make the first purchases of land in what became Crestwood Industrial Park. Brown and Meuller played major roles in convincing the Committee of One Hundred to purchase the 600+ acres destined to be central in Wilkes-Barre's industrial development for fifteen years beginning in 1952. Twelve years later, as the successful development at Crestwood Industrial Park necessitated plans for additional sites for industrial expansion, PP&L provided a ten-year, interest-free loan to permit the acquisition of a site in Hanover Township that eventually was developed as Hanover Estates Industrial Park.[30]

In addition to its assistance in raising funds, developing industrial sites, and recruiting new industries, PP&L produced a lengthy series of annual reports on industrial development. These sources survive intact for the period between 1940 and 1987, except for a few years during World War II, and provide extensive data on firms that located in the company's service region, including the prior and new locations of operations and their anticipated new employment and payroll. The reports also record major layoffs and firm closings, thus tracing job losses in the utility's service region.[31]

The existence of the PP&L annual reports makes it possible to tell the story of economic development in the anthracite region in a systematic way, moving beyond the anecdotal and self-promoting publications of the local industrial development groups themselves. The PP&L's lists include more than 1,250 firms that operated in the region between 1940 and 1995. An analysis of the experiences of these firms over time permits an evaluation of the success and limitations of industrial development efforts. Most studies of reindustrialization campaigns have been limited to describing the original job projections offered to justify low-interest loans, relocation assistance, or funds to assist firms with training new workforces.[32] To move beyond boosterism, it is vital to know how many jobs were actually created and how long these sought-after industrial firms operated in their new homes.

The number of new or expanding firms in the anthracite region recorded by PP&L increased steadily between the 1940s and the 1960s, growing from an average of twenty-five firms a year to almost fifty firms annually.[33] In addition, the proportion of these firms receiving local and state financial assistance increased steadily over time. In the 1940s almost 24 percent of firms listed by PP&L received financial assistance. By the 1950s and 1960s, that proportion exceeded 28 percent. In absolute terms, the number of assisted firms increased from 44 in the 1940s to 129 in the 1960s. Since 1970 the absolute number of new firms has declined, but the share receiving assistance has continued to grow. Fully 40 percent of firms listed by PP&L as starting up business or expanding operations in the anthracite region between 1970 and 1987 received some sort of state or local financial assistance. Increasing numbers of communities in the postwar decades had local industrial development organizations in operation, and the founding of the state PIDA program in 1956 further supported their efforts.

Local recruitment and state assistance led to an influx of industry, with garment and metalworking trades predominating among the new firms beginning or expanding operations in the region after 1940. More than a third of new firms noted by PP&L were garment factories; another 19 percent were in the metal trades. Food and beverage and wood products were the next two leading groups, each accounting for just under 6 percent of new firms. Whereas almost exclusively male jobs were lost with the closing of the mines, new firms moving into the region brought significant new employment opportunities for women. PP&L noted the numbers of male and female jobs projected to result from the operations they surveyed and women were expected to fill more than 41 percent of the new positions. Among garment firms, the proportion was greater than 83 percent, but these figures were offset by the predominance of men in wood products and metal trades firms. Still, local chambers of commerce sought to recruit employers of skilled, male workers and they typically had to settle for lower-paying industries with significant proportions of female workers. Men interested in unionized, higher-paying work, such as that found in auto plants and steel mills, would commonly have to commute on a weekly basis or migrate outside the region to find employment. Those who stayed close to home had to accept what came their way.[34]

The typical new firm in the region between 1940 and 1970 employed on average 108 workers and survived for seventeen years after starting up or expanding its operations. Recruited and expanding firms tracked by PP&L employed somewhat more than 80,000 workers all told, a total comparable to the number of mining jobs lost in the postwar years.

Overall, just over 30 percent of all these firms received some sort of assistance from a chamber of commerce, a local industrial authority, or the Commonwealth of Pennsylvania. How did assisted firms differ from those that did not receive such financial support? Assisted firms were considerably larger than unassisted ones, employing on average 185 workers, compared to 66 for firms that did not receive assistance. Assisted firms also exhibited significant growth in the size of their workforce over time. In their first five years of operation, assisted firms employed 140 workers on average, figures that rose to 199 and then to 224 in each of the next five-year intervals. Unassisted firms began with 60 employees on average, a figure that increased only slightly, to 72, after the first decade of operations. Finally, as might be expected from these figures, assisted firms survived in business considerably longer than did unassisted ones, almost 22 years on average compared to 14 years.

These differences in size of operation and length of life were not a function of a firm's start-up date, county location, or industrial sector. Within firms started in a particular decade, assisted firms outlasted unassisted ones. Within product groups, assisted firms were larger than unassisted ones. In specific counties both of these differences held up strikingly.

Assistance to firms contributed to their longevity and to the number of jobs they created. But how did this performance compare to original expectations? For example, firms may have exaggerated the number of jobs they anticipated creating in order to justify lucrative subsidies for their relocation or expansion. Such was the case for a number of high-profile assistance projects. The first postwar firm recruited to Hazleton, Electric Auto-Lite, provided only a fourth of the number of jobs it had initially promised. Disappointment with the performance of Electric Auto-Lite led local leaders in Hazleton to revamp their assistance program with the founding of CAN DO and to offer only lease/sale arrangements on shell buildings and to cease providing direct subsidies to recruited firms.[35]

On the whole, however, firm performance in terms of job creation did not replicate the disappointing experience of Electric Auto-Lite. Among assisted firms, the average number of workers actually employed stood at 185 compared to original projections that averaged just over 152. Thus, assisted firms eventually exceeded expectations by more than 20 percent. These figures do not represent the immediate experience upon relocation or expansion, but indicate that over the lifetime of assisted firms their original projections were not out of line. Assisted firms basically delivered on their initial promises.

The contributions of state and local assistance programs are particularly visible for those firms that relocated into the anthracite region during the 1940–1970 period. PP&L records note 103 firms that relocated into the region during this time. About 40 percent of these businesses came from New York City, 20 percent from other areas in Pennsylvania, with the remainder from elsewhere in New York, New Jersey, and other states.[36] Most striking about relocating firms is the high proportion that received assistance—almost 72 percent. Among local firms expanding their operations, the comparable figure was less than 23 percent.[37] Moreover, assisted relocating firms were larger and survived longer than those that made the move without assistance. Assisted firms averaged 201 workers, compared to 121 for unassisted firms. Assisted relocating firms remained in business almost 22 years, 7.5 years longer than the average for unassisted ones. Local industrial development organizations sought to recruit large firms from outside the region, and their efforts bore fruit.

Focusing on the three largest industrial development organizations in the anthracite region in the postwar decades—the Scranton Lackawanna Industrial Building Company (SLIBCO), Hazleton's Community Area New Development Organization (CAN DO), and the Greater Wilkes-Barre Industrial Fund (GWBIF)—reinforces the overall portrait provided by the PP&L reports. These organizations led the anthracite region in terms of securing mortgage support from the Pennsylvania Industrial Development Authority, accounting for almost 130 firms assisted by the Authority. All three organizations recruited roughly similar mixes of firms, with textile, garment, and metalworking firms accounting for 50 to 60 percent of their assisted firms. They were larger than was typical for new or expanding firms in the region, averaging 165 workers at the outset and 308 workers at firms that survived more than ten years. Of the three communities, Wilkes-Barre fared best, with assisted firms continuing in business on average for almost twenty-five years, roughly 25 percent longer than in Scranton or Hazleton. Comparisons among the three communities reveal that they competed for a single pool of potential recruits, and the similarities among them outweigh any evident differences in terms of the performance of recruited firms over time.

These findings point to an important fact about economic development efforts in the Pennsylvania anthracite region in the postwar decades: Assistance to individual firms did make a difference in the employment picture. Assisted firms did not just take their subsidies and run, but stayed on, grew significantly, and contributed to the economic life of the communities where they set up business. Although there were notable examples of businesses that failed to meet expectations or moved out of the region almost as quickly as they came, these firms were exceptions to the pattern that emerges from the broader analysis.

Shifting from evidence on specific firms that moved into the anthracite region to broader aggregate data on the regional economy as a whole confirms these initial findings. The three counties most dependent on anthracite employment—Lackawanna, Luzerne, and Schuylkill—bottomed out, as far as overall employment was concerned, in 1960. At that date the overall labor force in the three counties totaled more than 179,000. By 1970 that figure stood at 206,000, and in 1980 it exceeded 247,000. Over a period of two decades, then, the three-county labor force grew by 38 percent. In that same interval, employment in anthracite in the three counties declined from almost 13,700 to less than 3,000. The garment industry, formerly a mainstay of the regional economy, also declined, losing 14,000 employees, about a third of its workforce between 1960 and 1980.

What economic sectors in the anthracite counties showed gains between 1960 and 1980, offsetting the losses in the anthracite and garment industries? The service sector showed the most dramatic growth, more than 200 percent over the twenty-year period. Trade held its own with growth of almost 35 percent; in contrast, manufacturing employment declined by 3 percent over the period. While manufacturing had accounted for 51 percent of the three-county labor force in 1960, by 1980 its share was less than 36 percent.[38] Still, the regional economy had shown remarkable resilience and rebounded substantially from the low point in 1960. Chambers of commerce had been active, and localities and the state had focused resources on the region, and there were results to show for these combined efforts.

Pennsylvania's and other states' reindustrialization efforts were complemented in the 1950s and 1960s by federal legislation aimed at the redevelopment of what were termed distressed areas. Democratic Senator Paul Douglas of Illinois—joined by the anthracite-region congressman Daniel J. Flood in the House of Representatives—spearheaded the federal effort by sponsoring the Depressed Areas Act in 1955. Douglas proposed an array of approaches that included loans for construction of industrial buildings, much like Pennsylvania's PIDA program, as well as federal grants and loans for infrastructure. The Republican administration of Dwight D. Eisenhower countered with a much more modest alternative that deemphasized the federal role in dealing with industrial decline. Twice in the 1950s Congress passed area redevelopment legislation, only to find the bills vetoed by Eisenhower.[39]

Appalachia in general, and Pennsylvania in particular, played pivotal roles in the emerging federal debate. Advocates for the anthracite region had been making their case for federal assistance for some time. In March 1954, members of the Northeast Pennsylvania Industrial Development Commission wrote an open letter to senators and congressmen describing the region's plight and urging federal aid.[40] Between

January and April 1956, the Subcommittee on Labor of the Senate's Committee of Labor and Public Welfare held hearings on Douglas's area redevelopment legislation. Pennsylvanians were prominent among those making statements before the committee. In testifying before the subcommittee, Congressman Flood joined citizens and officials from the anthracite communities of Carbondale, Georgetown, Moosic, Plymouth, Pottsville, Scranton, and Wilkes-Barre. Min Matheson, district manager for the International Ladies' Garment Workers' Union in the Wyoming Valley, joined by a variety of UMWA and Pennsylvania AFL-CIO (American Federation of Labor–Congress of Industrial Organizations) officials, also spoke. The subcommittee held one set of hearings in Wilkes-Barre, where Pennsylvania's experience with industrial development initiatives received a respectful hearing.[41] Subsequent hearings in 1959 before a subcommittee of the Senate Committee on Banking and Currency heard testimony from Congressman Flood, both Pennsylvania senators, and Governor David Lawrence. Senator Hugh Scott spoke for all when he argued that "the local communities are running out of money," and "that chronic unemployment and underemployment . . . is a matter of national concern." From Pennsylvania came a single voice calling for federal action.[42]

With the election of John F. Kennedy in 1960 the legislative terrain in Washington shifted dramatically. Kennedy had been a cosponsor of Douglas's area redevelopment legislation in the Senate and his primary campaigning in West Virginia had further alerted him to the needs of depressed areas. He had met with Appalachian governors and designated the Area Redevelopment Act a top priority on assuming the presidency. By May 1961 both houses of Congress had passed, and Kennedy signed, legislation establishing an Area Redevelopment Administration (ARA) in the Commerce Department. The new agency had an appropriation of $390 million to be spent over a four-year period, much to be utilized in revolving loan funds for distribution in regions with high and persistent rates of unemployment. The ARA's first and only administrator was William L. Batt, formerly Secretary of Labor and Industry in Pennsylvania during the period in which the PIDA program was implemented.[43]

The ARA, while representing a departure from the federal inaction of the 1950s, had a task well beyond its modest means. It soon had responsibility for funding programs in about a thousand counties that met fairly loose eligibility requirements, and what the agency could do for any specific locale was quite limited. Aid to the anthracite region included a loan to establish a mine tour in an abandoned mine in Ashland, loans to a variety of employers in the region, and support to test methods of controlling culm bank fires.[44] Assistance was often indirect, as the terms of the authorizing legislation required that the ARA provide funding to other agencies for program implementation. Moreover, its lifetime was short, and in 1965 the passage of legislation that launched Lyndon Johnson's Great Society programs undercut the mandate of the ARA, and Congress did not renew the program when it expired in 1965. While it was in operation, however, the ARA did channel a large share of its funding to the depressed Appalachian region, and federal support continued to flow from a variety of programs into the Pennsylvania anthracite region in succeeding years.[45]

In the early 1960s the United States rediscovered poverty in the midst of affluence, and Appalachia epitomized a region and a population that had been left behind during the nation's postwar economic boom. Here was an area suffering economic decline, in both its bituminous and anthracite regions, while the rest of the national economy was booming. The federal government did not directly address the need for new industrial employment as the coal economy declined, but it did recognize the connection between improved economic infrastructure and economic development in the Appalachian region. The appearance in 1962 of Michael Harrington's exposé, *The Other America*, reflected the increasing awareness of poverty in the midst of an affluent society.[46] In this and other studies, the Appalachian region stood out as lagging behind the postwar economic development that characterized the nation as a whole. A weak agricultural sector, declining coal mining, high outmigration, and few prospects for improvement characterized Appalachia, north and south.

The ARA's first efforts focused on Appalachia, responding to contemporary studies and to a variety of state initiatives. Counties in northern Georgia and eastern Kentucky promoted regional planning efforts, and eventually the Council of Appalachian Governors met to share common concerns and mount a lobbying effort to secure federal recognition of their problems. After his election in 1960, Kennedy established the President's Appalachian Regional Commission to study regional problems and propose legislative solutions. After Kennedy's assassination, the Commission filed its report, and in early 1965, following the reelection of Lyndon Johnson, Congress passed the Appalachian Regional Development Act that placed these efforts on a more solid foundation by creating a "new independent agency to coordinate state and federal actions" regarding Appalachia.[47]

The Appalachian Regional Commission (ARC), established by the legislation, gave governors of Appalachian states a major role in setting priorities and establishing policies to be supported with federal funds. Compromises reached during the legislative process led to a primary emphasis on highway construction, with additional support for programs related to environmental conservation, vocational education, and health care. The earlier, and more controversial, focus on federal support for economic development receded as Congress translated into legislation the recommendations of the report of President Kennedy's commission.

A crucial focus of ARC resulted from the creation of local development districts (LDDs), multicounty units within Appalachian states that served to promote local planning and input into the process of policy making. The Commission left it to the states to pass enabling legislation that established the specific LDDs within their boundaries and offered funding for administrative support that proved an inducement to state action. By 1972 more than fifty LDDs were in operation, a number that eventually reached seventy-two across the 410 counties in 13 states eligible for support from ARC. These districts, in turn, took on increasing planning and coordination responsibilities under subsequent federal legislation relating to intergovernmental cooperation and environmental policy.[48]

While federal and local highway construction took the lion's share of ARC funding, other programmatic support varied across Appalachia. Within the anthracite region, the Economic Development Council of Northeastern Pennsylvania (EDCNP) served as the local development district that organized research, coordinated input from various public officials and agencies, and generally established local priorities for ARC funding.[49] Environmental issues ranked high for the anthracite region in the early years of ARC, with substantial funds going to fight underground mine fires and to backfill abandoned mines to reduce the threat of mine subsidence. For several years running, ARC funds were used to extinguish smoldering fires in Carbondale, Throop, and South Scranton, all in the northern anthracite field. Strip-mine restoration projects were also concentrated in the anthracite region, with major projects in Scranton, Wilkes-Barre, and Conyngham Township.[50] Federal programs complemented the recruitment efforts of area industrial development organizations and the statewide PIDA program. The federal government relied on local and state officials and avoided initiatives that smacked of central planning or a national industrial policy. Federal programs—particularly the construction of the two interstate highways (I-80 and I-81) that crossed in the region—contributed to reindustrialization with the decline of anthracite mining, but they were calculated to avoid controversy or stepping on the toes of local power brokers. They offered a lowest-common-denominator approach to industrial development, not even using the term, and proceeded with programs that were broadly acceptable on the national level.

Both major programs, ARA and ARC, in the end, made only modest, indirect contributions to economic redevelopment in the anthracite region. Of much greater significance to the regional economy was a more traditional federal influence—pork-barrel politics. Anthracite congressmen were masters at channeling federal dollars—for whatever purposes—into their districts. Daniel J. Flood, who represented the Wilkes-Barre district from 1945 to 1980, had few peers in this regard. Flood was largely responsible for siting a Veterans Administration hospital in Wilkes-Barre and a major Army repair depot in Tobyhanna. Under a 1962 amend-

Smoke from underground fire, Carbondale, 1949. Photograph by George Harvan.

ment sponsored by Flood, the Army exported anthracite to West Germany to heat its military bases there. As vice chairman of the Defense Appropriations Subcommittee in the House, he was in a position to direct a great deal of military spending to his district. Indicted on bribery and corruption charges, Flood resigned from the House in 1980 and received a year's probation for his offenses. Meanwhile, Joseph McDade, who represented Scranton in Congress for thirty-six years, picked up where Flood left off. He served as the ranking Republican on the House Appropriations Committee, and after 1985 the most senior Republican member of its Defense Subcommittee. Adept at pork-barrel politics, McDade funneled defense dollars to his district as well as energy and water projects, parks and recreation funds, and support from the Small Business Administration. In 1992, at the height of McDade's power, federal projects valued at $420 million were underway in his congressional district.[51] In his winter 1992 constituent newsletter, McDade detailed recent appropriations that benefited the Scranton district.

Within the defense budget, he listed eleven distinct contracts or grants to private firms, military bases, and area institutions of higher education. In addition, he reminded constituents that he had recently been elected vice chairman of the House Appropriations Committee, providing himself a "strong voice on all 13 appropriations subcommittees." From that position of seniority, he claimed credit for numerous projects affecting his district: eight projects worth $16 million related to the Interior Department; four worth $1.8 million related to agriculture; six worth $19.4 million related to energy and water; and three worth almost $4 million for housing and environmental programs.[52] These earmarked appropriations contributed far more to the anthracite region after the mines closed than did the targeted programs through ARA and ARC.

While the federal programs provided resources for infrastructure and general economic assistance to the anthracite region, local groups and the Commonwealth of Pennsylvania maintained their focus on industrial recruitment that had been in place for a decade before the passage of the Appalachian Regional Development Act. While the economic redevelopment program in the anthracite region appears to have been quite successful, some contemporaries offered other, less positive evaluations. In 1979 the Pennsylvania Office of Budget and Administration criticized the PIDA program, the state's main contribution to the redevelopment effort. The evaluation offered a series of statistical tests, examining the relationship between PIDA assistance at the county level and a variety of measures of economic well-being. It posed a number of related analytic questions: What counties received PIDA assistance? Did aid target areas of high unemployment, counties in need of economic development, as intended in the original legislation? Which industries received PIDA assistance? Were they ones with above- or below-average wage rates? above- or below-average rates of job growth? Would assisted businesses have located in Pennsylvania without these subsidies?[53]

In addressing these questions investigators used county-level aggregate data from across the entire state rather than data from specific assisted firms. Their findings were striking. First, levels of employment in a given county did not correspond with levels of PIDA assistance. Counties receiving substantial PIDA financial assistance had slightly lower average unemployment in the mid-1970s than was true for counties that were not receiving such funding, though the difference was not statistically significant. Reinforcing this finding, the study reported that the mean per capita income in counties receiving some PIDA aid in the mid-1970s was $5,400, while in counties not receiving such assistance, the mean figure was only $5,000.[54] Looking at economic development and growth, the assessment concluded that there was "a strong tendency for large investments to be made in counties that are highly developed."[55] Despite the intent of the enabling legislation, the 1979 study found that well-heeled, well-organized areas of the state, which had strong industrial recruitment efforts, garnered a disproportionate share of PIDA funds. Conversely, areas most in need of economic development benefited least.

How did a program intended to assist depressed economic areas when it was established in 1956 become an unfocused assistance program for business throughout the state by 1976? Initially, quite strict eligibility criteria channeled PIDA assistance to businesses in counties with above-average levels of unemployment. However, in 1967 that legislation was amended to ease the eligibility restrictions, so that businesses in virtually all of the state's counties could now qualify for low-interest PIDA loans. State legislators from counties originally excluded from the benefits of the PIDA program apparently joined together to enact the change. Legislators intent on getting a share of subsidies for their constituents shifted the focus of the program. The unfocused nature of the PIDA program constituted the core of the critique offered by the 1979 report.[56]

Shifting its attention from the county- to the industry-level, the assessment pointed to similar problems. It found that PIDA support tended to go to industry sectors—the garment industry in particular—where average earnings and productivity were below national and state averages. This focus meant, of course, that PIDA assistance was tending to drive down average wage levels within the state. In a more positive conclusion, however, the study found that PIDA assistance was directed at industrial sectors of somewhat higher than average job growth.[57]

The 1979 assessment stresses the uneven impact of the state's major economic development effort. In terms of PIDA financial assistance, there were clear winners and losers. Communities and counties that had strong economic development programs, led by well-organized commercial groups, were able to take advantage of the PIDA program to support their efforts. Areas of strong economic growth found their efforts reinforced by state-financed low-interest loans. Counties with higher levels of unemployment and lower per capita income often did not have such well-organized efforts and did not qualify for PIDA assistance. It is ironic that a reasonable conclusion to draw from the 1979 assessment is that the main impact of this major state assistance program was that the rich got richer and the poor got poorer—just the reverse of the legislators' original intentions in passing the PIDA legislation in 1956.[58]

Thus while firm-level data based on the experience of businesses in the PP&L service area paint a picture that appears positive, and aggregate evidence for the three major anthracite counties shows gains in the number of jobs in the anthracite region between 1960 and 1980, still, the 1979 analysis of the PIDA program suggests a less sanguine view of the impact of assistance efforts. Given the ambiguity of the quantitative evidence, it may be useful to consider how reindustrialization efforts drew on and were perceived by residents of the anthracite region. Local businessmen and chamber of commerce officials invariably headed up these campaigns, which raises the question: Can they be viewed as genuinely broad-based community undertakings? Did working-class residents of declining mining communities support these efforts undertaken in their names? How satisfied were they with developments in their home communities?

The role of the local community in reindustrialization efforts is particularly visible in the Hazleton case, because CAN DO published accounts of the process in 1974 and 1991. In addition, the CAN DO offices in Hazleton house scrapbook collections that provide extensive contemporary documentation of their campaigns. Moreover, a cultural anthropologist from the University of Pennsylvania, Dan Rose, did fieldwork in Hazleton in the 1970s and published his own analysis in 1981. Rose relied extensively on local published sources but did interviewing as well, and his independent exploration of local efforts offers a counterweight to the local boosterism so evident in CAN DO's own accounts. Finally, oral interviews with anthracite region residents in the 1990s reveal strong feelings about the bond campaigns that played such a major role in the reindustrialization efforts forty years earlier.[59]

Hazleton had notable success in recruiting new employers during anthracite's decline, perhaps more so than larger communities such as Scranton and Wilkes-Barre. Local organizers relied on extensive community fundraising to support the activities of both the Hazleton Industrial Development Corporation and CAN DO. Chamber of commerce leaders sought to raise money in both the local business community and the broader Hazleton community. Recalling these early efforts some years later, the authors of CAN DO's in-house history summarized the dominant ethos of the mid-1950s: "it was deter-

mined that all of the people should become active participants. It had to be a true community effort."[60] Reviewing the successive campaigns and examining cooperation and conflict in them permits a fuller understanding of the character of this community effort.

Reliance on community fundraising characterized Hazleton efforts in the postwar period. In 1947, needing $500,000 to build a plant for Electric Auto-Lite, leaders succeeded in raising $650,000 from a combination of business and individual contributors, leaving a good margin for seed money for future recruitment.[61] In 1955, leaders again sought local support, this time to fund the purchase of land for an industrial park. To reach out to the community's working class, they created a "Dime-A-Week Fund," encouraging small donations. A Lunch Pail Committee placed lunch pails in local businesses to encourage small contributions, thus drawing on a powerful symbol for working people in Hazleton. Moreover, they approached all local civic and fraternal organizations with their One Hundred Percent Committee, trying to encourage contributions from all members of these groups of a dime a week, or just over five dollars a year. Articles in local newspapers listed the names of contributing individuals and publicized those organizations that succeeded in getting 100 percent participation of their members. In all, the chamber raised $14,000 from more than 5,000 donors, an average of three dollars per person. The funds were more than enough to purchase a five hundred–acre tract outside the city at ten dollars an acre, which became the location of Valmont Industrial Park.[62]

With land in hand, the group next needed to raise a much larger sum of money to install basic utilities on the site and build an initial shell building for a prospective recruit. In early 1956 leaders once again looked to the community. From local businessmen and professionals, they sought contributions to a revolving fund; from Hazleton's working class, they sought the purchase of fifteen-year, 3-percent bonds in $100 denominations. They offered interest and eventual repayment, hoping to motivate residents to "invest" in the community's and their own future. A handbook, a "Primer for Giving," posed the question that leaders felt would stimulate contributions and the purchase of bonds. "Want

Jobs for Men?" the cover headline asked. The pamphlet went on to describe the campaign, making it clear that Hazleton residents were in competition with their neighbors from "Scranton, Wilkes-Barre, Pottsville, Pittston, Nesquehoning, Shenandoah, Shamokin—and even Freeland and Tamaqua." Hazleton needed to be able to offer modern buildings to prospective employers, and the pamphlet claimed that the community had lost "at least TEN good, solid male-employing industries in the last three years" for lack of such buildings.[63]

CAN DO organized local civic and fraternal organizations to sell the bonds and offered to each organization a representative on the board of directors of the fund for each $25,000 raised. In this way, chamber leaders fostered competition among local groups to see who could raise the most money. The drive was hugely successful, raising $200,000 in cash contributions and $540,000 in bonds over a four-week period. With these funds in hand, CAN DO held its first public organizational meeting in June 1956, proceeded to erect its first shell building, and recruited its first tenant, the General Foam Company. Coming into operation just after the start-up of the state PIDA program aimed at recruiting industry into economically depressed areas, CAN DO was able to offer prospective recruits 100 percent financing for their buildings.[64]

To keep their efforts going, CAN DO needed to do extensive fundraising. Improvements at the Valmont Industrial Park were permanent investments, and the organization needed to provide a low-interest second mortgage loan for 20 percent of the initial investment for any recruited firms to secure state assistance. The initial $740,000, though impressive, would only begin the effort at securing adequate new employment to replace the mining jobs lost in these years. A second campaign in 1959 raised $835,000, and a third campaign in 1963 netted another $710,000. In the latter effort, organizers sought either cash contributions or the surrendering of bonds purchased by earlier contributors. In other words, local residents who had made investments in the first two drives were asked to turn those investments into contributions in 1963. As a result of the three drives between 1956 and 1963, CAN DO had a revolving

CAN DO fundraising at garment factory in Hazleton, ca. 1956. Courtesy of CAN DO, Hazleton.

Entrance sign, Valmont Industrial Park, Hazleton, 1960. Courtesy of CAN DO, Hazleton.

fund of more than $2 million for continuing activity. Moreover, in 1967 the state of Pennsylvania reduced the required contribution from industrial development authorities from 20 to 10 percent of funded projects and increased the state's contribution for qualifying industrial development projects. CAN DO's revolving fund could stretch much further with this change in the rules.[65]

Newspaper coverage of the second fundraising campaign reflected the depth of community support. Representatives of local organizations met in late January 1959 to mobilize support for the campaign and formed the Hazleton Council of Civic Clubs. In March thirty local labor leaders met at the Hotel Altamont to hear CAN DO officers outline the upcoming campaign. The stories read like CAN DO press releases, repeating verbatim the exhortations made by CAN DO leaders. Local leaders of the United Mine Workers of America and the carpenters' union added their statements of support. As was typical of local press accounts, the story named all the labor leaders and their union affiliations, permitting working-class readers to see that their unions backed the campaign.[66]

Civic and fraternal organizations were intensely involved in the fundraising efforts. The Lions Club, the Rotary Club, the Kiwanis Club, and the Unico Club all were featured in newspaper coverage after the successful completion of the 1963 fundraising campaign.[67] Clubs received representation on the CAN DO board of directors in proportion to the funds they raised,

and in the second campaign, leading fundraising groups had the right to designate names for streets in Valmont Industrial Park. While chambers of commerce dominated reindustrialization efforts in other cities in the anthracite region, in Hazleton broader community groups were deeply involved in the recruitment effort. Still, the fundraising and recruitment activities were not entirely free of conflict. Despite the active, formal support of local trade unions in the campaigns, local workers had their complaints. In one instance, workers at Superior Sleeprite, a firm recruited by the Hazleton Chamber of Commerce, but not a tenant in Valmont Industrial Park, asked the head of CAN DO to "mediate a settlement" for a strike. The workers made concessions in the course of mediation only to learn shortly after the settlement that the company was closing its plant permanently. CAN DO officials were certainly not responsible for the firm's hasty departure, but as a mediator between business and labor in Hazleton, CAN DO got some egg on its face when assisted companies closed down abruptly.[68]

The former Sleeprite workers became supporters of CAN DO's efforts when a fair number of them received jobs with a newly recruited company, Highway Trailer, that moved into the area just as the Sleeprite operations closed down. This was not always the case, however, as CAN DO's 1994 history, *Upon the Shoulders of Giants*, makes clear. That account offers an anecdote about unsuccessful fundraising at one local factory:

[Clifford L.] Jones [CAN DO executive director] and several CAN DO solicitors were visiting dress factories where women comprised most of the workforce. Jones, a naïve thirty year old, felt he was an outstanding failure at one plant because he was reviled and cursed. He couldn't believe that the women used cruder language than many of the men who had served in the armed forces. The solicitors were actually cursed out of the plant, without even one pledge card being signed.

Thus, while union officers were wholeheartedly behind the fundraising campaigns, their memberships sometimes resented the repeated calls for donations and bond purchases.[69]

Although the evidence on the contemporary attitudes of working people toward CAN DO's efforts is very limited, subsequent interviews from the Panther Valley, twenty miles south of Hazleton, provide insights into the limitations that working-class residents of the anthracite region recall about the fundraising campaigns. Nesquehoning, Lansford, Summit Hill, Coaldale, and Tamaqua were less well-heeled and less mobilized as communities than Scranton, Wilkes-Barre, and Hazleton, and redevelopment initiatives in the valley were fraught with disappointments and conflict.

Calls for bringing new employment opportunities to the Panther Valley emerged in a concerted fashion during World War II. The war brought increases in coal production, but better times did not quell anxieties about long-term job prospects, especially for returning servicemen, given the bleak future for anthracite and the Lehigh Navigation Coal Company, historically the economic lifeblood of the valley. In February 1944, John R. Watkins, owner and editor of Lansford's *Evening Record*, pleaded with local business leaders, and directly with LNC president J. B. Warriner, to work to bring "sizable and permanent industry" to the district. In response, Warriner offered $25,000 to launch a recruitment campaign and promised to donate company land for factory sites. Watkins's prodding led to the formation of a Panther Valley Post War Planning Commission that included representatives from each of the valley's communities.[70]

Jobs for returning veterans became a number one issue with the war's end as ex-servicemen and others organized demonstrations in the five communities.[71] In March of 1946, local civic leaders recharged their original redevelopment initiative and recast it as the Panther Valley Industrial Commission.[72] A year later, after intensive recruitment efforts, the Commission issued a flurry of heartening announcements. First, the Condenser Service and Engineering Company of Hoboken, New Jersey, had agreed to open a branch plant in Hauto, just north of the valley, providing upward of 2,000 jobs; the Commission would lead a campaign to raise $500,000 to subsidize factory construction. Second, the Bundy Tubing Company of Detroit would establish a facility at Hometown, northwest of the valley, with anticipated employment of 700; the Commission sought $360,000 of donations from local businesses and residents to defray building costs. Third, the Commission announced that deals were in the works to recruit a fabric printing firm to Nesquehoning and a company from Cincinnati to a yet-to-be-determined location.[73]

With high ambitions the Panther Valley Industrial Commission embarked on campaigns to raise funds.[74] Hopes, however, were soon dashed. The Hoboken firm withdrew from the deal when it appeared that only half of the community subsidy could be raised.[75] Negotiations broke off quickly with the fabric printing company and the firm from Cincinnati.[76] The opening of the Bundy Tubing Company in 1949 remained the single accomplishment of the first burst of recruitment, though with 450 people employed, not the projected 700. Much to the dismay of local promoters and Bundy executives, the firm was hit almost immediately by a strike. Newly hired Panther Valley workers quickly learned that they were being paid forty cents less an hour than workers in the home plant in Detroit and promptly walked out.[77]

Redevelopment efforts lagged in the early 1950s. In October 1952, the Panther Valley Industrial Commission announced the successful recruitment of Robert M. Green & Sons, a manufacturer of stainless steel hospital equipment. The Commission agreed to provide $95,000 from its depleted treasury to enable the firm to locate in Nesquehoning. Representatives from the town, expressing anger at the placing of Bundy Tubing outside the valley proper, lobbied intensely for their community.[78] With great fan-fare, Green & Sons opened for business in an impressive facility in September 1953.[79] Unfortunately, the company declared bankruptcy fifteen months later.[80] An aluminum company would eventually occupy the plant.

With little to show and fundraising bogged down, the Commission faced mounting criticism. In October 1954, a dissident group of civic leaders established a rival entity, the Panther Valley Business and Professional Men's Committee for New Industry, and launched a separate campaign.[81] The new group succeeded in recruiting a replacement for Green & Sons and soliciting funds to facilitate the expansion of a local shoe company.[82] Upon the creation of the state's PIDA program in 1956, Committee leaders also moved to have themselves, not the old Commission, designated as the official development agency for the area with eligibility for state aid. They then formed the Carbon-Schuylkill Industrial Development Corporation, according to PIDA stipulations, and in January 1957 organized meetings in the Panther Valley where residents from Nesquehoning, Lansford, Summit Hill, and Coaldale, but not Tamaqua, voted and gave their seal of approval to the Corporation. Tamaqua had not figured in early discussions of recruitment, and the town opted not to participate in the new redevelopment efforts of their neighbors to the east.[83]

In the spring of 1958, the Carbon-Schuylkill Industrial Development Corporation unveiled a $500,000 fundraising campaign, S.O.S., or Save Our Selves, aimed at the construction of a major industrial park in Hauto ("a last ditch project" to halt the exodus of families as board members of the Corporation described it to reporters).[84] Girl Scouts knocked on doors to raise funds, women ran bake sales, garment workers at Rosenau Brothers collectively contributed over $25,000 of their earnings, local businesses gave generously as well, and with a seeming $600,000 in hand by July of 1959, the Corporation held a trumpeted groundbreaking for the industrial park.[85] Unfortunately, the S.O.S. campaign lost its sheen within a few years. In January 1961, factory buildings at the Hauto complex remained partially occupied or empty, and negotiations with various recruited firms had broken

down. S.O.S. leaders admitted that less than 60 percent of the pledges had been collected and that the Corporation had limited means for further subsidies.[86]

The road to redevelopment in the Panther Valley was a rocky one—with false starts, dented dreams, and divisiveness. The record, though, is not entirely bleak. PP&L reports indicate that between 1946 and 1987 more than fifty firms either relocated to the valley or expanded their operations there.[87] These companies share much in common with other businesses in the anthracite region enumerated by PP&L. Garment firms represent 40 percent of the total, manufacturers of metal products another 18 percent. The Panther Valley companies were smaller than elsewhere, averaging between 60 and 70 employees, and they remained on average 14 years after relocation or expansion. Local champions of redevelopment thus could point to achievements, but residents of the valley held less positive opinions, expressing misgivings, disappointment, and even anger with the fundraising campaigns of the 1940s and 1950s and the kinds of jobs created.

Workers in the Panther Valley felt pressured to contribute their hard-earned wages. As Steve Pecha of Nesquehoning recalled, "We pledged. If our wages were thirteen dollars that day, we pledged thirteen dollars toward a fund to create industry." Mary Matrician contributed much more to one such campaign and remembered bitterly the outcome of her generosity:

> They started an S.O.S. drive at the time we were in the shoe factory. . . . We were supposed to contribute towards building up industry and everybody pledged what they would give towards it. Most of us pledged a hundred dollars. In a number of years we were to receive a bond. They were collecting it off our paycheck. They were taking so much per pay. I thought, "Oh, what the heck, I'll get double taken off, and get done paying sooner." So when it was time to go and see how much more I owed, I went to the bank and asked how much I had in on my S.O.S. . . . And they told me forty dollars. I said, "Oh no, I'm supposed to have about eighty dollars." And they said, "No you don't." They weren't turning [all] our money in . . . at the shoe factory. We were supposed to get a hundred dollar bond at the

end. We never got that. We didn't get all our wages either when they closed up. They went bankrupt.[88]

Matrician believed that only half of her contributions were ever properly credited. Furthermore, she never received the bond that would have permitted her to reclaim her contributions, with interest, some years later. Other contributors fared better, and some, like Pecha, worked for a business that received direct assistance to set up a factory in his hometown. But even though he had benefited directly, Pecha expressed resentment about the assistance offered to incoming businesses. His attitude was common in a number of the interviews: "We gave them ten years, tax-free, we put the building up for them. . . . They made profits and then when it came time to pay taxes, they moved out to other towns . . . other states—like Tennessee took a big industry away from us."[89]

Rapid turnover among assisted firms was only one source of dissatisfaction among area residents. Low pay and lack of union representation bothered Grant Gangaware when he worked for a lamp factory in the Panther Valley community of Lansford. Majestic Lamp Manufacturing, the recipient of a low-interest loan of $60,000 under the PIDA program, channeled through the Carbon-Schuylkill Industrial Development Corporation, remained in business for more than twenty years, employing thirty-five to forty workers on average in that period. Gangaware, who had worked in the forestry department of Lehigh Navigation Coal before it closed down, did not like the work or the low pay at Majestic and did not remain there long. He recalled,

We had to train six weeks at no pay. You collected unemployment compensation while you were working for training. . . . [After the training period ended,] we worked for a dollar and a quarter an hour with this outfit. And we had quite a few arguments there. That was a nonunion plant at first and they were trying to put different unions into it. And there was a lot of disagreement on which one would come in so we finally got the International Brotherhood of Electrical Workers to take command of it. But we still stayed at the low rate, we weren't by no means paid electrical workers' wages.[90]

Many of the firms that entered the anthracite region hired workers at or near the minimum wage. One former miner thought this was not simply a coincidence: "Any factory that come in here, it was rumored, I can't say this for sure, but, the companies would ask, 'Well, what should we pay them?' and the one businessman from Lansford would say, 'Give them a dollar an hour, they'll work for it.' Which we did, because we were glad to get something. But everything that come in left. . . . All low-paid and stayed awhile and left. . . . When wages would begin to climb, evidently they . . . buzzed off."[91] Anthracite residents resented the minimum-wage level earnings and the transience of the firms that entered the region.

Whatever the dynamics, there were a lot of unemployed former mineworkers in the region, and wages were extremely low. As Gangaware's wife, Irene, recalled of the 1950s and 1960s, "I swear to God, I thought everybody in this whole wide world got paid fifty dollars a week. I did, I really did."[92] And men who had worked previously in the mines, making perhaps $15 to $20 a day, did not feel great when their earnings fell to the minimum wage of $1.25 an hour, or $10 a day. Between the low wages and the need to fight the unionization battles all over again, there was considerable bitterness on workers' part.

Some of the area's residents resented the efforts to recruit new businesses because they perceived them as begging. Steve Pecha looked back on the efforts with a decidedly negative view: "It was dog eats dog between different towns and states—offering different incentives to them. And the one with the least incentives got knocked in the head and the others with the good incentives got the industry. And who was holding the bag but us poor suckers."[93] Asked whether the local community had benefited from these efforts, Pecha offered an emphatic no.

These comments express a sense of a fundamental contradiction between the sacrifices that workers made to contribute funds to the recruitment campaigns to bring new industries into the anthracite region and the near minimum wages that the new firms paid. And to add insult to injury, after paying substandard wages for a period of time, many of the recruited firms left once their tax breaks expired. Many former mineworkers expressed the belief that they had been had.

A second complaint emerges in many of the interviews, focused on the earlier period when anthracite mining dominated the region. Those who felt that the redevelopment efforts had been a failure often attributed that failure to the years of economic and political domination by the region's coal companies. In numerous interviews, residents repeated stories about how coal companies had consistently opposed efforts to bring in other industries—industries that might have offered jobs to miners and competed with the mines for the local labor force. Mike Sabron told a story repeated by others: "Ford wanted to put a plant over at Hauto. Well, the coal companies won't let 'em because all the men would probably leave the mines and go to work for the Ford Company." William Wrightson, who grew up in Blakely, near Scranton, had a similar story: "I remember Ford wanted to put a place in up there, and many different places, RCA, but they just wouldn't let them in . . . they would never let them, because they knew the people would get out of the mines, and that was their mainstay." Mary Mogilski had a different story, but her interpretation was the same:

Now, my son tried to write to this, is it Fortune 500 [company]. He tried to get industry here. Since he became [a county] commissioner, he was very much interested in trying to keep the younger people here in Panther Valley. And he tried. He wrote to so many different places to bring some industry here. The reason why we couldn't get other industry here in this valley, they blamed the Lehigh Coal & Navigation Company for that. They wouldn't allow them to come in. Their excuse was that due to the railroad, they didn't have access to come here. But . . . the way I was told, . . . the coal company wanted the miners to work in the mines. They didn't want them to get easier jobs elsewhere.[94]

Area residents saw a direct connection between the coal companies' opposition to other industry coming into the region and the limitations of later recruiting efforts. John Lazar was most articulate about the region's difficulties:

You know, years ago, Ford wanted to build a plant. . . . And the coal company fought them left

and right, 'cause they said they would take away the workers. People would quit working in the coal mines and go work in the auto plant. And they wanted to buy land, and of course the coal company back then owned most of the land here, they wouldn't sell them the land. So they said, "well hey, if that's your opposition, we're leaving." Then after the mines closed down, they opened this industrial park over in Hauto. And they would make deals with the companies to come in. We'll give you ten years and no taxes and this [and] that. And then when the tenth year was up and it came time for the company to pay taxes back to the area, they closed up. So the guys were out of work again.

Ziggie Whitecavage lived in Shenandoah, but he told pretty much the same story: "the coal companies dominated everything up there. They wouldn't let no other industry in around that."[95] Whether there had ever been a realistic prospect of bringing in high-wage, unionized work is hard to tell, but area residents clearly blamed the coal companies for the failure to recruit high-wage jobs and for the low-wage, nonunionized jobs that came with the industrial development effort in the anthracite region.

Local residents, in recalling their years after the mines closed, also commented on the insecurity of their employment in the postanthracite period. Robert Daniels had experiences that were shared by others. After the Panther Valley mines closed, he worked for nine years in the shipping department of a recently relocated aluminum firm in Nesquehoning that made storm windows and doors. The Lehigh Aluminum Company had relocated from Ohio and purchased the 65,000-square-foot community-constructed building formerly occupied by Robert M. Green & Sons. It planned to employ 500 men with an annual payroll of $1.6 million. Hopes were high, but employment peaked at 313 in 1959, the firm folded in 1965, and Daniels was again looking for work. He found work at another metalworking firm, this time commuting forty-five minutes each way to Bethlehem. That work lasted for sixteen years, when that firm also went out of business. By then Daniels was sixty-one years old and relatively fortunate that the firm was able to provide him a modest pension a year later. As Daniels wryly summarized his work career: "It seems like I had a history of closing down businesses! That's the way it goes." It was a perspective that a good many anthracite miners would have shared.[96]

Panther Valley oral histories reveal the existence of multiple, conflicting narratives concerning the reindustrialization effort in the anthracite region in the post–World War II decades. The viewpoints conflicted with one another in public debates and also within individuals. The dominant, public narrative, articulated by chamber of commerce leaders and elected officials, described the region as needing jobs and the local industrial development organizations and their fundraising campaigns successfully recruiting new companies and holding onto existing firms. Many working-class residents accepted this narrative and contributed to support SLIBCO, CAN DO, and the other organizations that spearheaded these efforts. At the same time, the oral histories offer a counter, popular narrative that reveals the resentment that built up over decades of industrial development efforts that did little to stop economic decline. We see in the official and vernacular versions of this history—to use John Bodnar's apt terms—that contemporary views of industrial redevelopment efforts were contested. There was an ongoing struggle over historical memory, over how anthracite residents viewed these efforts. In chamber of commerce publications, in the statements of elected officials, and in the mainstream local press, the official version predominates. In personal interviews, though, numerous accounts offer an opposing version that expressed considerable resentment about the perceived unfairness of the overall process. And to complicate the picture still further, individuals sometimes held onto both of these versions of the region's reindustrialization efforts. These alternative perspectives reflect the "conflicted cultural memory" that Bodnar has found in oral histories of Indiana communities in this same period.[97]

While anthracite-region residents framed their criticisms of low wages, nonunion conditions, and the insecurity of employment based on their individual experiences with the reindustrialization effort, their arguments have much in common with those offered by professional economists operating within a radically different frame of reference. Consider, for instance, the perspectives of Melvin L. Burstein

and Arthur J. Rolnick, economists at the Federal Reserve Bank of Minneapolis, who have decried an industrial policy that they term "The Economic War Among the States." They see the tax incentives, mortgage subsidies, and individual job training programs offered to lure companies to specific locations as a terrible waste of taxpayer resources. They have called for a cessation of interstate competition for jobs and investment through the enactment of federal legislation that would tax any benefits corporations would garner in this process. They argue that only targeted, migratory businesses benefit from the present system and that concerns for equity and economic efficiency point toward the broader form of business assistance that they recommend.[98]

What these economists call the "Economic War Among the States" had its counterpart in the anthracite region in the war among the towns and cities, among the three dozen industrial development corporations competing for a limited number of industrial prospects, as each community tried to improve its own tax base and provide jobs for its own residents. And of course the anthracite-region industrial development corporations competed with similar entities across the state for limited PIDA loans to subsidize plant construction and expansion. The criticism that the 1979 state report had made of the PIDA program was precisely that the subsidies were no longer promoting economic development in the neediest areas of the state but rather going to the most well-heeled counties in the state that were able to raise the funds that permitted them to come out on top in this competition.

Criticism of competition among towns and states in reindustrialization campaigns and of the faithlessness of corporate beneficiaries of this competition is not limited to reminiscences from the Pennsylvania anthracite region. A recent newspaper account of the experience of the Maytag Corporation in Galesburg, Illinois, is strikingly similar to the complaints offered by Panther Valley residents. In Galesburg, "after a decade of tax breaks and union concessions" aimed at keeping Maytag in town, the appliance firm closed operations in September 2004 and opened a new factory in Mexico. The local district attorney is considering a lawsuit to recover what he views as "excess tax breaks" he claims that the company took during the ten-year period. Similar conflicts over tax breaks and union concessions in Iowa, New York, and Florida suggest that the difficulties evident in industrial development efforts in the anthracite region in the postwar decades continue in the first decade of the twenty-first century.[99]

An alternative policy emerged in the 1990s, though few states adopted this strategy. The state of Oregon refused to participate in the economic war on the terms that most states did. Policy makers in Oregon, for instance, did not compete to attract low-wage jobs, arguing that such jobs actually cost the state more than the taxes it received on the increased wages and spending. Instead, Oregon chose to spend its resources to create "an educated work force, a well-developed infrastructure, [and] a diversified economic base," on the assumption that these resources would serve well corporations already in the state or others considering relocating. By making more generalized investments in human capital and economic infrastructure, such an industrial policy avoids creating winners and losers that are so prominent in the "Economic War Among the States" and in the version that filtered down to the anthracite region of Pennsylvania after the mines closed. It seeks rather to create a win-win situation in which all communities and all companies benefit.[100]

Viewing the evidence as a whole, reindustrialization efforts had a mixed impact on the anthracite region. While chamber of commerce leaders sought jobs for unemployed males, the companies that they attracted more often offered insecure, low-wage jobs for men and women. Certain communities were better positioned than others to recruit new firms, yet even in those instances, residents expressed resentment at the concessions that were required to secure the low-end employment opportunities. While the reindustrialization campaigns relied on community support, residents often came to feel that they had been taken advantage of and that community values had been violated in the process.

Much of this resentment stemmed no doubt from the interplay of the various actors in the drama of redevelopment. The anthracite railroads that had driven development in the late-nineteenth and early twentieth centuries—and had profited enormously from the toil of the region's mineworkers—were conspicuously absent during the region's concluding decline. Stock speculators, not the area's residents, benefited from the anthracite conglomerates' liquidation. Moreover, the United Mine Workers of America provided no effective leadership in the final crisis; as the anthracite industry evaporated in the 1950s and 1960s, officials of the UMWA were more concerned with keeping their own nest well feathered. Although federal and state government officials and agencies significantly propped efforts to strengthen the region's infrastructure and to retain and lure manufacturers, none of these actors offered a comprehensive regional plan for renewal.

Redevelopment in the anthracite region after World War II can thus be seen as both a glass half full and a glass half empty. The campaigns to attract new businesses bore positive results for the leading communities in the region—Scranton, Wilkes-Barre, and Hazleton. Employment in Lackawanna and Luzerne counties did rebound from low points, and the population in the anthracite region steadied after its precipitous decline during the 1950s. But even as the population of the anthracite region stabilized and a new economy evolved to replace the earlier dependence on anthracite coal, area residents felt the loss keenly. Residents in recalling the reindustrialization efforts felt that there had to have been a better way. Whatever the ultimate judgment, residents of the anthracite region were still left largely in the lurch, and they had to rely primarily on families, neighbors, and their own resourcefulness in coping as the mines closed. Personal responses of the generation of working men and women who bore the brunt of the region's misfortunes are as integral to the story of economic decline as the organized responses of capital, labor, communities, and government.

CHAPTER 6

Personal Responses to Decline
Fathers and Mothers, 1945–1990

MIKE SABRON, KNOWN TO HIS buddies as "Crow," was forty years old in the spring of 1954 when Lehigh Navigation Coal (LNC) closed all the underground mines in the Panther Valley after sustaining operations there for more than 130 years. Sabron had begun work for LNC right after high school, working first in the company's bagging plant. In 1940 he moved into the mines and worked as a mule driver, carting waste rock dislodged in driving tunnels for new underground gangways. After later stints dumping and loading coal cars, he began work as a contract miner in 1947, the elite job he held when the mines closed in 1954.

Like the other four thousand mineworkers laid off by LNC in May 1954, Sabron was not sure what to do. He had been married for eighteen years and had five children, so he needed work to support his family. With some of his mining buddies he went to Linden, New Jersey, and got work at the General Motors plant there. The group commuted together, boarding near the factory during the week and returning to their homes in the Panther Valley every weekend. He hated the auto work: "on that assembly line . . . that was rough. . . . you had to move all the time, and I just wasn't used to fast work like that." After six months, Mike was rehired to work in the No. 9 mine, now leased by the Panther Valley Mining Company. As he commented when interviewed years later, "Oh, the guys were glad to get back [to the mines]," even

though they continued in operation only intermittently for another five years.[1] Sabron's recollections capture the overwhelming love of mining among those displaced by the closing of the mines.

Sabron's wife, Edith, was thirty-six when the mines closed, and she too responded to anthracite's decline. She first went to work as a presser in a garment factory in the east end of Lansford for four years. After that she worked as a waitress for a stretch, supplementing her husband's irregular income as anthracite mining in the valley declined steadily. Like other fathers and mothers of their generation, Mike and Edith Sabron refashioned their lives with the mines' closing, but they drew on the values they had nurtured while coming of age in the anthracite region.

The Sabrons' experiences were not unique, but Mike was unusual among Panther Valley miners as he continued to work underground until 1972, mining for the Lanscoal Company, still in the No. 9 mine where he had begun work underground in 1940. Few miners in Sabron's generation managed to work underground for three decades. He was drawn back to the Panther Valley, and though he might have earned more money and had more security as an autoworker in New Jersey, he preferred life in the anthracite region, even with the uncertainty and intermittent work of the mines in their declining years.[2]

The Sabrons' story encapsulates much of the response of their generation to anthracite's decline in the post–World War II years, including the economic hardships, job possibilities at a distance from the region, and the ultimate attachment to family, friends, and familiar settings close to home. The institutional responses of capital, labor, and communities to the decline of the anthracite industry have dominated preceding chapters, but the personal perspectives of families and individuals were equally important.

Two kinds of sources permit the reconstruction of the postwar lives of the generation of men and women who experienced the closing of the mines in the anthracite region. Decennial federal censuses provide an objective view of changing patterns of employment and retirement, while contemporary testimony and subsequent oral histories illuminate subjective responses to changing circumstances. Together these sources complement one another and offer multiple perspectives on personal responses to economic decline.

Anthracite's final decline began in earnest in 1950. During the ensuing decade, anthracite employment declined from 77,000 to 20,000. Out-migration took a heavy toll in the 1950s as the working-age population in the three principal anthracite counties—Lackawanna, Luzerne, and Schuylkill—declined by more than 15 percent.[3] Individual responses to the dramatic decline of the region's principal industry after 1950 varied by generation. Older residents faced one set of choices; young people coming of age, another set. We begin with the responses of the fathers and mothers of the older generation.

◆

Who continued to work in and around the mines as the decline resumed in earnest after the war? The 1950 federal census offers a window onto postwar developments; although anthracite was sputtering after World War II, it had not yet begun its precipitous decline.[4] The census reveals that the regional economy still had jobs for many adult men and women.

Men and women in the anthracite region, as in postwar America more generally, were situated quite differently in the regional economy in 1950. Men were much more likely than women to be gainfully employed. More than 86 percent

of men between the ages of 18 and 65 worked, while for women the comparable figure was less than 29 percent. Just as important, men and women worked in different sectors of the economy. Mining in 1950 still employed more than 23 percent of the region's workingmen, with construction, railroads, and agriculture also accounting for significant numbers. The generally disadvantaged position of women in the regional labor force resulted in the crowding of employed women into a limited number of occupations. Fully 66 percent of working women were employed in the three leading sectors: apparel and textiles (36 percent), clerical (19 percent), and sales work (11 percent). Men had more occupational choices in 1950, contributing to the higher earnings they enjoyed.[5]

This occupational sex-segregation was particularly evident in anthracite mining. Men comprised 99 percent of anthracite mineworkers in 1950, with less than 1 percent of the workforce consisting of women employed in clerical or cleaning jobs in the offices of mining companies. Miners and mine laborers composed over 75 percent of the mine workforce, with the remainder spread among a wide range of skilled and semiskilled positions. Since there had been little hiring of new workers as the mines had declined in recent years, the miners were advanced in age: almost 52 percent were between thirty-five and fifty-four years of age and 72 percent were heads of households. The mean age for heads of households among male mineworkers in 1950 was forty-four.[6]

The view of mineworkers that emerges from the 1950 census is reinforced by statistics for the workforce of LNC in the Panther Valley. Focusing on mineworkers who were laid off by the company between March 30 and April 30 of 1954, the group most directly affected by the closing of the company's mines, provides a remarkably similar picture. More than four thousand miners lost their jobs at this time; their average age was forty-five, and they had been working for the company for a period that averaged twenty-two years. These were the men whose mining careers halted abruptly with the closing of the mines, and their age and experience profiles were no doubt similar to those of experienced miners across the anthracite region in the decade of the 1950s.[7]

As men lost their jobs in the region's mines, their daughters and wives increasingly sought work in area garment shops. Among anthracite-region women in 1950 the leading occupational group consisted of needleworkers. Almost 28 percent of employed women in 1950 worked as operatives in firms making apparel and accessories. They were distinctly younger than male miners as a group, with two-thirds ranging between fifteen and thirty-four years of age. These women garment workers divided almost evenly into two groups: Slightly more than half were never married, while 49.5 percent were married or had been married. Among the married women the average age was thirty-five years. These married garment workers along with the homemaker wives of miners were the mothers of this generational cohort.

In discussing personal responses to the closing of the mines and the ravaging of the local economy in the anthracite region, three strategies stand out: (1) migration out of the region, (2) commuting to jobs outside of the region, and (3) remaining in the region and seeking new work near home (a group we refer to as "persisters"). Exploring these strategies affords a broad view of the range of individual responses to economic decline. What becomes apparent from the outset is the gendered impact of economic decline on the older generation. Fathers and mothers were differently situated when the mines closed and their responses were necessarily distinct. As men and women reconstructed their lives in the wake of industrial decline they also reshaped gender relations within their families and communities.

Migration transformed the anthracite region in the postwar decades as the region's young people left in droves. Successive census snapshots reveal the character of out-migration and the composition of the out-migrant stream.[8] The population of the five-county area that comprised the anthracite region fell unevenly during the post–World War II period. The region's residents numbered 1,025,000 in 1950, and the population fell 11, 3, 0, and 4 percent, respec-

Lansford Sportswear workroom, W. Bertsch St., Lansford, 1956. Photograph by George Harvan.

tively, over the next four decades, until in 1990 it numbered only 853,000, a figure roughly 17 percent below the postwar high.[9] The decline was concentrated in the 1950s, as net out-migration in that decade accounted for almost two-thirds of all out-migration during the forty-year period.

Leaving the region appealed to some groups of residents more than others. The influences of decade and age on out-migration from the region are apparent in table 6 (see appendix 1), which shows in detail migration patterns during each decade after 1950 in Lackawanna County, home to the northernmost anthracite deposits and the region's largest city, Scranton. The bulk of out-migration from Lackawanna County occurred in the 1950s, when roughly 14 percent of county residents left over the ten-year period. In succeeding decades out-migration ranged from one-third to about an eighth of that level. Moreover, in each decade out-migration was concentrated among fifteen- to twenty-four-year-olds. The younger generation just coming of age was much more likely to leave the region than were their older neighbors. For each decade out-migration in this age group was at least twice as great as that in any other age group, ranging from a low of about 12 percent in the 1970s to a high of 29 percent in the 1950s. Among all ages, out-migration was relatively evenly distributed along gender lines. In two decades more women left the region than men; in two others, men were more likely to migrate. Still male-female differences were minimal.

At first glance the migration enumerated here does not seem striking. After all, the population of Lackawanna County declined only 17 percent over a forty-year period. But these numbers need to be placed in context. Pennsylvania gained in population by 1.4 million, or 13 percent, over this forty-year period, and for the nation as a whole, the corresponding increase was 64 percent.[10] The anthracite region's decline was evident in the 1950s when Pennsylvania's overall population grew by almost 8 percent, while Lackawanna County declined by 9 percent. Anthracite-region population loss stands out, in particular, when compared to other hard-hit labor markets across the nation. A comparison of population change for eighteen depressed labor markets across the country between 1940 and 1950 reveals that the Scranton metropolitan

area suffered by far the greatest population loss.[11]

Contemporaries expressed great concern about this out-migration during the 1950s, especially the impact of the loss of the region's young people. Congressman Daniel J. Flood noted in 1956 that his district was losing residents at the rate of 220 per month; he also commented that the region's most "notable export . . . was 'high school graduates.'" Min Matheson, district manager of the International Ladies' Garment Workers' Union (ILGWU) in Wilkes-Barre, remarked: "Young men are compelled to leave home in search of work or remain here in the morally destructive atmosphere of enforced idleness." These population losses had repercussions across all dimensions of life in the anthracite region. School populations shrank dramatically. In Schuylkill County, enrollments dropped by 9 percent between 1948 and 1953. In consequence, tax revenues for school districts fell at even greater rates than enrollments, and school boards had to cut operations drastically. The out-migration of young people, a consequence of the closing of the mines, reinforced the region's economic decline and undermined the economic prospects of those who remained there.[12]

The fathers and mothers who were working and raising children in the anthracite region as the mines closed included both out-migrants and persisters. Men and women in this age cohort left the anthracite region in significant numbers in the decade of the 1950s and then slowly filtered back in succeeding decades. As table 6 (in appendix 1) indicates, between 10 and 11 percent of men and women between the ages of 35 and 44 left Lackawanna County in the 1950s. For the 1960s net out-migration was slightly negative for both men and women in the 45–54 age group; in other words, return migrants slightly exceeded out-migrants in that decade. The same pattern held for the 1970s for the cohort, now age 55–64, except that net in-migration accelerated in that decade, amounting to almost 4 percent of males in that age group and 6 percent of females. In the 1980s the population for 65- to 74-year-olds in Lackawanna County did not change, with modest net in-migration of women and similar out-migration for men roughly canceling each other out. Totaling the separate decadal flows for men and women reveals a net loss of 8.5 percent of men in this age cohort over the forty-year period, but a net loss of only 0.6 percent for women. The gender difference reflects the sharp decline in men's jobs in the region with the closing of the mines, compared to a distinctly better job picture for women in the same period. This gender differential contributed to a growing proportion of women in this age cohort as time passed.

Migrants, by the nature of their choice to pick up stakes and leave, disappear from subsequent censuses in their original hometowns. Still, contemporary accounts and interviews forty years later provide clues as to the destinations of anthracite-region migrants in the 1950s and 1960s. Migrants typically did not move far from their mining hometowns; common destinations included Bridgeport, Connecticut, northern New Jersey, the greater Philadelphia metropolitan area (especially Levittown and Fairless Hills), and Wilmington, Delaware. Migrants were drawn to areas with strong manufacturing sectors and often found jobs in steel or automobile manufacturing within a radius 100–150 miles from their hometowns.[13] Unemployed miners were cautious as they anticipated migration. The process began with job hunting, and if that was successful it was typically followed by a period of commuting. Only after these first two steps did most husbands and wives decide to confirm the choice and move their families out of the anthracite region. Oral history interviews confirm this stepwise process.

The process of job hunting emerged as a collective male endeavor that allowed former mineworkers to maintain the close relations with fellow workers that had characterized their working lives in the mines. Unemployed men sat in bars and clubs exchanging tips on job possibilities in places near and far. They also traveled together in search of work and remained loyal to each other. Joe Rodak described how he and three buddies roamed through industrial districts in Philadelphia, knocking on factory doors. Joe had training in electronics and received an offer at the Link-Belt Company, which he refused when his friends were turned down. They later secured employment together at a General Electric plant (helped by a guard who recognized them as fellow "coal crackers").[14] Some-

times a father and son would look for jobs together, as in the case of Mike and James Melovich, hired at U.S. Steel's Fairless Works in October 1953. The Meloviches left their family in Hazleton and slept in their car for a week or two, using their first paycheck to set themselves up boarding in Morrisville. Later, Mike Melovich recruited other of his mining buddies to join him at the Fairless Works.[15]

When these job-hunting trips were successful, former mineworkers often carpooled and commuted on a weekly basis to jobs in the Philadelphia area or northern New Jersey—then a two- to four-hour drive from their homes. Men drove downstate together to the Fairless Works northeast of Philadelphia. Similar groups of area men commuted to northern New Jersey and worked at General Motors in Linden, Johns Manville in Manville, or Phelps Dodge in Elizabeth. These areas were home to major industrial corporations, with large and growing workforces, strong unions, and expanding suburban developments. Journalists, economists, and sociologists writing about the growing affluence of the U.S. working class had these workers and communities in mind. These were areas to which former mineworkers commuted and sometimes migrated after the mines closed.[16]

In the oral interviews, the recollections of Mary Painter offer the most positive view of migration. Mary recounted her family's experience migrating from Tamaqua, in the Panther Valley, to Fairless Hills, near Philadelphia. Mary and Tom Painter married in 1950. Mary's first marriage had ended in divorce and she brought two children into her second. The new family lived in Tamaqua, at the southwestern edge of the Panther Valley, and Tom earned good pay as the operator of the rotary dump at the top of the Tamaqua breaker of LNC. It was very unhealthy work, exposing him constantly to clouds of coal dust, but compensating him more because of that. Tom had worked for the company for fourteen years, and his father and several older male relatives had all worked for the mining giant that at its peak employed some eight thousand men across the Panther Valley.

After the company closed its underground operations in May 1954, Tom learned of jobs at the U.S. Steel Fairless Works just northeast of Philadelphia while working part-time as a bartender. Tom commuted the ninety-five miles from Tamaqua to Fairless Hills and lived in a boardinghouse for his first months on the job, returning to his wife and family in coal country for stretches of two or three days when he had time off from the mill. Finally, in December 1954, the family bought a house in Fairless Hills and Mary and the children moved permanently to join Tom. Mary's parents had died and only Tom's mother remained as a family link back in the Panther Valley.

Tom and Mary Painter were younger than the typical married mineworker and wife. When the mines closed in 1954, they were 31 and 32 respectively. The period of commuting lasted only six months for Tom, and the Painters made a relatively quick and smooth transition between the Panther Valley and Fairless Hills. As Mary recollected, they moved quickly, without misgivings, and barely looked back. Tom worked in steel for twenty-five years, retiring relatively early at fifty-seven because of growing health problems associated with black lung. Mary worked for more than twenty years as a medical secretary and immersed herself in voluntary activities in the community. She worked in the library, on a local advisory school board, and for the Little League, and she enjoyed the rich fare that theater and opera in the Philadelphia area had to offer. The cultural and community life of Fairless Hills stood in stark contrast with the more limited opportunities in the Panther Valley, and Mary Painter embraced the new possibilities with enthusiasm.[17]

Migrants like Tom and Mary Painter found more secure employment outside the anthracite region and a very different community life than they had known in their working-class mining hometowns. Those who moved to the Fairless Hills area most frequently resided in the massive Levittown development that encompassed parts of four townships in surrounding Bucks County. Between 1952 and 1958 more than 17,000 families moved into the area as the Fairless Works of U.S. Steel went into operation.[18]

Migrants commented in their interviews about the changes they experienced leaving the mining towns in which they had grown up. First, there was a much higher standard of living in Fairless Hills than they had experienced back in anthracite country. Mary McHugh's comments

about a relative's house in Fairless Hills, when she first visited the area, captured a response that was widely shared: "When I saw that house, I thought it was a palace after coming from little row homes in Centralia. I thought that was a palace. I wanted so badly to move down here. It was like night and day over where we came from." Another Fairless Hills migrant, Paul Melovich, had a similar reaction: "This is like heaven now" is how he described the new housing offered by the Danhurst Corporation, a subsidiary of U.S. Steel. He also commented on the affordability of the housing for former coal miners. A down payment of one hundred dollars was all he needed to buy a three-bedroom home with mortgage payments of less than $90 a month.[19]

A second major difference between the working-class suburbs in which anthracite migrants settled and their old hometowns was the declining place of ethnicity in their new communities. In the late nineteenth and early twentieth centuries, immigrants established strong ethno-religious institutions in anthracite communities and their descendants sustained those ties. On Sunday, August 1, 1954, for example, just weeks after LNC closed its mines, five thousand area residents gathered at an outing of the Italian-American Association of Carbon, Schuylkill, and Luzerne counties, while another ten thousand attended a Polish Day celebration.[20] In the same period the St. David's Society of Schuylkill and Carbon Counties held Welsh Day festivities and the Panther Valley Irish-American Association sponsored their own annual gathering.[21] In addition to participating in the activities of these broader organizations, residents of the Panther Valley continued to attend their small ethnic churches.

This reliance on ethnic churches and a strong ethnic associational life did not hold in the communities in southeastern Pennsylvania and northern New Jersey to which migrants from the anthracite region moved. Joseph McHugh captured the shift clearly: "Here [in Fairless Hills] it's one parish for all parishioners. But in Shenandoah there's an Irish church, a Greek church, two Polish churches, one Lithuanian church, a Slovak church with fifteen people. Each nationality had their own church." Similarly, Joe Rodak described the ethnic diversity in his hometown, noting Polish, Lithuanian, Russian, and Italian churches in Mount Carmel, something lacking in the working-class suburbs to which anthracite migrants moved.[22] Theresa Pavlocak experienced the same change. After attending St. Michael's, the Slovak church in Lansford, she and her family joined St. Matthew's Church in Edison, New Jersey, "the church around the corner." Helen Mordock summarized the change that virtually all anthracite migrants experienced: "Down here [in Morrisville] . . . the Catholic [Church consists of] all different nationalities. Where up there [Freeland] they had eight, nine, ten churches. And every [ethnic group] had their own church."[23] This move from the ethnic church to the neighborhood church recast the religious dimension of migrants' lives as they moved from coal country to working-class suburb.

In addition to the new ethnic diversity within Catholic churches, there was much more religious diversity in northern New Jersey and the Levittown area than migrants had known formerly. Catholics in the anthracite region in the early 1950s comprised about two-thirds of the population; in Levittown, by way of contrast, only 39 percent were Catholic, with Protestants outnumbering Catholics slightly.[24] Catholics had suddenly become a religious minority in their new communities.

There was also much greater occupational and class diversity in the Levittown–Fairless Hills area than in the anthracite region. U.S. Steel never dominated employment in the suburban communities built around the Fairless Works the way the coal companies dominated the anthracite region. In the Panther Valley, for instance, more than 60 percent of the fathers of high school graduates in the 1946–1960 period were mineworkers; in Levittown, adults held more varied jobs. Their ranks included steelmen, shipyard workers, local service providers, professionals, and white-collar commuters employed in Philadelphia and Trenton offices.[25]

Levittown and Fairless Hills proved much less closely knit and homogeneous than towns in the anthracite region, and migrants participated in non-ethnic community voluntary organizations and cultural activities that had no counterparts in coal country. As one journalist summarized in a contemporary account of "The

Mass-Produced Suburbs": "Nearly everyone belongs to organizations and, generally speaking, tries to be actively involved."[26] Among anthracite migrants, Mary Painter commented that she worked in the town library and served on an advisory board for her children's school. Participating in various cross-class organizations in Fairless Hills reminded Mary that members of miners' families would never have gone to a golf club or a woman's club back in Tamaqua. These were organizations reserved for upper-class residents in the anthracite region, but they functioned more broadly as community institutions in Fairless Hills. Moreover, these class connections crossed the thresholds of people's homes in ways that would never have happened in the more firmly class-stratified mining communities. Neighbors of different class backgrounds visited each other and their children played in one another's homes. As one former Levittowner commented, "It was hard to look down on somebody in Levittown, because you all lived in the same houses. . . . It didn't matter whether your friends' fathers were doctors or lawyers or well-educated or not. It didn't matter because you all lived in the same house and you knew how much it cost."[27] Sociological studies of the new working-class suburbs confirmed the oral testimony: The new communities displayed a heterogeneity and an aura of classlessness that were distinctly absent from the mining communities in which these migrants were born and raised.[28]

The ethnic and religious diversity, and the very newness of Levittown and Fairless Hills, rendered these communities more anonymous in character than was the case in the close-knit, largely parochial, mining communities. William Wrightson spoke to the difference: "Blakely, you know, small town—everybody knew everybody else. I live in Levittown all these years and sometimes you don't even know people two doors up the street." Joe Rodak came from a much larger mining town, but made basically the same point: "Mount Carmel is a nice little town . . . Everyone knew everyone because it was a town that was one mile square. And. . . . you recognize almost everybody." Gloria Rehill reinforced these comments: "In Wilkes-Barre, when you met somebody, you always knew somebody that knew somebody that knew somebody that knew something about the people you were dealing with. Here it was all strangers. There was no family ties, no knowing anything about anybody."[29] In the new suburban communities emerging in the mid-1950s, people did not know your parents and your sisters and had not gone to school with you as they had back in Lansford or Shenandoah or Blakely. In Levittown or Fairless Hills about 20 percent of residents came from "upstate," the contemporary term in Bucks County for the anthracite region, while more than half came from metropolitan Philadelphia and another large share from other U.S. Steel operations in Pittsburgh or Gary, Indiana. Similarly, in northern New Jersey and other destinations, anthracite migrants comprised only a small proportion of residents in their new communities.[30]

Many migrants to new communities coped with the dislocations by maintaining close connections to their hometowns back in the anthracite region. They typically moved to places 75–100 miles away and returned repeatedly to visit parents, siblings, other relatives and friends, to attend local reunions, and to camp and enjoy the mountain outdoors.[31] Some, like Don Hunsinger, held onto family properties and used them throughout their lives as vacation retreats. Paul Melovich went back home every weekend the first year he was in Morrisville, near Fairless Hills. "I would stand at the bank in Morrisville, the main street coming out of the mill, and they'd see me, they'd wave," and he'd get a ride back up to Hazleton to visit friends or see his girlfriend. Visiting back home frequently became a way of life as recalled by Joe Rodak, who moved from Hazleton to Philadelphia. "Everybody that left the coal regions to go somewhere else were going up for the weekends. 'You going up for the weekend?' 'Yeah, I'm going up.' That was what we said."[32] Why did Joe travel back and forth practically every weekend for more than a decade after his move to Philadelphia? For the "fresh air, camaraderie, people you knew," he noted. "The city was so strange to everybody." No doubt, there were some who never looked back, but interviews with out-migrants reveal the persisting importance of hometown ties.

In the out-migration process, commuting often was the first step. However, that was not always the case; commuting might provide displaced miners and their families an alternative

to leaving the region permanently. Mike and Mary Vitek were among the significant numbers of fathers and mothers who maintained their residence in the Panther Valley after LNC ceased underground mining. While they, too, depended on work opportunities beyond anthracite, they clung more strongly to their roots in the region. Commuting rather than migration was their response to economic decline. Mike Vitek responded to the mine closings by heading to New Jersey to look for work. His wife's brother lived in Manville, New Jersey, and helped Mike find employment at Johns Mansville, adding asbestos to roofing material. Mike commuted to work, heading to New Jersey on Mondays with some Panther Valley buddies, living in a boardinghouse during the week, and returning home to his family each Friday.[33]

Mining had been a male occupation, and commuting sustained the separate male world. When long-distance commuting was involved, the men boarded together during the work week, returning to their families in the anthracite region on weekends. The demands of weekly commuting made for trying times. Men found themselves separated from their families from early Monday morning until late on Friday. Life in the rooming houses could be unsettling. John Pavuk recalled living in Manville with eight men in a room furnished with four single beds as each boarder replaced another coming off successive shifts at a local factory. Mary Vitek also noted that after paying the boarding house rent and buying gas for his car, her husband returned home to the anthracite region on weekends with little of his paycheck left.[34]

Commuting separated husbands and wives and often undercut earlier accommodations that couples had made to the intermittent nature of mine employment in the 1950s. Mary Vitek had worked at a local department store in the last years that the mines were operating, but she did not work during the period that her husband commuted to New Jersey. She had her hands full raising five children, from an infant of a few months to a fourteen-year-old, when Mike first began to commute. Mike commuted long-distance for a couple of years, but rather than moving with his family to New Jersey, he kept looking for work closer to home, finally landing a job at Mack Truck in nearby Allentown. Once his long-distance commuting was over, Mary found work at a local garment factory. Mike continued to work at Mack Truck for almost eighteen years and commuted daily (but now only forty miles each way), permitting him to spend more time with his growing family. Mike finally took early retirement at the age of sixty, his stamina shot by years of exposure to coal dust in the mines and paint fumes at Mack Truck. Mary worked for a number of years after Mike's retirement, but in 1983 at the age of sixty-two she too called it quits.

Mike and Mary Vitek were among the lucky ones in the anthracite region after the mine closings because each was able to work eighteen years and to qualify for a pension on retirement. Mike worked steadily at Mack Truck, at well-paying, unionized work, and Mary, though she worked for several garment firms, qualified for a union pension as well. The Viteks survived a period of commuting without giving up their ties to their hometown. When interviewed in 1995, they were living just a couple of doors down from the house to which Mary had moved when she was eleven years old. They had achieved secure retirements even in the face of the region's evident economic decline.[35]

Interviews revealed quite a few families like the Viteks who managed to shorten the commute and continue living in their mining community, with the husband often working just outside the region. The more fortunate job-seekers found work closer to home and commuted on a daily rather than a weekly basis. Like Mike Vitek, Bob Sabol commuted for thirty-one years from his home in Coaldale to Western Electric in Allentown, though he vividly recalled the perils of winter driving through the mountainous area. Mike Knies would stay at a YMCA in Reading some days during the winter to avoid the drive home in bad weather after his night shift work, but still he slept at home virtually every day. Robert Daniels recalled "sleep[ing] half the way home" in his carpool after a grueling day at Bethlehem Fabricators, but most of the time the commuting distance was manageable.[36]

With so many men from the anthracite region commuting on either a weekly or daily basis and working for wages that were much lower than they had earned in the mines, both domestic duties and wage earning fell to their wives,

who remained in their hometowns. Women were prepared for the increased necessity of wage-work because even before the mines closed family survival relied on women's employment outside the home. This generation of married women in the region worked for wages, if intermittently, throughout their lives. Miners made good money when production of anthracite was high. But in the postwar years there were numerous stretches when miners worked only two or three days a week, and wives needed to supplement their husbands' fallen incomes. Women had always secured work during mine strikes and when men suffered accidents. While married women were reluctant to leave young children to other caretakers, large families forced some women to work even when their husbands were fully employed. Thus when the mines closed, there was nothing new in women becoming wage earners, despite prevailing sentiments that decreed the home as women's proper place.[37] Lillian Verona had been in and out of employment in local garment factories, and she remembered: "I was working part time, 'cause this was a common thing, to work part time in any factory." Theresa Mogilski lent support to this perspective commenting, "you know, most of the women had been working already at the time." The closing of the mines simply confirmed and underlined the crucial contributions that married working women made to the economic support of their families. As one contemporary newspaper article noted, "It's the women of the hard coal country—not its rugged men—who are bringing home the bacon . . . these days."[38]

Even as married women in the anthracite region responded to the need to work outside the home, they remained committed to their work as homemakers. As Mary S. noted, even though she worked full-time, "my house was perfect"—referring to the meals she put on the table and the cleanliness of her home. For working wives whose husbands commuted on a weekly basis after the mines closed, the multiple responsibilities could become burdensome, but they were also a source of pride. Mary Daniels, in speaking of the tasks taken on by women while their husbands were away during the era of economic decline, imparted a heroism to them: "the women, they were the ones that kept this valley going."[39]

◆

Married men and women thus made concerted efforts to remain in the anthracite region, not following their former neighbors to new communities. Those who stayed faced a changing region, obviously economically but also demographically. Out-migration dramatically reshaped the age distribution of the population in the anthracite counties each decade after 1950. The changing proportions of young and old in the region across the forty-year period reveal the demographic shift (see table 7 in appendix 1). The share that children and young adults contributed to the population of the anthracite region declined steadily from almost 55 percent in 1950 to just over 40 percent in 1990. At the same time the share of residents over sixty-five more than tripled, increasing from 5 to 17 percent over the period. As fertility declined and young adults departed steadily across the forty-year period, the elderly, past their years of employment, remained in the anthracite region, supported by social security and black-lung compensation payments. In many cases, elderly former residents returned to the anthracite region after retiring from jobs elsewhere in Pennsylvania or in adjoining states. By 1990, Pennsylvania had the second highest share of retirees of all states in the United States, despite the fact that its winter weather would hardly mark the state as a retirement haven.[40] As the share of young people in the anthracite region declined and the proportion of elderly increased, the region's population aged significantly. In the 1950 sample the median age for residents was 26; forty years later, that figure had increased to 37. This development reflected the region's economic decline and made it more difficult for employers and public officials who undertook efforts to revitalize the region's economy. The aging of the region's population was simultaneously a cause and consequence of the region's economic decline.

Sketching finely the changing economic landscape of the anthracite region between 1950 and 1990 illuminates the choices made by those men and women who remained in the region even with economic decline. Anthracite dominated male employment in the region in the immediate postwar years, but that domination did not last long. In 1950, more than 23 percent of all working men found jobs in and around the region's mines. By 1970 the mines employed only 3 percent of males in the region, and by 1990 the proportion was down to 1 percent. At that latter date, wholesale and retail trade employed more than 18 percent to lead all sectors, while durable and nondurable manufacturing also employed substantial numbers. Construction had increased in relative significance over forty years, while transportation (railroads in particular) had shrunk considerably. The leading sectors of forty years earlier played a less important role in the regional economy, as male employment was more dispersed across occupations than it had been at midcentury.[41]

Women's employment in the anthracite region was consistently more concentrated than men's. In 1950, nondurable manufacturing (principally textile and garment work) employed almost 44 percent of all working women in the region. Trade and professional services (chiefly teaching and nursing) also had substantial shares. Altogether these three sectors accounted for almost 79 percent of all women's jobs in the region. Forty years later, the decline of textile and garment work had taken a toll on the region, and by 1990 professional services provided the leading sector for female employment with over 30 percent of jobs. Trade now held second place with almost 25 percent of women's jobs, while nondurable manufacturing was third with about 16 percent. Altogether these three sectors accounted for more than 71 percent of women's jobs, down from the earlier concentration.[42]

While these sectoral shifts were occurring in the regional economy, there was another more general change taking place, a trend that the region shared with the nation in these decades: a dramatic increase in women's employment relative to that of men.[43] Overall employment in the region increased only slightly over the forty-year interval. Yet men and women were affected quite differently by the economic changes of the period. In 1950, 86 percent of men in the age group 18 to 65 were employed, compared to only 29 percent of women; by 1990 labor force participation for men in this shrinking age group exceeded 93 percent, but for women the comparable proportion now amounted to 80 percent.[44] The earlier dominance of the anthracite industry had supported a primarily male workforce; its decline ushered in an era with a more even distribution of employment across gender lines.

While more women in the anthracite region were finding their way into the labor force in the postwar decades, persisting men tended to move out of the labor force. Focusing on male mineworkers in 1950, it is possible to infer family and work patterns by tracing this age cohort through succeeding censuses.[45] The main change evident between 1950 and 1970 for men in this age group was an increase in the proportion not in the labor force from 3 percent to almost 30 percent.[46] Although the men in the group were all 55 to 65 in 1970, almost a third had given up looking for work. As the employment of men in this cohort declined over two decades, the proportion of employed married women in these families almost doubled, going from 20 percent to more than 37 percent. Although we do not have comparable figures for 1950, by 1970 wives of men in the 55–65 age group were earning about 20 percent of their families' income while husbands contributed just under 70 percent.

Over the next decade, declining proportions of men and women in this age cohort continued to work, but the decline was far sharper for men than women. By 1980, about 17 percent each of men and women in the age group 65 to 75 continued to be actively employed. The change over time is even more striking in a thirty-year perspective. In 1950, in the age group 35–45, men had been five times as likely to be employed as women; in 1970 that ratio for men and women between 55 and 65 had declined to less than 2:1; by 1980 men and women between 65 and 75 were equally likely to be employed.

Over time, as the members of this cohort aged, social security and retirement income became the primary means of family support.[47] Only 5 percent of wives and 7 percent of husbands in the 75–85 age group continued to work in 1990, and their wage and salary incomes accounted for less than 6 percent of family income. On the other hand, more than 92 percent of husbands and 86 percent of wives were receiving social security. Social security and retirement income accounted for more than 64 percent of family income by 1990, up from 40 percent a decade earlier, and 4 percent twenty

years earlier. Judging from these successive snapshots, anthracite-region families in this age cohort saw their incomes rise over the forty-year period. In 1950 the average total income of families in which the male household head ranged between 35 and 45 years of age was just under $3,400; twenty years later families with a male head 55 to 65 had incomes of almost $8,800, and by 1990 the comparable figure was slightly more than $20,750. Taking into account the rising cost of living over these decades makes the gain in real income less dramatic. Adjusting for inflation indicates that real family income (in constant 1950 dollars) went from $3,400 to $3,823 over the forty-year period.[48]

Ironically, even this modestly rising real income of anthracite-region families resulted largely from the high incidence of black lung among former mineworkers. Black-lung compensation payments after 1970 were roughly equivalent to the social security income that former mineworkers could expect, so those former mineworkers whose black lung was crippling, rather than lethal, received roughly twice the social security income that they would have received had they been healthier.[49] The higher income for survivors of black lung was counterbalanced, of course, by the high rates of death among former mineworkers from decades of exposure to coal dust underground and in the breakers.[50] One study of retired anthracite miners estimated that 80 percent (12,000 out of 15,000) of retirees receiving pensions from the Anthracite Health and Welfare Fund also received black-lung benefits. The circumstances suggest the anomalous conclusion that to have enough income to live decently during retirement a former mineworker had to be crippled by black lung—quite a convoluted and sad logic.[51]

A family perspective of employment patterns and joint income of married couples is a necessary complement to consideration of the wages and entitlements of individual fathers and mothers. Demographic and income information from the decennial federal censuses make possible this approach.[52] The mean age of married anthracite mineworkers who resided with their wives in 1950 was 42 (with the middle third ranging between 36 and 46); their wives averaged 39 years of age. One of seven wives was employed, and those who were working contributed substantially to the family's total income. These employed wives averaged $930 in wages annually, compared to $2,530 for their mineworker husbands.

Women's contributions to these now older families increased markedly in the twenty years after 1950. More than 35 percent of wives whose husbands were between 56 and 66 in 1970 (the age range that included the middle third of married mineworkers twenty years earlier) found employment—up from 14 percent for this cohort in 1950. Moreover, as more wives found employment, the proportion of their husbands who were working declined from 95 to 65 percent. And lastly, wives' earnings increased relative to those of their husbands. While in 1950 they had earned on average 37 percent of their husbands' wages; by 1970 that proportion had increased to 72 percent. These census snapshots demonstrate that anthracite region "fathers and mothers" were moving away from the strictly patriarchal family economy that had characterized the region at anthracite's height and toward a more equal sharing of economic responsibilities. The census picture is reinforced by many of the personal histories of men and women of this generation.

Tom and Ella Strohl were shaped by the patriarchal value system that characterized the world of their immigrant parents and grandparents. They felt close ties to their ethnic, working-class origins and gave no thought to moving when the mines closed. Still, they found themselves breaking new ground as they coped with the economic crisis of the postwar years.

Tom and Ella were both born in the Panther Valley. Her father was a contract miner, and his father ran a mule stable for the coal company. They married in 1940. Tom worked underground in the mines; Ella stayed home and raised their two daughters. Tom was forty when the LNC mines closed in May 1954, and Ella was thirty-five. They'd been married twelve years, and their daughters were ten and six.

Times were tough in the Panther Valley as its largest employer closed its operations. After six months, however, local operators leased mines from the company and managed to produce anthracite coal intermittently for another five years. In February 1960, though, the last of these operators acknowledged defeat and closed down.[53] Tom had seniority and managed to get work in the mines during these last years, even though employment shrank drastically between 1954 and 1960.

When the mines closed for good, Tom began looking for other work, insisting that his wife remain at home, that "her job was raising the kids." He first took work as a carpenter building suburban track homes near Allentown, about an hour from his hometown of Nesquehoning. He and a number of mining buddies worked and carpooled together. On the job he was under pressure to complete one home and move on to the next, to cut corners to speed the work and save on materials. After a stretch, he succeeded in finding a job closer to home and worked as a maintenance man in a paint pigment factory in neighboring Lansford until he was injured in a fall from a ladder and took disability retirement at the age of fifty-two.[54]

At that point Ella was forty-seven, and she began working at Cassie's Sportswear, an apparel firm just up the street from their home. She was able to come home at midday and share lunch with her now-retired husband. She worked as a trimmer in the factory for fourteen years, retiring at sixty-one. In addition to earnings that increased from $1.25 to $4.00 per hour over the period, she qualified for a pension of $31 a month. Both her earnings and the pension added significantly to her family's income, initially limited to her husband's unemployment benefits and later to a combination of social security and black-lung payments.[55]

On the surface, Tom and Ella's work histories tell a straightforward story of a family supported by the husband's earnings until he could work no longer and a wife's stepping in at that point to make up for the loss of his regular income. And their oral histories followed this narrative line without revealing much apparent conflict. It was only when their daughter, Ruth Strohl Ansbach, offered her perspective on the impact of the mine closings on the family that a more complex story appeared. "I can remember them arguing," she recalled in a 1993 interview. "I can remember her saying, 'I want to get a job,' because there was nothing coming in. [He replied,] 'No, no, no, you can't get a job, I'll get something.'" Despite this opposition, her mother did take a job. Her daughter described her mother coming back saying to her father, "'I'm working,' [she said.] He didn't like it too much . . . but I think he kind of figured there would be no money coming in until he would get his Black Lung [compensation]. They were collecting unemployment, but at that time then it wasn't enough either, you still had other bills and stuff to pay. But I can remember my sister and I saying, 'Just go, just go and get the job, don't worry about what he says!'" The family struggle over traditional patriarchal relations worked itself out, with Ella having the strength to stand up to her husband and her daughters encouraging her despite their father's opposition.[56]

The changing gender relations evident in the Strohls' stories reshaped the lives of many fathers and mothers during the postwar decades after the mines closed. Their different experiences as migrants, commuters, or persisters all worked to undermine the strong patriarchal relations that had characterized working-class life during the heyday of anthracite mining and contributed to more egalitarian relations between men and women.

For families of commuting men, changes took place immediately. For women, the separateness of the home front intensified with the economic crisis. Wives of commuting husbands in effect became single parents, almost entirely responsible for the raising of children. Mary Jasinski's husband commuted on a weekly basis for seven years, meaning that he spent little time with his children, being absent during the week and fatigued on weekends. In this context, Mary did virtually all the parenting and, as her daughter reminded her, even took the children fishing. Irene Gangaware's husband, Grant, was on the road for fifteen years. He admitted to playing a minimal role in his children's upbringing: "I didn't spend the time with my kids, raising my children, it was my wife that really done that." Mike Knies commuted on a daily basis, but he worked the night shift (for twenty-eight years), driving to Reading right after his children came home from school and returning in the morning just about the time that the new school day began. He, too, lamented the limited time spent with his son and daughter during their growing years. These comments invariably reflect the

price that commuting outside the region exacted. Neither husbands nor wives spoke positively about the experience of weekly commuting. One perspective that emerged repeatedly was the importance of mothers' contributions to their children's upbringing. As Paul Melovich commented about his mother, "She is the block; she is definitely the strong one."[57]

Thus, with regard to childrearing, the economic crisis of the anthracite region reinforced the traditional distinction between men and women's family responsibilities. In other aspects of home life, however, new patterns of gender relations emerged as women worked more regularly outside the home. Both men and women offered stories commenting on the reversal of gender responsibilities. In the immediate aftermath of the closing of the mines, some men remained unemployed for substantial periods of time. They stayed at home taking on homemaking tasks; here a blurring of boundaries occurred rather than strict gender distinctions. Mary Kupec described the new situation in her household: "He [my husband] was at home. There was no work—you couldn't get anything. So then I went to work full time . . . and he babysat. . . . The man became a woman, and the woman a man." Mary Jasinski remembered a similar development in her household: "I never had a babysitter [when my husband was unemployed]. No, every time I worked, he was home." Grace Ferrari similarly recalled her husband's domestic work with some amusement. "When he wasn't working, he washed the clothes. My neighbor next door would say, 'Boy, he hangs the clothes funny.' I said, 'I don't care how he hangs them, as long as he washes them.'" Thus, working women encouraged their unemployed husbands' sharing of household domestic work.[58]

The assumption of domestic duties, however, was not always easy for men. Sarah Fibac recalled that when her unemployed husband "was hanging clothes they used to call him sissy, and why is he doing it, that his wife should be doing it." Bitterness crept into men's lives. Years later Joe Orsulak depicted a frequent scene in mining communities with the closing of the mines— idle men driving the family car to pick up their working wives after a day's toil in the garment factories:

In those days when the men lost their jobs the women went to work. But most of them had cars at that time already. And they would line up down at Rosenau Brothers waiting for the wives so as to pick them up at work. That was their main duty of the day. . . . [H]ere they had to go down there, and she made the money, and he had to bring her home and that's all he did. That was degrading, sure, it was. That's the way they saw it.[59]

Married men found themselves performing work traditionally done by their wives, and for some the adjustment was not easy.

Orsulak recalled these feelings forty years after the fact, but contemporary sources confirm the perspective he described. Min Matheson, ILGWU district manager for the Wyoming Valley, offered testimony before a Senate committee on area redevelopment discussing the economic crisis in the region. She commented that the increased employment of women in garment factories was not necessarily "a cause for rejoicing." Her point was that many of her union's members were not eager to go out to work, but did so "only because the men in their homes . . . are unemployed and are at home doing the housework, the shopping and tending to the children—a complete reversal of the normal course of family life." Quoting a garment worker interviewed in the course of a *Business Week* story on the anthracite region, Matheson reported, "It's one thing to have an independent income if your husband is working, but it is no fun being the breadwinner." Several times, Matheson reported that the role reversals necessitated by male unemployment were "demoralizing" for the family. She reported increased drinking and suicide among the consequences of this state of affairs.[60]

Other testimony reinforced Matheson's perspective. Wilkes-Barre area congressman Daniel J. Flood testified before a Senate committee in 1959:

My men are in the kitchen. Do not tell me that is where they belong. . . . They are the babysitters. They wash the dishes, they are preparing the meals, they sweep the front porch, they are wearing the aprons, sir—aprons. Thousands and dozens of

thousands of the best workers in the world are housekeepers.[61]

The repeated reference to "aprons" emphasized the abnormality of this "feminization" of coal miners' lives in the aftermath of the mine closings. Another witness at the redevelopment hearings, Peter McMahan of Georgetown, in Flood's district, shared the view that unemployment undermined miners' masculinity. He described his encounters with housebound men in no uncertain terms: "Many times I knock on doors, and the man comes to the door with an apron on. I have seen them come to the door with a dirty diaper in his hand. I am speaking the truth." His language underscored the change that he witnessed. He emphasized that he was "speaking the truth," and not exaggerating. Not liking what he saw in miners' homes, McMahan supported area development legislation on the grounds that it would help turn these circumstances around: "What we need are industries in this valley that will put the father to work and leave the mother [to] stay home and . . . take care of the children." Contemporary opinion, male and female, from both leaders and rank-and-file, confirmed the recollections of Joe Orsulak on this score.[62]

Although we might view these numerous complaints in the 1950s as the first, halting steps toward more egalitarian gender relations, men and women during the period of anthracite's decline did not welcome such changes. To pick up one's wife at the factory gates was viewed as "degrading" as Orsulak recalled, and as "a complete reversal of the normal course of family life," as Matheson testified. Images of working-class men opening the front door wearing an apron or holding a soiled diaper struck contemporaries as evidence of a world out of joint. Men and women agreed that these stories pointed to the problems they faced in the midst of anthracite's decline.

Contemporary testimony before area development hearings and subsequent oral interviews reveal that the economic collapse of the anthracite region generated tension in the household as well as the mutual efforts of mothers and fathers to cope with crisis. Some husbands refused to allow their wives to work, no matter how desperate the situation. Recall that Tom Strohl felt his wife's job was "raising the kids." "I wouldn't leave [her] go to work," he commented. Even though his earnings in a series of jobs were close to minimum wage, he insisted that his wife stay at home. In her interview Ella Strohl glossed over her husband's opposition to her work outside the home, but she did recall the difficulties of stretching her husband's limited income. Only when an accident forced her husband's early retirement did Ella Strohl begin working in a nearby garment factory. Eleanor Yelito similarly remembered her father's refusal to entertain the notion of her mother's working. "A woman belongs at home," she remembered him repeating.[63]

As women worked more and men had increasing difficulty finding employment, earnings and control of family income surfaced as another area of conflict in some families. Women sometimes earned more money than their husbands did—and not just when the men were unemployed. Handling of family budgets was an issue that could appear in good times and bad, but the commuting of husbands added to the difficulties. Irene Gangaware remembered that when her husband was on the road for months at a time, she made sure that his company sent his paycheck directly to her. He lived on a company expense allowance. When he demanded once that he receive his check, she argued back, "'The day I don't see your paycheck, that's the day you go out the door with the old luggage.' Because I wasn't drinking his money; I wasn't gambling the money. It was our money. . . . [W]e never had yours and mine."[64] Irene Gangaware's strong words encapsulate the changed relations between men and women that emerged with economic decline.

Two stories encapsulate the changing ways that men and women handled family budgets as the mines were closing down, revealing important shifts in family dynamics between the Depression and post–World War II era. Lillian Verona, daughter of a miner, remembered a mealtime conversation from her childhood in the 1930s: "And I even said to my mother, 'Why can't I drink this milk?' [Her mother replied,] 'Because that's for your father. That's for your father. He's the one that works; he's the one that has to eat.'" Men's advantages at the dinner

table were part of their dominance over broader family expenditures. Verona recalled this element of her father's control within the family: "My father was a gambler and a drinker, too. Whenever he worked, he would just give [my mother] so much money and the rest he kept for himself."[65] As these recollections suggest, patriarchy had a strong material base in the years before World War II. Women stayed at home and made the basic decisions about running their households, but they did so within a framework in which husbands often unilaterally decided how much of their income to contribute to their families' support.

That practice became less common after World War II. Mike and Margaret Mikovich recalled the very different way they handled the family budget in the late 1940s. This was a slack period in the mines, and Mike was typically bringing home only four or five days' pay for each two-week payroll period. Mike described the care with which the couple paid their bills: "when payday would come we'd sit down and we'd have the bills on the table, the bills that we owe, and then we'd say, 'Well, we could pay this guy or we can't pay this one or we pay that one.' That's the way we got through." As if to underscore that the "we" was to be taken literally, Margaret added, "at that time we both sat down and did it."[66] The line between the patriarchal approach in the Verona household and the shared responsibility of the Mikoviches could hardly have been drawn more sharply.

Even in families that continued to maintain a relatively rigid gender division of labor, by the 1950s women were taking increasing responsibility for managing family budgets. In the interviews with Tom and Ella Strohl, for instance, it was Ella who detailed how the family coped with its irregular income when the mines closed:

We let the rent go, we let the butcher go, but we explained to them what happened and all. Then when they [her husband and other mineworkers] went back to work again, we were six months due on our rent. We went down to talk to [our landlord] and [when Tom] went back to work we were paying her twice a month, thirty-five dollars. [Once] . . . to catch up and the other [time to pay part of what] . . . we owed. So in a couple months we straighten[ed] her out. Then . . . we did the same thing with the butcher.[67]

Ella Strohl's detailed recall of how the family coped suggests that she was probably the one who had to stretch her husband's limited income and make extra payments to the landlord and the butcher to catch up on credit that had been extended when the mines were not operating.

In this retelling, the Strohls seem to have been on the same page when it came to Ella's responsibility for overseeing family expenditures. In other stories, women were more assertive in claiming their rights to manage the family economy. Paul Melovich remembered mothers in his Hazleton neighborhood accompanying their husbands to work on paydays. At this date, in the early 1950s, Melovich's father was working in Wilkes-Barre, and normally he would carpool with friends to drive to the mine where they were working. But on paydays, wives would drive along with their husbands. As Melovich recalled, "I went up to Wilkes-Barre with my mother on the weekends when he'd get paid, because when the men would go, the wives would go along to collect, to make sure [of] the check." The men would pay the bar tabs that had grown since the last payday, and their wives would ensure that "the rest of the money came home with them." As Melovich described his mother's role in this repeated drama, "they'd pay the bar tab. . . . And then she took all the money. That was her territory."[68] What a contrast between Paul Melovich's description of his mother and Lillian Verona's recollection of her mother's Depression-era comments about her father—"He's the one that works; he's the one that has to eat." In the postwar period, wives of anthracite miners increasingly assumed the right to control the family purse strings, as these examples confirm.

Another factor that eroded the traditional patriarchy of anthracite-region families was the increasing importance of women's earnings in older families after the mines closed. The evidence for husbands' and wives' earnings drawn from postwar census samples provides an initial indication of this change, but oral history recollections offer a fuller understanding of how men and women experienced and understood these developments. Time and again anthracite-region men were forced to retire early due to health problems traceable to their years in the mines, while their wives continued to work in garment factories. Men and women discussed the aging process in their oral interviews, and their words provide insights into their attitudes toward the strategies they adopted in response to economic decline.

Mike and Margaret Mikovich from Lansford spoke directly to this issue. Mike was fifty-eight when health problems led to his early retirement in 1974. He had not worked long enough in the mines or at any of the jobs he had held in the post-mining years to qualify for a pension. But the couple pieced together enough to support themselves. As Mike recalled, "between my Black Lung [compensation] and my wife's wages we were able to manage" until social security kicked in when Mike turned sixty-five. Similarly, when Tom Strohl had to retire when he was fifty-two, his wife worked in a garment factory for fourteen years. By the time she retired, her husband was sixty-six, just old enough to qualify for monthly retirement benefits from social security to supplement his black-lung payments. And for another sixteen years, the couple enjoyed her $31 monthly pension, a lasting, if modest, legacy of Ella Strohl's willingness to break with her husband's traditional expectations.[69] That pension exceeded by a dollar a month what retired anthracite miners received from the United Mine Workers of America in this period.[70]

Gender relations shifted in the anthracite region for the men and women who stayed. Change and conflict also marked the lives of married couples who migrated. Mary Painter recalled several instances, for example, where women refused to move after their husbands had been worn down by years of commuting or resettled grudgingly and then summarily moved back alone to the anthracite region. "We had one friend," she related. "He was a guard at U.S. Steel [in Fairless Hills], and he came home one day, and the house was empty." His wife had left and taken their children back upstate with her. This couple had not discussed the relative merits of their alternative choices, but the wife's decisive action carried the day and led her husband to give up his new job and return to the uncertain prospects of their anthracite hometown.[71]

Mary Painter, however, told this story in the light of her own. When the mines closed, she and her husband had jointly decided that he would work in Fairless Hills and that if all went well she and the children would follow. Don Hunsinger told a similar story about his wife's part in family decision making. He had been injured in an accident in the mines, and after the accident, his wife insisted, "You're not going to work down there! You're not going to work down there." Hunsinger became an autoworker in northern New Jersey, and after less than two months his wife and two sons joined him in a rented home in Rahway.[72]

Resentments and tension surfaced between men and women during the economic collapse of the anthracite region—the crisis giving women more leverage in their dealings with their husbands. Yet, as interviews with Mary Painter, Don Hunsinger, and others attest, hard times also required and encouraged joint decision making. Thus a unilateral world of male authority dissolved in cases of both marital discord and agreement.

These anecdotes speak to the ways that the fathers and mothers who coped with the mine closings had to negotiate a number of transitions as the regional crisis deepened beginning in the 1950s. First, of course, they had to accommodate to the loss of mining jobs and the need to

find work, either in the Panther Valley or by commuting or migrating to jobs at some distance from their homes. These circumstances set the anthracite region's fathers and mothers apart from those Americans who benefited from the nation's postwar economic prosperity. But, like all Americans of their generation, they also had to cope with advancing age and the need and desire to retire. Typically in their late thirties or early forties when the mines closed, their first challenge was to accommodate to the changed economic circumstances and support their families. Still, they had a second transition to face—from work to retirement—and struggled to make ends meet in retirement. Examining this final transition addresses the lasting impact of the closing of the mines.

Individual oral histories provide numerous stories about how former miners and garment workers negotiated the transition from work to retirement. Before drawing on the personal accounts, though, it will be helpful to establish a broader demographic context. By 1970 fathers and mothers who faced anthracite's postwar decline were primarily in their late fifties and early sixties, and retirement was no doubt on their minds. Examining the evidence for male and female labor-force participation in the anthracite-region public-use sample for 1970 reveals patterns of behavior for men and women of this

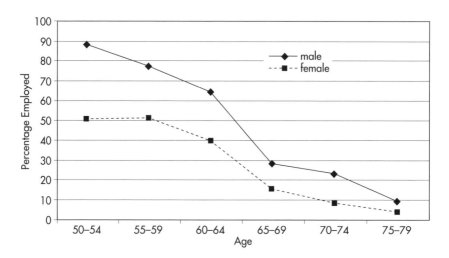

Figure 1. Labor force participation in the anthracite region, 1970, by age and gender. *Source*: IPUMS, 1970 (for U.S. and anthracite region), Minnesota Population Center, University of Minnesota.

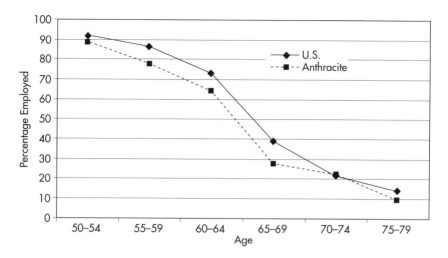

Figure 2. Male labor force participation, 1970, U.S. and anthracite region. *Source*: IPUMS, 1970 (for U.S. and anthracite region), Minnesota Population Center, University of Minnesota.

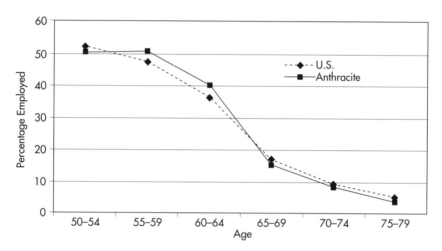

Figure 3. Female labor force participation, 1970, U.S. and anthracite region. *Source*: IPUMS, 1970 (for U.S. and anthracite region), Minnesota Population Center, University of Minnesota.

generation. Figure 1 offers a view of the steady movement of men and women in the anthracite region out of the paid labor force after fifty years of age. The steepening slope of the graphs for men and women beginning with the 60–64 age group reflects the quickening pace of retirements after that age. For anthracite residents between 60 and 64, almost 65 percent of men and 40 percent of women continued to be employed in 1970. For the 65–69 age group, however, those shares fall dramatically, and less than

30 percent of men and less than 20 percent of women continued to be employed, proportions less than half the figures for those five years younger. Men and women in the anthracite region in these years performed hard, low-paying work, and once they became eligible for social security, they did not tarry long in retiring. Moreover, the former mineworkers in this population commonly suffered from black lung and they simply could not continue to exert themselves as they had done earlier in their lives.

Physically and mentally, men and women in the anthracite region were ready to move to a new stage in their lives—retirement.

Comparing the work and retirement patterns for older anthracite-region residents with those for a similar national sample in 1970 is instructive (see figures 2 and 3).[73] Males in the coal region between ages 50 and 54 worked in roughly the same proportion as men in their national age cohort. However, men between 55 and 70 worked distinctly less than was true for their counterparts across the nation. In the 55–59 year age group, for instance, the labor force participation rate for men in the national sample was 8.9 percent greater than that for the anthracite region. For the next two age groups, the comparable gaps were 8.5 percent and 11.1 percent, with anthracite men once again less likely to work than all men nationally. As figure 3 demonstrates, however, anthracite women's work patterns were more similar to national patterns. Changes in the proportions of employed women parallel one another across the age spread, typically differing by only 2–3 percentage points for a given age group, sometimes with anthracite women working more, sometimes with the figure for the national sample being higher. In a regional economy still reeling from the shock of the mine closings, jobs for men were simply not available in numbers comparable to those in the national labor force. The high levels of unemployment characteristic of a depressed regional economy were reflected in the lives of the men in this age cohort as many of them gave up looking for work and retired earlier than they would otherwise have done. These older men paid a high price for the decline of the anthracite economy. Anthracite-region women made up for some of this lost family income, however, and worked in proportions roughly comparable to those for the nation as a whole.

What stands out for the men in this population, whether they migrated or remained in the anthracite region, is the way that deteriorating health played a key role in the decision to retire. In interview after interview, this influence proved central in men's stories. Tom Strohl was forced to retire at fifty-two when he was injured in a fall from a ladder while working in a paint pigment factory in Lansford. He saw a doctor, who focused less on the actual injury than on Strohl's broader health problems. After examining x-rays of his lungs, he told him, "Tommy, if you want [to live] a little while, get out of there." Strohl lived another thirty years after he retired, though in the last years, his black lung condition worsened and he required an oxygen tank at times to breathe.[74]

Others had similar stories. Grant Gangaware retired at sixty-three, when a heart attack and lung problems forced him to quit. Although it took repeated efforts, he eventually began receiving black-lung compensation. Even those who migrated from the anthracite region could not avoid mining's health complications. Ziggie Whitecavage, for example, began working at fourteen in mining and subsequently in shipbuilding before going into the military during World War II. After the war, he worked for almost twenty years at U.S. Steel in Fairless Hills, but at sixty-two he had to retire. "My miner's asthma started bothering me and that was hard for me to breathe." As soon as he was approved for compensation payments, he retired. Joe Marouchoc told an almost identical story. He had pieced together a series of jobs in small mines around the area after Lehigh Navigation Coal shut down, but then he applied for black-lung compensation and quit work when his approval came through. Like Gangaware and Whitecavage, Marouchoc was in his early sixties.[75]

While men and women in the anthracite region tended to retire at roughly the same ages, two patterns of retirement proved about equally common. For one group, wives often worked several years after their husbands had retired. For instance, Ella Strohl worked fourteen years during her husband's retirement; Irene Gangaware worked ten years after Grant retired; Margaret Mikovich worked for thirteen additional years; and Mary Whitecavage worked ten years after her husband Ziggie retired. For a second group, husbands and wives retired at about the same time, to enjoy their new freedom together. Mary Phillips anticipated her husband's retirement by one year; Helen and John Mordock retired at the same time; and Mary Jasinski retired "shortly after" her husband retired in 1988.[76] The stories reflect the new importance of wage work for women after the mines closed. Some of these women retired at the same time as their husbands; for others, their husbands' health problems required that they keep working to support the family. Both the frequency with which older women were earning wages and the varied timings of their retirements represented distinct departures from practices in their parents' generation.

As fathers and mothers made decisions and moved into the retirement stage of their lives, their children and grandchildren took on increasing importance in their lives. And just as their shared wage earning represented new departures, so their attitudes toward their children revealed new ideals. Those who remained in the region in the face of economic decline developed strikingly different expectations for their children and grandchildren than those of their parents when they had been growing up.

During the Depression, for instance, children frequently felt parental pressure to quit school at early ages and work to contribute to their families' immediate needs—a common practice a generation earlier for many of the fathers and mothers who described their school years and early adulthood in their interviews. Irene Gangaware had hoped to train to be a nurse, but her family could not afford the cost—$300 for three years—so she went to work in a local clothing factory instead. For Lillian Verona, the story was much the same. Despite having graduated sixth in her high school class, she went right to work in a local factory.[77] In the post–World War II period, however, men and women in this cohort began to view their children's futures in new ways. They took more interest in the education of the new generation than their parents had. Increasingly, many parents and grandparents saw education and migration as keys to children's long-term economic success—a perspective that differed markedly from the framework in which they had come of age.[78]

Theresa Pavlocak spoke for many of her neighbors, evincing this new attitude toward education in her generation:

Most of the girls are educated; they're nurses or whatever. . . . The boys are educated; they go to Marian High or Lehigh; all the mothers wanted their children educated. When the children got educated, now there's grandmothers, like myself and

all. . . . I have girlfriends, they go all over America because they have four or five children living away from here, they're going to California . . . Texas . . . Florida now.

Pavlocak began her commentary noting that "all the mothers wanted their children educated." As a consequence of that education, many of the children of this generation were able to leave the Panther Valley and find better-paying and more fulfilling work. She and her neighbors, then, spent a good deal of their time visiting successful children and grandchildren living all over the United States. They worked to educate their children and helped them leave the region so that they might find economic success.[79] In the face of difficult times, they remained in the area even though with reduced economic opportunities. In turn, they placed their hopes in their children and took great pride in their educational and occupational achievements.

But even as children struck out in new directions with their parents' blessings, fathers and mothers remained strongly attached to the anthracite region. Strong community ties often proved more important to members of the older generation than the economic insecurity of their hometowns. Even for those who had migrated in response to the closing of the mines, the attraction of their hometowns remained strong and they visited often. So strong was the connection that some out-migrants returned permanently to the region to start new chapters of their lives. Theresa Pavlocak and her second husband lived twenty-two years in Edison, New Jersey. But when a son died in Vietnam, they buried him in Summit Hill and returned to the Panther Valley. As she recalled, her husband said, "We're going back home." Mary and Stanley Jasinski also found it hard to leave the Panther Valley permanently. Stanley commuted weekly to New Jersey for seven years and worked for a paper company in Spottswood. Mary and the children moved to Spottswood and lived with him for nine months. During that time they even picked out a home that was under construction. But before they moved into it, they got cold feet. Mary and the children moved back to Summit Hill, and as soon as Stanley found a job, he rejoined his family permanently.[80]

Other migrants, nearing the ends of their careers, often took the occasion of retirement to settle back in familiar territory, as Irene Gangaware's parents did, returning from Newark, New Jersey to Lansford. Similarly, John Mordock, though he and his wife decided to remain in the Fairless Hills area, recalled a number of friends who retired from U.S. Steel and moved back to Hazleton or Wilkes-Barre.[81] Between those who persisted in the anthracite region continuously after 1950, those who commuted for periods of time but decided not to settle outside of the area, and those who returned there after living and working for a stretch of years in more prosperous areas, a large share of the fa-thers and mothers continued to call the region their home. Mike Sabron spoke for many in his generation when he said, "When I think back . . . I could have stayed in Linden [New Jersey] when I worked in Linden, but I didn't like it. I just loved, right here. [Later,] my kids wanted me to go and live with them. No, I want to die here, baby. I worked hard for this."[82] The stories of these fathers and mothers after the closing of the mines are a testament to the power of place and community in people's lives. The testimonies speak poignantly to the ways their individual identities were indelibly connected to the anthracite region.

CHAPTER 7

Personal Responses to Decline
Sons and Daughters, 1945–1990

FORMER MINEWORKERS AND their wives remained in the anthracite region after 1950, even at the cost of stagnating family income. Increased wages from wives' garment work, black-lung compensation payments, and social security income permitted older families to make a go of it even in difficult times. And the close familial, church, and ethnic ties of mining communities often proved more important to members of the older generation than the evident lack of economic opportunity in their hometowns.

Their children, however, had different perspectives on the region's economic collapse. Coming of age in the midst of the crisis, sons and daughters were much more likely to leave the region than their parents. They had little experience in or attachment to the area's mines and held no hope that the mines might reopen. In addition, they were making a transition in their lives from high school to college or work. For both these reasons they were more open than their parents to resettling in areas with better employment prospects. As evidence of the weaker attachment to the hometowns in which they grew up, a large minority—29 percent of young men and women enumerated in 1950 in Lackawanna County, who were 15–24 at the beginning of the decade—had migrated out of the county by 1960 (see table 6 in appendix 1). Some left the region for military service or to attend institutions of higher education; others

sought better-paying jobs available in northern New Jersey or the Philadelphia region, two common destinations for anthracite-region migrants. Most young out-migrants left the region permanently, but a significant minority returned to the area over time.

One man whose life story reflects much that went on in the lives of the young migrants in the postwar period is James Coon, who grew up in Girardville in Schuylkill County, some miles southwest of the Panther Valley. Son of a mineworker who lost a leg in a mine accident at the age of nineteen, Coon quit school in the tenth grade in 1952 and migrated to Bristol, Pennsylvania, with two older brothers, looking for work. There he found a job on a Delaware River tugboat. The brothers soon brought their parents down to live with them and reconstituted their family in southeastern Pennsylvania. Coon then worked at Pacific Steel for a couple of years and finally shifted to U.S. Steel's Fairless Works in 1955. Except for a brief stint in the Marines, Coon worked for U.S. Steel until he retired in 1988. Always a strong supporter of his union, the United Steel Workers of America, Coon began working for the union in 1988 and when interviewed in 1995 he was serving as president of the Fairless Hills branch of the Steelworkers Organization of Active Retirees.[1]

This postwar generation of sons and daughters broke new ground; most did not follow in the footsteps of their parents' or grandparents' generations. They did not recapitulate the "traditional" experience of sons and daughters growing up in coal country. Labor in the region's breakers and silk mills had been common for the generation of youngsters who came of age before World War I. In the 1920s and 1930s, child labor virtually disappeared, but sons and daughters in mining families either quit high school to enter the workforce or sought employment immediately upon graduation. Higher education was out of the question for children from working-class families in these earlier generations, and the pressure to earn money to assist parents was intense. Horizons were limited, and children typically followed well-worn paths to mine or factory work close to home. Historians have commented on the generational continuity that had characterized the transition from youth to adulthood in the patriarchal world of anthracite at its apogee. Donald Miller and Richard Sharpless thus wrote, "In the deep snows of winter, fathers carried small boys to breakers on their backs in the predawn darkness," and "boys learned early that they would follow their fathers into the mines." John Bodnar concurred with this picture: "familial obligations directed wage sharing and entry into work. . . . Daughters and sons grew to emulate their parents."[2]

For the parental generation, occupational profiles had differed markedly along lines of gender. In mining communities in 1920, more than half of young working women were employed in garment or textile factories, while another 17 percent found sales jobs. Among young men, more than 75 percent took jobs in and around the mines, with garment or textile work accounting for another 22 percent.[3] There were distinct occupational tracks for sons and daughters at that date, and the regional economy was able to absorb the rising generation.

Such was decidedly not the case for young people after World War II. Census snapshots for the anthracite region between 1950 and 1990 provide a useful view of the experiences of members of this younger generation. One change evident in tracing this age cohort across four decades is its declining share as a percentage of the regional population. In 1950, 15- to 24-year-olds comprised 17 percent of the population of the anthracite region; by 1970 this age cohort (now 35 to 44 years old) represented only

about 10 percent of the region's population. As time passed, the proportion of males in the cohort declined notably—from 49 percent in 1950 to less than 43 percent in 1990. Higher rates of out-migration and mortality thinned male ranks more than those of females across these decades.

For both men and women in this younger generation, changing work possibilities after 1950 took them in occupational directions that were a departure from the world of work their grandparents and parents had known. Mining never proved to be as important an occupation for young men coming of age after World War II as it had been for earlier generations. While 14 percent of employed men in the 15 to 24 age range held jobs in the anthracite industry in 1950, that proportion declined sharply in the next two decades, fluctuating at levels below 2 percent in 1970 and after.[4] These figures contrast with those for 1920, when more than 65 percent of employed 15- to 24-year-old males in mining communities worked in anthracite. With the decline of anthracite mining, young men in the post–World War II years found considerable employment in wholesale and retail trade and various manufacturing occupations. Women, in contrast, were found principally in operative positions in the garment and textile industries early in the postwar period. The employment of anthracite-region women in the apparel and accessories sector, however, declined steadily between 1950 and 1990, both in absolute numbers and as a proportion of overall female employment. In 1950 some 15,600 anthracite-region women between the ages of 15 and 24 were employed in this sector; by 1990, the number of women in this cohort (55 to 64 at this later date) employed in apparel and accessories had declined to 7,400.[5]

The decline in the number of garment workers in the daughters' generation was part of a substantial contraction in the apparel and accessories sector in the anthracite region over the four decades. Between 1950 and 1990 the number of female garment workers in the region declined by 46 percent. The erosion in garment employment in the anthracite region accelerated in the early and mid-1970s as U.S. firms increasingly outsourced work abroad and imported women's clothing. Employment in the apparel industry in the Scranton–Wilkes-Barre metropolitan area declined from more than 26,000 in 1972 to 8,400 twenty years later. The workforce plummeted still further in the 1990s, so that garment employment in the metropolitan center stood at only 1,300 in 1999. The garment workforce had declined by 95 percent in less than three decades.[6]

Moreover, as the garment workforce shrank, it also aged dramatically. The generation of the daughters was not being replaced by a younger generation as time passed. In 1950 the mean age of women garment workers was slightly over 30; by 1990, it exceeded 49. Workers 15–24 years of age comprised more than 40 percent of the female workforce in 1950. All told, 89 percent of female garment workers in the region at that date were under 45. By 1990, however, only 36 percent of female garment workers were under 45. What a difference forty years made. In its dying years, a moribund anthracite-region garment industry relied on a shrinking, aging female workforce and was hard pressed to replace these women as they retired.

These figures indicate that sons and daughters in the Pennsylvania anthracite region experienced successive waves of economic decline: first, anthracite mining shrank dramatically in the 1950s; later, beginning in the 1970s, garment jobs began moving overseas and women had to look for alternative employment, just as their fathers, husbands, and brothers had done earlier.[7] A third wave affected a significant share of the sons but occurred outside the anthracite region: Those members of the generation who migrated either to southeastern Pennsylvania or northern New Jersey were hit by massive layoffs due to downsizing in the mid-1980s at U.S. Steel's Fairless Hills plant and the General Motors assembly plant in Linden, New Jersey.[8] The decline of industrial employment that characterized the U.S. economy in the last half of the twentieth took a continuing heavy toll on the generation of the sons and daughters who came of age as the mines were closing down.

The labor-force participation rates of women in the anthracite region generally grew in the post–World War II period—as elsewhere in the United States—but the rates varied by stages in the life course and in different ways than for men. In 1950, some 39 percent of women be-

tween 15 and 24 were in the paid labor force, compared to 52 percent of men in the age group (see Figure 4). This modest differential grew, though, as women married and took on family responsibilities. By 1970, when daughters in this group were between 35 and 44 years of age, this differential had grown. What had been a 13 percent gender differential in 1950 stood at 42 percent only twenty years later. After 1970, though, as the demands of raising children for this cohort lessened and increasing numbers of older married women sought paid employment, the employment gap narrowed. By 1980 the male-female differential had declined to 26 percent, and in 1990 the comparable figure stood at 22 percent. Figure 4 shows that the employment gap between men and women in the anthracite region grew in the period 1950–1970 and shrank in the two succeeding decades and that men's and women's labor force participation peaked at different points in their respective life cycles.[9]

The gender differences in terms of employment for this generation of sons and daughters reflected national patterns in the postwar decades (see figure 5). For the nation as a whole the gender gap in employment for this age cohort increased to its maximum in 1960 when cohort members ranged between 25 and 34 years of age. At that point in the life cycle some 95 percent of males, but less than 40 percent of females were working. Steadily this gap narrowed over succeeding decades until 1990, when cohort members in a national sample were 55–64 years of age and the difference between the proportions of men and women employed was less than 25 percent, a figure close to the pattern evident in the anthracite region at this date.

While there were similarities in the life-cycle employment experiences of this age group in the anthracite region and a comparable national age cohort, significant differences are also evident. For every decade, male employment is distinctly higher for the national sample than for the comparable age group in the anthracite region. There is a twenty-year stretch, between 1960 and 1980, when on a national basis, men in this age group were working steadily with more than 90 percent employed in each of the three census years. Among comparably aged male workers in the anthracite region, employment exceeded 90 percent for the group only once, in 1970, when

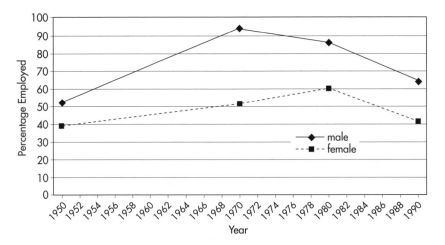

Figure 4. Proportions of age cohort in paid labor force, 1950–1990, by gender, anthracite region (cohort members 15–24 in 1950). *Source*: IPUMS, 1950, 1970, 1980, and 1990 (for anthracite region), Minnesota Population Center, University of Minnesota.

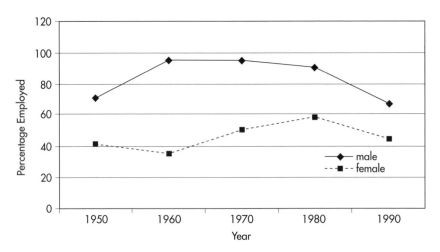

Figure 5. Proportions of age cohort in paid labor force, 1950–1990, by gender, United States (cohort members 15–24 in 1950). *Source*: IPUMS, 1950–1990 (for U.S), Minnesota Population Center, University of Minnesota.

cohort members ranged between 35 and 44 years of age. Figures 4 and 5 reveal the impact of economic decline on this generation of men who entered the workforce immediately after World War II.[10]

In contrast to the picture for men evident here, the labor-force participation rates of women in this age cohort in the anthracite region mirrored national trends. Between 1950 and 1990 employment for women in both groups ranged between a low of 40 percent,

when women fell in the 15–24 age range, to a high of 60 percent for women between 45 and 54. The proportions of employed anthracite-region women in this age cohort almost perfectly match those for women in the United States as a whole in these decades (compare the female portions of figures 4 and 5). Both men and women had to cope with declining jobs in the mines, but anthracite-region men in this age group faced much harsher employment prospects than did women.

Analysis of the experiences of young people coming of age in the late 1940s and 1950s—with the advanced collapse of anthracite coal mining—is sharpened by tracing the lives of specific individuals over time. Questionnaire research on almost 600 graduates of seven high schools in the Panther Valley, including four public schools—Lansford, Nesquehoning, Coaldale, and Summit Hill—and three parochial schools—Marian High, and its two predecessors, St. Ann's and St. Mary's—between 1946 and 1960 provides rich evidence.[11] The questionnaire responses, like the oral history interviews for the older generation, speak to the dynamics of outmigration during the postwar years. Fully 81 percent of the responding graduates moved from the Panther Valley in search of educational or employment opportunities during the most precipitous period of decline.[12] Many would ultimately return to the Panther Valley, but by the mid-1990s 68 percent of these alumni still lived elsewhere.[13]

These midcentury high school graduates are not a random sample, but they are well representative of members of the Panther Valley communities in which they grew up. Equally divided between men and women, more than 90 percent were born and raised in the communities in which they attended high school.[14] At the time that they graduated from high school more than 60 percent of their fathers were employed in anthracite mining; two-thirds of their mothers were housewives who did not work outside their homes, and almost half of the mothers who were employed found work in area garment factories. Their religious affiliations also mirrored local patterns. Just over 60 percent were Roman Catholic, another 9 percent were Russian or Eastern Orthodox or Byzantine (Greek) Catholic, while just under 30 percent were members of various Protestant denominations.[15] The graduates represent the Panther Valley–born grandchildren of the southern, eastern, and central European immigrants who came to work in the valley's mines between 1900 and 1920 and the children of the "fathers and mothers" of the post–World War II period.

Coming of age in a period of economic crisis and seeing so few economic opportunities, a significant share of these young people, not surprisingly, continued their schooling after gradu-

ation. More than 66 percent of these high school graduates had additional higher education or training in skilled trades, with 69 percent of males and 63 percent of females continuing their education after high school.[16] Given gendered differences in occupational patterns for their fathers and mothers, this modest difference in the educational attainment of sons and daughters is striking. Only 6 percent more men than women received some sort of further training or higher education after graduation, and this difference was not statistically significant.[17]

This pursuit of higher education distinguished sons and daughters sharply from their parents. Only about 2 percent of men and women in their parents' generation in the region (35–44 in 1950) had attended college.[18] Among their sons and daughters, however, at least among those who completed high school, 40 percent graduated from four-year colleges or universities or three-year nursing training programs. The greater educational opportunities for sons and daughters was yet another difference between their world and that of their parents' generation.

The overall similarity in the numbers of sons and daughters continuing their education after high school masks a gendered element in higher education for this generation. Almost 52 percent of men attended a college or university, compared to 21 percent of women. The differences in college attendance can be attributed to the different track for higher education taken by women, with 27 percent of them attending hospital nursing programs.[19] While 111 men earned bachelor's degrees compared to only 48 women, another 66 women earned registered nurse (R.N.) degrees. When both those who completed college and those who earned R.N. degrees are included, women slightly outnumbered men, 114 to 111. Because men were more likely to complete four-year college degrees, male graduates typically had more years of higher education, but the difference was less than half a year on average. Overall, the similarities in the education attainment of sons and daughters are more striking than the differences. More than 39 percent of these Panther Valley graduates earned either nursing or college degrees, remarkably high proportions for a working-class population in the postwar period.[20]

What other factors influenced a graduate's likelihood to pursue some from of higher education? Two factors stand out: father's occupation and mother's employment. Most strongly associated with higher education was father's occupation: about 61 percent of graduates whose fathers worked in the mines had some higher education; fully 76 percent of those whose fathers worked at other occupations—such as white-collar employees—did so. The second factor most strongly correlated with higher education was mother's employment. Graduates (whether male or female) were more likely to go on to higher education if their mothers were employed outside the home at the time of their high school graduation than if they were not, the percentage difference amounting to almost 11 percent.[21] The evidence does not explain this difference, but a family's extra income when both parents were employed made it easier for them to contemplate the loss of a child's income that higher education entailed. Or perhaps working mothers had more ambition for themselves and their children, an influence reflected in their children's choices after graduation.

This final relationship had an interesting gendered component. Both sons and daughters were more likely to seek higher education if their mothers were working, but mother's employment had a greater impact on the higher education of their daughters. Among sons, an additional 5 percent of those whose mothers were employed outside the home went on to higher education in comparison to sons whose mothers were housewives. Among daughters, the comparable figure was almost 17 percent, a far greater differential. The earnings and examples of working-class mothers helped launch their daughters into more middle-class lives.

Questionnaire responses provide important overall evidence on the higher education of sons and daughters as the mines closed, but individual cases belie generalizations. Edward Herron, for example, a 1949 graduate of Lansford High School, eventually graduated from Penn State University after a four-year tour of duty in the Navy. Herron was the son of a coal miner, and his mother was a housewife. Both his grandfathers and his father were miners, and "almost all of his 14 uncles worked in or around coal

mines." He came from a mining family through and through. Two factors played crucial roles in Herron's path to college and out of the Panther Valley—the Korean War and the influence of his father. Herron explained his life course after high school:

The best thing that ever happened to me was the Korean War. One of my friends died in Korea but it was my ticket out of the Valley and to college. While my father didn't send me to college he did something I will always remember. He would not let me work in or around the mines.[22]

The story that Herron told recurs in oral history testimonies. After the 1920s many seasoned miners did not want their sons working in the mines. In an interview with the historian John Bodnar, Vincent Znaniecki noted how his father during the Great Depression strongly discouraged him and his brothers from becoming miners: "He never wanted us to go in the mines if we could help it. He figured that the best place for us would be to get into some business, and he encouraged it very much."[23] Like Znaniecki, Edward Herron, given his father's adamant opposition, never considered the mines as an option. Limited initially to work in local stores, first as a stock clerk and then as a shoe salesman, he took the opportunity three years after graduation to enlist in the Navy and that choice proved his "ticket out of the Valley."

Herron's experience makes a connection that was apparent for many sons in the postwar generation—the importance of military service as an economic option in hard times and as a source of financial support for higher education. Joe Rodak, who graduated from high school in Mount Carmel in 1946, commented directly on the latter point:

I think we had about ninety boys in our graduating class, and I can remember sending off two busloads to the service, going to say goodbye to them from the enlistment center—right after graduation. I know of one busload of about forty went to the Navy, and a whole bunch of them went to the Marine Corps. Of course, the war was now over, but that was a good option because it was someplace where you could extend your education, and get paid for it. And then they had the G.I. bill of rights.

That was a big selling point. . . . because you knew when you got out, you were going to be able to go to school, and extend your education.

Rodak himself later attended LaSalle College in Philadelphia, drawing on the benefits that the G.I. Bill provided.[24]

While many sons in this generation went to college, like Herron and Rodak, supported principally by the G.I. Bill, other children of mineworkers did receive substantial assistance from their parents. David DiFebo, a graduate of St. Mary's in Coaldale, commented that his father worked as much overtime as he could get during World War II and that his savings "funded several college educations," including his own. John Evans, another miner's son, could not afford the cost of college, but told a similar story, of how his parents sent him to a local business school, "after my Dad was awarded 'Black lung' payments."[25] In both cases, parents' sacrifices permitted sons and daughters to gain educations that had eluded them.

Education, in turn, was strongly associated with another development that set sons and daughters apart from their parents—their decisions to leave the anthracite region for good. While modest numbers of the older generation migrated from the Panther Valley with the closing of the mines, fully 81 percent of midcentury high school graduates left their home communities.[26] Sixty-eight percent of the group migrated. Another 13 percent left their hometowns for other than college or military service for an average of thirteen years and then returned. While it is hard to claim that the Panther Valley was representative of the anthracite region generally, evidence for Mt. Carmel High School, in neighboring Northumberland County, is strikingly similar. A survey conducted in 1981, at the fifteenth reunion of the class of 1966, revealed that 50 percent of the alumni already lived outside the anthracite region and that another 25 percent expected to move. These numbers are consistent with those for the Panther Valley.[27]

Migration of anthracite-region high school graduates after World War II was strongly associated with higher education. Those who went on to more schooling were much more likely to leave the region than were those who only had high school diplomas. And the greater the number of years of higher education, the more likely the respondent was to leave the region permanently.[28] No other variable was as closely associated with permanent out-migration as higher education—not size of family of origin, father's or mother's occupation, gender, religion, school attended, marital status, military service, date of graduation, or occupational attainment.[29]

Migration proved a shared experience for postwar high school graduates, but they did not move far from home—notably like the migrating fathers and mothers of this period. Of those who migrated (other than for military assignments), some 59 percent moved to other towns in Pennsylvania and another 22 percent to the adjoining states of New York, New Jersey, Delaware, and Maryland. Only 19 percent migrated beyond a narrow band reaching roughly 100–150 miles from the Panther Valley.

The graduates of Panther Valley high schools who left the area with the postwar collapse of anthracite mining did not go far, and a sizable proportion subsequently returned to establish homes in the valley. Of the almost one-third of the graduates who resided in the Panther Valley in the mid-1990s, more than 40 percent were return migrants. Gender differences are evident in the return migration. More than 16 percent of the women graduates came back to the valley, compared to just under 14 percent of their male classmates; female graduates returned on average twelve years after they had left, two years sooner than their male counterparts. More important, men and women who returned home explained their actions in very different terms. Economic motivations predominated for men, 75 percent of whom returned for new jobs, or because they had lost jobs elsewhere, or were retiring back to their former homes. For women, family concerns and community ties predominated among the reasons they expressed. The need to care for an ill parent, a desire to be near family, homesickness, and marriage or divorce were the principal explanations women gave for their return.

Differences between men and women in regard to their reasons for returning to the Panther Valley are highlighted in other ways. Altogether, 54 percent of returnees were female. But women comprised 78 percent of those who said that they returned due to the death or illness of a parent; 86 percent of those returning to marry; and 67 percent of those returning after a divorce. Conversely, men comprised 60 percent of those who mentioned retirement in conjunction with returning to the Panther Valley and 64 percent of those who returned for job or business reasons.

Even when women returnees noted jobs or economic reasons, they commonly placed the situation in familial rather than career terms. Four of ten women who listed job reasons for their returns mentioned a husband's loss of his job, while none of the eighteen men motivated by jobs or economic conditions referred to a wife's employment. Both husbands and wives operated in contexts shaped by familial circumstances, but returning women mentioned their spouses' economic circumstances and familial concerns back home; men did so less frequently.

The questionnaire responses not only reveal how women and men explained their actions but also provide a view into the choices members of this generation made in the face of economic crisis. Like their mothers, women in this cohort went to work when their husbands were unemployed. Ann O'Connell, a 1952 graduate of Coaldale High School, offered remarks along these lines twice on her questionnaire: In 1959, when she moved from Allentown to Bound Brook, New Jersey, she commented, "no jobs available in region for husband." Then when her family moved from Bound Brook to Whitehall, Pennsylvania, in 1961, she noted: "Husband unemployed/my employment."[30] The gendered nature of the logic that men and women offered in describing the interplay of economic and familial motivations and their families' residential patterns is striking in the high school survey. In this respect the gender divide so characteristic of the lives of the generation of fathers and mothers in the anthracite region persisted in the lives of their children. Men continued to explain their residential choices principally in personal economic terms; women were much more likely to mention familial commitments.

While women privileged family obligations over personal careers, employment proved central in the decisions of both men and women graduates. Respondents recorded up to six jobs since graduation and all told male graduates

listed more than 1,200 positions, while females noted about 1,000. These jobs reveal how different the sons' and daughters' occupational choices were from those of their parents, the occupational changes they experienced over their careers, and the persistence of occupational segregation by gender.

At the outset, sons' occupational choices were dramatically different from those of their fathers. First, less than 3 percent of male graduates had first jobs in anthracite mining. The most common initial choice for men was military service, accounting for 22 percent. About 7 percent found work as laborers, 6 percent in sales jobs, and 9 percent in clerical or teaching positions. Over time military and blue-collar positions declined, and white-collar and managerial jobs grew significantly among the occupations reported by men. Many men took advantage of the G.I. Bill and completed college after their tours of military service.[31]

Women's jobs resembled the occupational choices of their mothers in some respects. Some forty-four women graduates found first jobs in the garment industry, long the leading industrial occupation of anthracite-region women. But that number was dwarfed by seventy-nine women whose first jobs were as nurses or in hospitals, sixty-nine who were employed in clerical work, and another twenty-two who worked as teachers. These last three occupations were much less common for their mothers' generation. Women who graduated from area high schools between 1946 and 1960 had occupational possibilities different from those of their male classmates, but also very different from those of their mothers.

As time passed white-collar occupations accounted for increasing shares of jobs for both men and women. Nursing, teaching, and clerical jobs stood out among women, while managerial and supervisory positions became increasingly common for the men in this cohort of high school graduates. For women, slightly more than half of last jobs were in the clerical and professional sectors. For men, proprietorship and professional occupations accounted for more than 15 percent of last positions, middle- and upper-level managerial posts totaled almost 24 percent, and white-collar occupations comprised another 38 percent.

Dividing sons and daughters into the three groups that emerge from the questionnaire responses—persisters, permanent migrants, and returnees—permits a more fine-tuned analysis of employment patterns. While the economic fortunes of their parents were uneven and often bleak, sons and daughters had more success in the years after the mines closed. The education that many of them received served them well in the changing national economy between 1950 and 1980. And their willingness to leave the anthracite region placed them in a better position than their parents to take advantage of economic opportunities that were emerging in this period.

Some 114 of the graduates of Panther Valley schools in the post–World War II period, 19 percent of respondents, remained in the Panther Valley all their lives and were more likely to work in low-paying blue-collar and service occupations than those who migrated permanently and occupied managerial, professional, or business positions. A few figures underscore this generalization. Among men who remained in the Panther Valley continuously after their graduations, 50 percent began their careers in blue collar or service occupations, while only 6 percent were employed initially in managerial, professional, or entrepreneurial occupations. At the end of these persisters' careers, however, 36 percent remained in blue-collar or service jobs and another 28 percent worked in high-level occupations. Over the course of their careers, then, roughly 20 percent of male high school graduates who spent their lives in the Panther Valley after the mines closed moved up into higher-paying middle-class occupations.

In contrast, young men who migrated permanently from the region had much higher-status occupational profiles. At their last jobs, 45 percent were employed in high-level occupations and only 16 percent in blue-collar or service jobs. Male migrants who subsequently returned to the Panther Valley typically found themselves in jobs that mirrored the occupational distribution of respondents who never left.[32] In other words, male out-migrants did much better economically than classmates who never left or subsequently returned to the Panther Valley. The economic world beyond the anthracite region offered an expanded set of economic opportunities, and the careers of different groups of high school graduates reflected that reality. Those who left the region did better than those who remained in the Panther Valley; and among those who migrated, those who did better economically tended to remain away permanently.

The occupational experiences of female high school graduates were similar to those of their male classmates. Once again, persisters were more likely to remain in service and blue-collar jobs across their careers than were permanent migrants; 27 percent of female persisters held these lower occupational positions compared to 8 percent among female out-migrants. Conversely, migrants were more likely to find higher-level jobs in managerial, professional, or entrepreneurial occupations than were those who stayed in the Panther Valley—with comparable proportions of 45 and 32 percent. Yet the occupational advantage for migrant women was less strong than for migrant men, and the occupational profiles of female returnees more closely resembled the patterns for migrants than for persisters—a distinct difference from the male experience.

A closer look at women's occupations helps to explain the similarities between women migrants and persisters. One occupational group stands out. Of the 283 female high school graduates, more than 80 had some nursing or medical training; 66, or more than 23 percent, earned R.N. degrees. Most striking about the residential patterns of nurses is how similar they were to patterns for all women graduates. They were slightly more likely to leave the Panther Valley than were all women graduates, but the difference was modest and not statistically significant. Many of the nursing graduates continued to live in the Panther Valley, first training and then working at area hospitals in Allentown, Coaldale, Bethlehem, and Palmerton, within a reasonable commute from their homes.

The expansion of medical services in Pennsylvania and elsewhere in the United States in the second half of the twentieth century proved a boon to Panther Valley female graduates. Opportunities in nursing expanded, and women responded eagerly.[33] Of the 66 women who earned R.N. degrees, 64 worked as nurses after graduation. Between their work as nurses and in supervisory medical positions, these Panther Valley daughters had nursing careers that had averaged 33 years in length at the time that they completed their questionnaires. Almost half continued to work in the nursing field in 1994, fully forty years on average since they had graduated from high school. In a period of economic uncertainty, these women had chosen higher education, which provided them with middle-class incomes and job security that had eluded their mothers' generation.

In contrast to nursing, most other higher-status occupations required out-migration from the anthracite region. This association held in two distinct respects. Not only did migrants tend to work in higher-status occupations than classmates who remained in the Panther Valley all their lives, but the spouses of migrants enjoyed considerably higher pay and status than did the spouses of persisters. More than 17 percent of spouses of migrants held professional occupations compared to less than 5 percent of spouses of persisters. Conversely almost 26 percent of spouses of persisters worked at the end of their careers in blue-collar occupations, compared to less than 10 percent of the spouses of migrants. These figures show that migration made an economic difference for the graduates of Panther Valley high schools in the postwar years and that migrants were able to turn their education and mobility into family economic gains relative to those who stayed in the depressed anthracite region in the postwar years.

The experiences of one daughter and one son among the high school graduates speak to the contrasting life patterns of persisters and migrants in the years after the Panther Valley's mines closed.[34] Lillian Verona was born in 1934, graduated from Summit Hill High School in 1952, married a former mineworker in 1955, and lived continuously thereafter in the Panther Valley.[35] James Ferrari was younger than Verona, graduating high school in 1960. After a period of service in the Air Force, Ferrari attended Penn State University, earning B.S. and M.B.A. degrees, and never returned to the Panther Valley.[36] Verona's and Ferrari's full life stories encapsulate the choices available to young adults when the mines closed.

Lillian Verona's father was a contract miner; her mother a housewife. She was the ninth of ten children in her family. Four of her older sisters quit school and went to work in New York City. Lillian helped by writing letters to these sisters for her mother, who could barely read and write. Lillian's mother thought that she should follow her sisters into wage work in New York, but Lillian was a very good student and was determined to finish high school. She graduated sixth in her high school class of sixty. She considered going into nursing training, but the lack of family encouragement or financial support quashed that possibility and Lillian went out to work.[37]

After a brief stint in office work, Lillian took a job in a local garment factory that paid better wages. She married in 1955 (at age 21) and her first child was born in 1956, though the new family only moved into their own home in 1957. Lillian worked intermittently in these years, as layoffs and strikes cut into her husband's earnings in a series of factory jobs. Moreover, local garment factories offered part-time work, thus allowing Lillian and other women of her generation to combine parenting and domestic responsibilities with wage earning.[38] Lillian went back to work when her first child was only six weeks old, but she worked five hours a day, a schedule that permitted her to balance work and home.[39] She and her husband handled their child-care needs by coordinating their respective shifts.

Work and pregnancies punctuated this period in Lillian's life, as cutbacks and closings of local companies meant that her husband never had any job security. The garment firms at which she worked typically paid piece wages to their women stitchers, and as workers became more productive, management set higher production quotas and lowered piece rates. After more than thirty years of employment, and with evident skill at her work, Lillian still earned only six to eight dollars an hour at the age of sixty.[40] She expected to be eligible for a union pension of $114 a month when she retired, but her earnings were so low and her retirement prospects so modest, that she had no plans to retire. "I told the boss the other day," she remarked, "[I'll retire] when I die. As far as I can see, I have no plans for the future or retiring. As long as I am able, I think I will work."[41] Life was clearly a struggle for those Panther Valley high school graduates who remained in the area in the face of economic decline.

James Ferrari's life followed a different path after graduating from Summit Hill High School in 1960. His father was a stone cutter at a local mortuary, but he was unemployed often when James was young. His mother was a garment worker whose income proved crucial to the family during the 1950s and 1960s. Theirs was a small family with only two children, James and a sister. One factor that must have influenced James's subsequent decisions was his father's insistence that he not go into the mines. The date of his graduation, 1960, further reduced that possibility. James left the Panther Valley after high school, and after a four-year stint in the Air Force, he began working at Mack Truck, moved to an apartment in Allentown, and married at age twenty-two a woman from his hometown of Summit Hill. Still, he was not satisfied with his prospects and looked to college to provide new opportunities. Drawing on the G.I. Bill, and with support from his wife working nights, he attended Penn State University, earning a B.S. degree in 1969 and an M.B.A. a year later. Successive jobs took him and his family to Indiana, New York, and New Jersey, before he returned to Pennsylvania, settling in Lebanon, near Harrisburg, in 1980. After earning his M.B.A., he had a variety of managerial positions with manufacturers and in 1994 took early retirement (at age fifty-two) from a heavy equipment manufacturing firm. Later he went back to work, and currently he is teaching business and marketing at Harrisburg Community College.[42]

James Ferrari had a measure of financial security that Lillian Verona lacked, perhaps in part because his wife, also a Summit Hill graduate, earned an R.N. degree and worked as a registered nurse. Ultimately she moved into hospital administration, managing a coronary care unit in a local hospital. James had no periods of unemployment over his years in the workforce, in stark contrast to the experience of Lillian Verona and her husband. Persisters and migrants, brought up in similar economic and social settings, faced very different economic environments in the years after their graduations.

Dividing respondents into persisters, migrants, and returnees helps in understanding the very disparate lives led by the Panther Valley high school graduates in the immediate post–World War II era. Still, it is difficult to place individuals or families uniquely in these categories. And individuals from very similar backgrounds sometimes acted quite differently. The lives of Grant and Irene Gangaware serve to complicate the collective portrait.

Grant and Irene Gangaware never moved from the Panther Valley after their marriage in 1952.[43] Grant worked in the forestry department of Lehigh Navigation Coal before the mines closed, and Irene worked in area garment factories for thirty-seven years to help support the family. After the mines closed, Grant tried to piece together income from a variety of jobs, but only when he went on the road, supervising the construction of floating roofs on oil storage tanks across the country, did he begin to earn a stable income. He lived away from home much of the year, and his willingness to do so made it possible for Irene and their two daughters to have a measure of security in Lansford.[44] A second factor helped the struggling family. Irene's parents moved to Newark, New Jersey, where her father worked as the superintendent of an apartment building, and they invited Grant and Irene to live in their Lansford house rent-free. Grant and Irene kept the house in good shape, paying for needed repairs, and they turned the house back over to her parents when they retired and returned to Lansford.[45] Her parents were returnees, but their decision to work away from the Panther Valley permitted Irene to be a persister. And her husband's willingness to work all over the country (should he be viewed as a migrant?) gave his family the economic security it would have lacked had he relied solely on work in the area. Irene Gangaware appears in this account as one who never left the Panther Valley, but her story is far more complex and interesting than that label suggests.

Although their economic fortunes varied markedly, Lillian Verona, James Ferrari, and Irene Gangaware had similar goals for their children. All three emphasized the importance of education. And their children absorbed the lesson. Three of Lillian's four children graduated from college, both of James's children did, and both of Irene's did as well. Lillian Verona expressed particular pride in the achievement of one of her daughters, who paid her own way through college. Even though Lillian could not contribute much financially to her daughter's education, she gave her ample encouragement.[46] Similarly, Irene Gangaware was determined that her daughters would have the benefit of education she had not received. Her family's poverty in 1946, when she graduated from St. Ann's High School, meant she could not afford the cost of nurse's training. What pleasure it gave her twenty-five years later to help her two daughters earn their R.N. degrees![47] Finally, all of the adult children in these three families have well-paying occupations and the security that eluded so many in their parents' generation. Persisters and migrants among the sons and daughters had quite different lives, but their expectations for their children were similar, and their children's fortunes do not seem to have been seriously limited by the economic crisis in the anthracite region in the postwar period.

The issue of the expectations that sons and daughters had for their children brings into question how their children's lives compared to their own. The focus thus far has been on the first parental generation—the "fathers and mothers"—and their "sons and daughters." The high school survey, however, illuminates a third generation—the children of these sons and daughters. What sort of lives did they lead after the mines closed?

The residence patterns of this third generation reveal that the choices of the Panther Valley graduates had a lasting influence on their children. Movement away from the Panther Valley continued in the next generation. While 32 percent of the high school graduates resided in the Panther Valley at the time that they completed the survey, only 17 percent of their children continued to reside there. Of the children of graduates who migrated, only 1 percent—10 of 962—resided in the Panther Valley. Children of returnees—high school graduates who had left but subsequently returned—were the most likely to remain in the area. Almost 41 percent did so. And among the children of those who never left, just under 35 percent made the same residential choice. These two final figures speak to the lasting appeal of the region even after the

almost complete disappearance of the anthracite industry. More than a third of the children whose parents continued to live in the Panther Valley in the mid-1990s chose to stay there as well. The rate of persistence of this third generation was actually slightly higher than for their parents' generation.

The appeal of the coal region, even when its economy could not provide the security or income available elsewhere in the United States, is apparent among the sons and daughters and their children—those who continued to call the region home even in the face of economic crisis. Thomas Shober, a 1956 graduate of Lansford High School, and his children offer an example of a family that found the draw of the Panther Valley stronger than economic possibilities elsewhere. Shober returned to Lansford after a four-year stint in the Air Force and married and settled in neighboring Nesquehoning, his wife's hometown. He turned down a promotion at one job when it would have required that he move to the Allentown area. He had been active in civic affairs, having served as a member of the Panther Valley School Board. His five children all graduated from the area high school whose program he helped to shape, and three went on to graduate from college. All five continued to live in the Panther Valley as of 1994, examples of children of persisters who made the Panther Valley their home even in the face of forty years of economic decline since the mines of the Lehigh Navigation Coal Company closed for good.[48]

While children in the third generation made residential choices that were influenced by those of their parents, in some respects the differences among members of the third generation were less dramatic than for their parents' generation. Examining the educational and occupational patterns for the third generation reveals a geographical convergence over time. The isolation that had earlier characterized the Panther Valley (and the anthracite region more generally) weakened markedly over time, and in educational and occupational terms, the region came to resemble more strongly the mainstream of U.S. society. Higher education and place of residence were strongly associated for the generation of the sons and daughters. A glaring educational gap existed between those who never left the anthracite region and those who migrated elsewhere. Only 18 percent of persisters earned college or nursing degrees compared to more than 44 percent of those who ever migrated from the Panther Valley. But the educational difference between the children of persisters and migrants was much narrower. Among the children of persisters, more than 47 percent earned college or R.N. degrees; among the children of migrants the proportion was higher, at 60 percent. Children's education, in other words, was a strong value for the postwar graduates regardless of where they lived after the mines closed. Their children, both those remaining in the Panther Valley and those living elsewhere, had increasing access to higher education and strong parental encouragement, and they took advantage of the opportunities.

Similarly, occupational differences between the children of Panther Valley persisters and the children of migrants were considerably muted in comparison to the differences between persisters and migrants in their parents' generation. Respondents to the survey questionnaires recorded the occupations of 1,260 of their children. Among the children of persisters, almost 28 percent reported occupations as professionals, proprietors, or managers; among children of migrants, the corresponding proportion was 35 percent. Conversely, the share of blue-collar workers was higher among the children of those who remained all their lives in the Panther Valley. In the older parental generation, the differences between the occupational attainments of persisters and migrants had been much more striking. Among the parents—those who graduated from Panther Valley high schools between 1946 and 1960—migrants enjoyed on average an 18 percent advantage in higher-status occupations over classmates who never left the Panther Valley. In other words, the occupational difference between migrants and persisters was more than twice as great for the generation of sons and daughters who came of age in the 1940s and 1950s as for their children. Just as the educational gap between persisters and migrants narrowed for their children, so, too, occupational differences were more muted for the next generation.

The declining geographical and generational differences reflect the ways that sons and daughters adapted to the changing world around them as the mines closed and as new economic opportunities emerged beyond the anthracite region's borders in the postwar decades. Changes in the lives of this generation of sons and daughters stand in even sharper relief when compared to the experiences and attitudes of their mothers and fathers. Ken and Ruth Strohl Ansbach spoke directly to these issues in discussing their lives and those of their parents between the 1950s and the 1980s.[49]

Ruth Strohl was born in 1948 and grew up in Nesquehoning, graduating high school in 1966. Ken Ansbach grew up in nearby Coaldale.[50] Their fathers both worked for Lehigh Navigation Coal, and Ruth recalled the crisis when the Panther Valley mines closed for the first time, when she was in second grade. Education and jobs were points of conflict for her and her father in her recollections of her high school experience and the years immediately after graduation. In her recollections she seemed surprised at how little thought she gave to her future as her high school years drew to a close. Speaking of her parents, she recalled, "I never remember them mentioning college or anything to us. I guess it was because they didn't know anything about it either. [My dad] quit school when he was how young, to work. But they never pushed us."[51]

Ruth took part-time work in a local garment factory just before her graduation and then began working full time that summer. While working in the factory, she learned about a vocational training program in a nearby town and with some friends looked into the possibility. She began to take courses in data processing, and as she put it, "my father took a fit. He did not like this at all, that I was going to school." After a year in the program, she found a job at a data-processing company fifteen miles from her home, and once again, her father objected. "When I told him where I got a job he near died. He thought that was terrible that I was going all the way to Hazleton."[52] Her father's world was confined basically to the Panther Valley, and he found threatening the idea of going out beyond that world for training or a job. His first job after the mines closed was building homes in the Allentown area, about forty-five minutes from the Panther Valley, but as soon as he was able, he found a job in neighboring Lansford to avoid the daily commute.[53] He thought that the boundaries of the Panther Valley were broad enough for himself and that they should have suited his daughter as well.

Ruth's husband Ken worked as a laborer at Bethlehem Steel and then at the New Jersey Zinc Company in nearby Palmerton, fitting in a three-year stint in the service a year after graduating from high school. While higher education was not part of his experience, he pointed to some differences between his generation and that of his parents. He was able to go "a lot further than . . . [his] father," in his view, first because he'd "always held a job," while his father had to cope with "everything always shutting down." He also viewed the closing of the mines as "a blessing," because it meant that he didn't have to "go underground." He recalled, with a sense of realism, that "if the mines would have stayed up, I would have been underground, I would have been working in the mines." And given the health hazards in the mines and the lack of future prospects, such a job would have been, in his view today, "a nail in a coffin."[54] The closing of the mines forced Ken Ansbach, and

others of his generation, to look farther afield and seek work at Bethlehem Steel, Mack Truck, and New Jersey Zinc, which all provided better pay and job security than did the mines in their last fitful years.[55]

The closing of the mines, a traumatic event in their parents' lives, thus opened up opportunities for many of the sons and daughters coming of age in the anthracite region in the postwar decades. The mine closings and the accompanying crisis in the local economy forced this generation to look in new directions, and in this process to give serious thought to higher education and to seek out jobs with more of a future than the mining and garment work that had dominated the economic landscape for their parents' generation. And although more than 80 percent of these sons and daughters migrated out of the anthracite region, strong continuities are evident across generations in terms of core values and beliefs. From their parents they absorbed an appreciation of higher education, a commitment to family, and an enduring connection to hometown roots. Even as the proportion that remained in the Panther Valley declined from one generation to the next, there remained the abiding influence of the generation of the fathers and mothers as their children's and grandchildren's lives took shape under changing circumstances.

CHAPTER 8

Legacies

THE RURAL CHARACTER OF THE
Pennsylvania anthracite region—the small
towns, sense of community, and rugged
terrain—kept many of its residents in place and
lured back others who had left. In coming to
grips with what they faced with the region's economic collapse, members of various communities deeply invested themselves in promoting the
histories of their hometowns. The public narratives they are fashioning emphasize the heroic
contributions of anthracite miners to the nation's
development and the vibrant communities that
the miners and their families established and
sustained. In these accounts, there is little anger
expressed toward the coal operators, the United
Mine Workers of America (UMWA), and governments for inadequate responses to economic
decline. Also absent from recent community
chronicling is the visible destruction of the natural surroundings.

The countryside that so beckons is grievously
scarred. The landscape and the built environment of the anthracite region have been casual-
ties of two centuries of mining's growth and decline. The land and the towns have repeatedly
struck outsiders with their grim bleakness. A vis-
itor to the area in 1922 offered the following description:

the anthracite industry stamps . . . communities
with characteristic features. Great culm piles; towering breakers, frequently with black dust rising

from them like smoke; streams discolored with silt;
roads, house lots, and fields broken by subsidence
and caving, combine to give a drab, bleak, forbidding setting for home life.[1]

A study of the borough of Shenandoah in
Schuylkill County, published in the same year,
reinforced much the same impression:

The streams are black with soot and there are black
piles of refuse and culm, and the men returning
from work wear masks of coal dust. Trees have
been cut down for mine timber so that only stumps
and scrubby bush, saplings, or misshaped trees are
left. The earth mixed with the slate and coal dust is
for the most part bare, and the few gardens, which
demonstrate that the ground can still be cultivated,
emphasize the general desolation. . . . The air is
usually filled with the sulphurous dust which blows
from the breakers and the coal cars. The noise of
the coal as it rushes down the breakers and of the
chugging of the mine fans and other machinery is
almost incessant.[2]

The early 1920s were good years for coal production, and dust and noise were fixtures of anthracite mining communities. A geographer
subsequently summarized the physical environment of mining communities at the industry's
apogee: "Most coal towns were drab rows of
drafty barn-red double houses . . . built along
grids of muddy roads that extended up steep

House fallen into Mine Cave at Mayfield, 1908. Courtesy of Charles Kumpas, Clarks Summit, Pa.

hills or through narrow gaps to avoid tying up the company's valuable coal lands."[3] This was the impression mining communities offered in prosperous times.

The collieries and mining communities changed, of course, over the succeeding decades of economic decline, but observers' comments suggest that there was little improvement in their appearance even with the drastic reduction of mining. Consider a view of one breaker and associated buildings in the middle of the Depression of the 1930s:

On the right is the Shenandoah City Colliery, also closed down. It was painted red once, but the coal has blackened it and now, with a thousand window lights smashed, the dark mountains of culm behind it, it is a wild ruin. There isn't a light or a puff of smoke. The great stacks are rusted, the engine room is shut up. On a spur track that runs up to the breaker are four empty Reading coal cars, but the tracks are rusted too.[4]

During the Depression decade, idled mines comprised a third to a half of the anthracite region's total capacity, but in the post–World War

II decades, mine closings affected a still greater share of operations. At first, mining companies abandoned their underground operations and shifted to strip mining. Eventually, the strip mines closed down, leaving gaping wounds in the landscape that were at once eyesores and also dangers for residents of surrounding communities.

Federal assistance, funneled into the anthracite region in recent decades under the Appalachian Regional Development Act, and state aid from the Department of Environmental Resources permitted the reclamation of a number of abandoned strip mines. It is an expensive proposition, however, to fill the massive pits that remain after decades of mining, and state and federal funds for this purpose have been decidedly modest. The Kelayres strip-mine project in Schuylkill County, for instance, involved filling a 261-acre pit that averaged some 300 feet in depth. The state undertook the project that was expected to take almost three years at a cost of $8.9 million. Given the magnitude of individual projects, there is no possibility that the state or federal government can restore more than a tiny fraction of the lands that had been despoiled by

Shenandoah: view of spires and houses as seen over coalyards on outskirts of town, 1938. Photograph by Sheldon Dick. Library of Congress, LC-USF33–02188.

surface mining in the anthracite region. The Abandoned Mine Land Program currently maintains a database that lists almost 5,000 problem sites in Pennsylvania viewed as high priority for remedial action. Of these sites only 326 have been treated, with another 78 funded for treatment. In all, only 8 percent of high-priority abandoned-mine problems have been addressed or funded.[5]

The existence of dangerous, abandoned mining sites, the lack of well-funded government reclamation programs, and continuing poverty in the region have led localities to seek novel solutions to their environmental and economic problems. Unable to attract high-paying jobs to the region and suffering from declining tax bases, anthracite communities have taken to accepting out-of-town and out-of-state waste, and they have used the trash to fill up abandoned underground mines and strip mines. A 1990 billboard advertised the business that had become the region's latest problem: "Waste Management of Scranton—Helping the World Dispose of Its Problems." And so the Empire dump in the Scranton suburb of Taylor accepted nine thousand tons of waste each day, "equivalent to . . . 22 percent of the waste produced in the state." Nearby Throop has its own thirty-million-

dollar landfill, and is now as dependent on trash as it was previously on anthracite coal. With taxes declining, income from the landfill provides 90 percent of the town's revenue. The new dumps opening in the Scranton area are inviting more health and environmental troubles in a county that already has fifty sites "listed by the EPA as waste sites of potential concern."[6]

Scranton is hardly an isolated case. Philadelphia Electric sends nineteen trucks a day, some 6,800 truckloads a year, to two landfill sites in former strip mines in Porter Township in Schuylkill County, carrying coal ash from two coal-powered generating plants. In 1995 the Girard Estate, owner of some 29,000 acres of coal lands in the southern anthracite field, negotiated a contract to establish a 145-acre landfill on its former coal holdings. The plan would bring in one hundred truckloads of demolition waste daily, more than six hundred thousand tons annually, with the sweetener of tonnage royalties to the Estate and to the two townships where the landfill would be located. Residents secured some 3,500 signatures on a petition opposing the landfill; activists recalled an earlier four-year struggle that had been required to close down a toxic waste pit on Girard Estate land that had forced the evacuation of local residents living near the facility. As one borough council member commented about the absentee officials of the Girard Estate, "They could care less about the towns. All they've ever cared about was the bottom line."[7]

Even in the face of these environmental problems, the lure of filling abandoned coal mines with others' waste remains strong. In the Panther Valley, home for 150 years to a dozen collieries and stripping operations, the Lehigh Coal & Navigation Company (LC&N), owner of the valley's coal lands, has a permit to dump coal ash to fill the Springdale strip mine at the western end of the valley.[8] The company has until the year 2008 to fill this giant pit that stretches across portions of Tamaqua, Coaldale, and Summit Hill. In early 2004 the company applied to the Pennsylvania Department of Environmental Protection (DEP) to continue the filling with a mixture of coal ash and muck to be dredged from the bottom of the Delaware River. In March 2004 the DEP approved the petition, citing stringent testing requirements for the river dredge before and after shipment to the landfill site. Further approvals will be needed before the new fill can be accepted; in the meantime, local representatives have proposed state legislation that would permit voters to decide the issue. LC&N reached an agreement with the Tamaqua Borough Council that would pay Tamaqua $800,000 annually should the company receive approvals to carry out the landfill operation; in subsequent elections several members of the council who had approved the agreement were voted out of office, and the local newspaper strongly opposed the agreement. While this latest landfill proposal has sparked notable controversy and the outcome is unclear, what is certain is that impoverished anthracite communities will continue to field propositions to accept other regions' waste, and it will be difficult to turn them down.[9]

In addition to the proliferation of landfill operations across the anthracite region in the past twenty years, there has been another "growth industry" in the region—prisons. The Schuylkill Economic Development Corporation (SEDCO) in Pottsville has been operating in the southern anthracite field for more than fifty years, recruiting prospective businesses, developing industrial parks, erecting shell buildings, and in general attempting to create jobs to replace those lost with the decline of the anthracite and garment industries since World War II. An article in the organization's summer 2002 newsletter, however, touted the importance of prisons to the local economy. Pointing to the federal prison in Minersville and two state facilities in nearby Mahanoy City and Frackville, SEDCO reported that the three prisons employed some 1,248 workers with an annual payroll approaching $63 million. Beyond the contribution of these payrolls to the local economy, the story noted that the prisons made purchases of $5.5 million a year from local businesses, supporting additional jobs. Finally, the story emphasized the prisons' contribution to the hospitality industry in the county, as families and friends visited inmates in the prisons, staying at area motels, and eating in area restaurants. The Web site of the Frackville Econo-Lodge lists fifteen area attractions, including "Frackville Prison . . . 1 mile," "Mahanoy Prison . . . 3 miles," and "Minerville [sic] Prison . . . 10 miles."[10]

Local communities compete for prisons in much the same way that they had sought industrial employers as mining declined. In a bidding competition, eleven counties in Pennsylvania submitted proposals to build prisons and lease them to the state; four sites were selected, and two of them—Frackville in Schuylkill County and Shamokin in Northumberland County—are in the anthracite region. "There is a pocket here that is prison-intensive," noted one local official. "It's primarily because aggressive public officials have been pursuing the opportunities they present."[11] These efforts underscore the limited success of industrial development campaigns in the southern anthracite field. Had more businesses found their way into the region, waste landfill sites and prisons would not seem so attractive. These developments in the 1990s provide a telling coda to a half century of industrial development campaigns.

As a result of the devastation caused by mining and its abandonment, the limited success of industrial redevelopment, and continuing environmental abuse in the post-anthracite years, the land and the communities of the region have never had a chance to recover. As one scholar, geographer Ben Marsh, commented in the late 1980s:

The land remains, and the people remain with it. Dozens of neat grey towns sit separated by piles of broken black rock—abandoned mine dumps and strip mines. The unstable mounds and brushy fields of this lunar landscape seem useless except to be mined again for coal still deeper in the ground. The towns are damaged, too, with downtowns too big and employment opportunities too limited for the current population.

Zeroing in on one community, Marsh described the urban landscape that remains today:

Shamokin looks to be a city, not a town, although its population is just 10,000. The housing is uniform row houses and duplexes . . . repeated block after block. Many houses are for sale or boarded up, but those that are occupied are in good repair, with neat yards and fresh paint or siding. Shamokin is a poor town, yet the business district is healthy and attractive. Independence Avenue, the main street, supports five banks, two department stores, an ornate movie theater, furniture stores, travel agencies, restaurants and bars, and professional offices. The bus stops briefly at a newsstand, then passes five fraternal organizations on its way out of town.[12]

Not all post-mining communities have fared as well. The main shopping district of Lansford, former home of Lehigh Navigation Coal, stretches for several blocks along Ridge Street. Once the lifeblood of a thriving community, Ridge Street's most prominent businesses in the 1920s included three banks and the J. C. Bright department store, where miners' families could purchase supplies on the book and have the expenses taken out of their biweekly pay. One long-time resident, Anna Meyers, contrasted Lansford in the mid-1990s with its former prosperity:

Oh my goodness, there used to be a lot of stores in Lansford. You'd go down Ridge Street, every store was taken. Now there's a lot of empty stores. . . . There used to be a shoe store . . . there are none there now. Sneakers you can buy, but . . . I don't think there is a shoe store in Lansford anymore.

There used to be two nice restaurants. There used to be a doggie place, they're all gone. The Porvazniks used to have a big flower shop; now they closed. They sold out to another couple now, but they're carrying Porvaznik's name.[13]

Robert W. Reichard offered a similar reminiscence about the vitality of Lansford at its peak. In addition to a bustling Ridge Street, he recalled that in "many of the neighborhoods there were grocery stores, hardware stores, butcher shops." And downtown "the stores stood shoulder to shoulder for several blocks." For entertainment, he noted "over 25 neighborhood bars, two movie houses, several pool rooms, veterans and fraternal clubs, and several undertaker parlors." With the mines operating, businesses prospered.[14]

Today, the scene is remarkably different. Walking east along the south side of West Ridge Street, in the heart of Lansford's shopping district, from Center to Coal to Tunnel Street, one passes an all-too-modest array of small storefronts, housing in turn:

Henry's Men's and Boys Shop
an empty storefront
Delrose Awards
Coaldale-Lansford-Summit Hill Sewer Authority
Lansford Appliance
Koch's Television Appliances
an empty storefront
Frank Gustus Barber Shop
HR/OSHA Visionaries
Bruch Eye Care Associates
an empty storefront
All Staffing, Inc. [a temporary employment agency]
Jet Company Data Service
Germaine's Gifts
two empty storefronts
a Pennsylvania state liquor store
Porvaznik's Flowers
Lansford Art and Ethnic Gallery
Frendak Signs
the former Ridge Street Grille—closed—for sale
an empty storefront
the former J. C. Bright Company building, now housing:

Valley Athletic Supply
The Lamp Outlet
Harleman's Family Restaurant
The Apple Tree Card and Gift Shoppe
The Medicine Shoppe
Suns Manufacturing, Inc.
The Monogram Shoppe
U.S. Air Force Recruiter
Crafts and Flea Market
Louis Daniels—Jewelry, Diamonds
Blue Ridge Cable TV
Lehighton Times-News
Fleet Bank
Marco's Pizza
Knittle Hardware—closed and empty
an empty storefront[15]

Echoing Anna Meyers's observations, nine of the thirty-eight stores along this stretch of Lansford's prime business district are empty; many of the others are small businesses open only part-time, permitting semi-retired owners to eke out a small supplement to social security income. Local businesses, which once supplied the needs of some 9,600 residents, now compete with more distant suburban malls for the much-reduced business of the 4,200 who remain in Lansford today. It is symptomatic of the marginality of Lansford in the twenty-first century U.S. economy that only three of the businesses located on this street—a pharmacy, a bank, and a newspaper—are parts of regional or national operations. The Gap is *not* in Lansford.

The current use of the J. C. Bright Company store speaks to Lansford's declension over seventy years. In its prime, the department store offered four stories of wares, from groceries in the basement, clothing, shoes, and household goods, to furniture on the third floor. Only the first floor is used today for commercial activity, with a small manufacturing operation on part of the second. The seven businesses and the Air Force recruiter that make up the mini-mall on the first floor may employ fifteen people, at most. The top two floors are mostly empty and used for storage. What a contrast to the past. A 1932 photograph of Bright's employees shows thirty-nine in all, arrayed in three neat rows in front of the store. Ten male managers sit importantly in suits and ties in the first row; twenty-three female clerical and sales workers stand in two rows behind their bosses, and six shirt-sleeved and overalled men round out the scene. Bright's dominated the downtown in its heyday, provided new white-collar employment, and did a thriving business, most of it "on the book." While not a company store, Bright's had a special arrangement with LC&N, and the company deducted store purchases before issuing its biweekly checks to mineworkers. The mini-mall today pales in comparison.[16]

Lansford, despite its unique history, is very

Empty storefront, formerly
Knittle Hardware, Ridge St.,
Lansford, 2004. Photograph
by Christine Harvan.

Empty storefront, formerly
Valley Footwear, Ridge St.,
Lansford, 2004. Photograph
by Ed Gildea.

much like other decaying towns in the anthracite region. Marsh describes Mount Carmel,
a somewhat larger former mining community
about forty miles west of Lansford, as a "fossil"
compared to "when it served three times as
many people." Like Lansford, Mount Carmel
has many more stores than the local population
can reasonably support. The oversized downtown persists, though, as "Many stores stay in
business because the elderly merchants own the
buildings for which there is no other use, and
they subsidize the stores with their labor because they, too, have little else to do." Family,
ethnic, and community networks keep these
stores and storekeepers in town, not the business or economic prospects.[17]

The decline of former mining communities is
evident throughout the region, but particularly
in its largest city, Scranton. The city's population
peaked in 1930 at more than 143,000 and shrank
steadily thereafter, as the Depression and then
the postwar mine closings took their toll. By
2000, only 76,000 residents remained in the city,
a 47-percent loss in the seventy-year period.[18]
With the closing of collieries and out-migration
of area residents, Scranton's tax base eroded
steadily, but it proved difficult to restrain city
spending within the bounds of limited revenues.
In January 1992 city officials applied "for designation as a distressed municipality" under legislation that would then make the city eligible for
considerable state assistance. The state Depart-

ment of Community Affairs (DCA) accepted the
petition, which placed the city under quasi-
receivership but also made it eligible for low-
interest loans to help meet the budget gap. The
DCA appointed the Pennsylvania Economy
League as the Recovery Plan Administrator,
charged to coordinate with the city government
that held ultimate authority and responsibility to
bring revenues and expenditures into balance.

The adoption of a Recovery Plan enabled
Scranton to secure a $2 million loan from the
state and authority to float $3.8 million of tax-
free municipal bonds. Still, three years of effort
did not materially correct the long-term imbal-
ance, and in January 1995 the state confirmed
the city's persistent "distress" and supervision of
its finances continued. The city's financial situation worsened when it signed successive
collective-bargaining agreements with municipal employee unions that violated the key terms
of the Recovery Plan. In October 1998 the state
sanctioned the city for its failure to implement a
fiscally responsible plan, cutting off considerable
state aid until the city once again came into compliance. City and state authorities remained at
loggerheads for another two years, but finally in
October 2000, the state lifted its sanctions and
released the funds it had held in escrow. Higher
taxes, greater control of overtime expenses,
shrinkage in the city payroll, and a variety of
one-time revenue measures and bookkeeping
changes had permitted the city to substantially

Bright's workforce, 1932, in front of store building. *Valley Gazette*, no. 251 (May 1993): 13.

cut into the shortfalls that had plagued its operations. Still, the city's distressed status remained and with it the continuing state supervision of its finances.[19]

While officials in Scranton struggled to maintain essential services in the face of a declining tax base, residents in Centralia in the southern anthracite field faced a more elemental challenge. In Centralia in May 1962 a routine burning of debris and garbage in the town landfill in an abandoned strip mine ignited an underground coal seam. Despite forty years of local, state, and federal efforts and the expenditure of millions of dollars, the fire continues to burn and spread. After nineteen years the federal Office of Surface Mines admitted that the fire might not be contained and extinguished, and in response to considerable local pressure the federal government began a buyout program to purchase local homes and provide residents funds to relocate. Most residents accepted the program, and the federal agency purchased their homes and bulldozed them; holdouts continued to live in town, however, and their row houses remained standing alone, their newly exposed sides reinforced with external brick buttresses. After another ten years, in 1992, state authorities ruled that the remaining town residents would have to relocate and that the state would exercise its right of eminent domain with the sixty or so residents who had thus far refused the buyout offers. In 1995 courts upheld the state's right to evict the remaining residents, vir-

tually all of them elderly retirees. Steadily the holdouts died or moved out, and by May 2004 the town, whose population had stood at 1,100 at the outbreak of the mine fire, was down to 12 residents.[20]

The mine fire has devastated the Centralia community, both physically and psychologically. As the fire spread through the underground mine workings, it emitted poisonous gases that made some homes uninhabitable and required constant, intrusive monitoring of remaining homes. Moreover, the burning of coal seams removed structural surface support and seriously undermined the ground in the vicinity of many homes and businesses. No one has been killed, but in February 1981 a twelve-year-old boy, curiously inspecting steam coming from a small hole in the ground, nearly slipped into a deep opening created by sudden mine subsidence that accompanied the fire and was saved only by hanging onto tree roots and screaming until an older cousin playing nearby rescued him.[21]

Differing responses to the spreading fire tore the Centralia community apart. For a long time many residents denied the seriousness of the fire, while a vocal minority tried to arouse their neighbors and demand effective action from local, state, and federal authorities. Once Congress appropriated funds to finance home buyouts, Centralians divided over whether to leave or stay in the face of mounting evidence that the fire could not be controlled. In a pattern that mirrored attitudes toward out-migration

Mine fire exhaust, Centralia, 2003. Photograph by Ed Dougert.

Route 61, south of Centralia, 2003, showing effects of Centralia mine fire. Photograph by Ed Dougert.

throughout the anthracite region, younger residents (especially those raising children) responded much more strongly to the crisis posed by the fire than did their older neighbors.[22] Younger residents sold out and left town, acting out of concern for their children, while older residents, many near or in retirement, felt the strong ethnic and community ties that made it so hard for other anthracite-region residents to consider moving in the face of economic decline. Thus an environmental crisis that might have brought the community together demanding action from state and federal agencies became a divisive force that turned neighbors into antagonists and even broke up marriages. A series of neighborhood meetings in the summer of 1983 energized opposing groups and led a state agency to sponsor a referendum in August that resulted in a vote of 345 to 200 in favor of relocation. Following this expression of community sentiment, federal legislation for a home buyout moved swiftly through Congress, sealing the fate of Centralia.[23]

Each of these anthracite-region communities offers a unique story, and yet the cumulative effect of the individual accounts offers a larger picture. The strangulation of Lansford's downtown business district, Scranton's designation as a "distressed" city, the abandonment of Centralia in the face of a smoldering mine fire, and the region's growing reliance on waste disposal and prisons are all evidence of the failures of the region's organized interests to respond effectively to the loss of its central industry—coal mining.

◆

Anthracite executives, the UMWA, and local, state, and federal governments all share responsibility for this collective failure. The cases of the Lehigh Coal & Navigation, Glen Alden, and Reading companies provide representative instances of corporate failure. In anthracite's prosperous years, the great majority of earnings from mining found their way into the coffers of the anthracite railroads that increasingly controlled the ownership and production of the region's coal mines. In the post–World War II years, as anthracite's decline became chronic, anthracite companies diversified. A process of decline that had internal roots accelerated with the entry of a new breed of outside investors with no commitment to either coal mining or the anthracite region. Eventually the holding companies that emerged in the 1950s and 1960s moved out of the mining of anthracite and divested themselves of their remaining coal lands. In the good years through 1920, New York and Philadelphia investors had reaped the lion's share of anthracite's bounty—but they had had a vested interest in the region's viability. In the final difficult times, by contrast, a new generation of outside investors sold the assets of the area's companies and employed tax shelters to launch their far-reaching conglomerates.

The dynamic for the UMWA was not the same as for anthracite owners and managers, but the union's abandonment of anthracite miners was as complete. Given the union's history and its purpose, its actions were particularly egregious. Capitalist firms are ultimately responsible to their owners and shareholders, so managers' commitment to profitability or share prices is not surprising. The UMWA, in contrast, existed for the sole purposes (in theory at least) of representing coal miners and defending their interests. The organizing of the anthracite region, after all, had been a key element in the emergence of the union as a labor powerhouse in the first decades of the twentieth century. Dues from the anthracite region had underwritten additional successful organizing in the nation's bituminous coalfields. But after World War I, as bituminous mining dwarfed anthracite output, it became increasingly clear that the UMWA was first and foremost a bituminous union. Separate contracts for bituminous and anthracite mineworkers and later separate health and welfare funds meant that workers in each branch of the industry would go their separate ways as coal mining accommodated to the changing energy environment in the United States.

The anthracite industry met a slow, drawn-out death, during which time the UMWA was occupied foremost with defending its own institutional interests. To John L. Lewis and successive UMWA presidents those interests consisted primarily in becoming a major economic player in the nation's soft-coal business. The creation of separate health and welfare funds for anthracite and bituminous miners and the milking of those funds for the economic benefit of the union as an institution, the consolidation of a centralized business unionism, and the destruction of union democracy were the central elements of the union's history in the immediate postwar years. By the time of the renewal of a democratic movement within the UMWA in the early 1970s, anthracite mining, for all intents and purposes, was dead.[24] The UMWA was largely "missing in action" during the difficult years of anthracite's final decline. Union leaders had periodically defended anthracite exports and opposed the St. Lawrence Seaway and the construction of natural gas pipelines, but they simply were not major players in the limited positive responses aimed at reconstructing the regional economy in the post-anthracite era. They made no serious effort to revive the industry, nor could they envision or contribute to a world after anthracite.

Governments, too, failed the region and its residents. Without responses from organized capital and labor to meet the deep needs of the area, with prevailing reliance on the mechanisms of the market economy, and divided, scattershot governmental initiatives, the formulation of coherent public policy never got off the ground. The federal government could have implemented a plan for the phasing out of anthracite mining, regional economic diversification, and meaningful assistance to the displaced, but federalism and the multitiered structure of U.S. government led the federal government to tread lightly in the redevelopment field. The Area Redevelopment Administration (ARA) did not have a clear mandate of its own, but worked through other agencies and never developed a coherent national agenda. Through the Appalachian Regional Commission, which built on the ARA program, the federal government focused on constructing the interstate highway system to strengthen transportation infrastructure, anticipating that such actions would integrate the Appalachian region generally and the anthracite region in particular into the larger national economy. In addition, the federal government provided funds for environmental initiatives and vocational education, though it left the setting of priorities to states and counties. None of these efforts began to address the economic crisis of the anthracite region.

The most significant programs that ultimately emerged were local and state efforts that fostered bidding wars among the region's town and cities in efforts to attract new business and to keep existing companies from leaving. Local chambers of commerce organized fundraising campaigns and sold bonds to provide subsidies to firms that promised new jobs in their communities. After a decade of exclusively local efforts, the state of Pennsylvania joined in, establishing the Pennsylvania Industrial Development Authority to distribute subsidized mortgage loans through area development agencies to firms moving into the state or expanding their operations. While these campaigns succeeded in attracting a variety of firms to replace the jobs lost in mining, the new employers typically paid wages close to the legal minimum. Oral history interviews suggest a good deal of resentment about the concessions made to obtain these new jobs and a sense that residents paid a heavy price and gained little in return. Finally, the continuing efforts today to develop waste landfill sites and attract prisons speak to the inadequacies of the industrial development campaigns.

Young people especially saw no promise in the reindustrialization effort and continued to migrate from the region. The experiences of the graduates of high schools in the Panther Valley between 1946 and 1960 reflected the first wave of departures of the region's young people, and the patterns evident then have continued in succeeding decades. As noted in a 1990 newspaper story, focusing on Hazleton, the region continued to lose many of its youth: "These bright, energetic students seem destined to follow in the footsteps of classes before them—out of high school, out of state and out of the economic package that Pennsylvania is trying to put together as it competes in the global arena for growth and prosperity." The article followed up this prognosis with supporting evidence, showing that Pennsylvania ranked forty-fourth among states in population growth (resulting in the decline in the size of its congressional delegation). Moreover, as the state's population stagnated, it aged dramatically, so that Pennsylvania ranked near the top among states in the proportion of residents over sixty-five. Certainly unemployment in the anthracite region no longer stood out as such a major problem, but much of the working-age population had left the region, and increasing numbers of residents had stopped looking for work.[25]

The failure of organized interests either to stem the decline of anthracite mining or to promote effective economic alternatives for the region is one conclusion that stands out in studying the decline of anthracite since World War I. The second theme is the varied face of economic decline. Its impact has differed by geography, gender, and generation. For example, the responses of organized miners during the Depression varied across the region's major fields, with equalization, dual unionism, and coal bootlegging offering distinct approaches to solving the problems facing mineworkers. How individual firms experienced decline also varied across anthracite's three fields. While the Glen Alden and LC&N companies mined, processed, and sold their own coal, the Philadelphia & Reading Coal & Iron Company (P&RC&I) in the western middle and southern fields operated through lessees, such as St. Clair Coal, adding complexity to the picture of decline. Finally, redevelopment efforts after World War II varied dramatically by community: well-heeled and -organized cities had great advantages in mobilizing resources to attract state subsidies and new business. Location shaped the character of economic decline in distinct ways.

Men and women variously experienced the final closing of anthracite mining in the post–World War II decades. Former miners commonly had to leave the region to find work—either as commuters or out-migrants—or accept declining wages in the less-skilled, insecure jobs that remained. In contrast, anthracite-region women of this same generation enjoyed increasing employment opportunities in the growing garment industry of the 1950s and 1960s. As women made increasing contributions to family income, relations between men and women and their respective responsibilities in their new households changed.

Just as gender shaped the impact of economic decline, so too did generation. Older workers already in the labor force were affected one way by anthracite's demise, while their children, just completing their educations and anticipating lifetimes of employment, had different experiences. While older residents tended to commute or find work in the region, fully 80 percent of the younger generation migrated away from their communities of origin. The younger generation took advantage of the greater possibilities for higher education in the postwar decades, and this education permitted them to achieve economic success far beyond that of their parents. Still, the return migration of a significant proportion of the generation of the sons and daughters and the persistence in the region of a distinct fraction of their children provide evidence of the continuing attraction of the anthracite region to succeeding generations even in the face of regional economic collapse.

A third theme that emerges is how economic development and decline are inextricably intertwined. Forces that drove the rise of anthracite also shaped its descent. Geology influenced patterns of investment, settlement, mining technologies, and labor relations during anthracite's nineteenth-century ascent as well as its subsequent demise in the twentieth century. The greatest reserves of anthracite, for example, remain in the area's southern coalfields, but the pitched, thin veins of coal there undercut the profitability of its mining and processing in today's energy markets. The region was always dependent on outside finance capital. In the early period outside investors poured enormous capital into the region; in the last half century, they dismantled anthracite's productive capacity. The state similarly chartered development in the early era and offered fractured assistance in downward times. The contributions of the UMWA are as ambiguous. On the one hand, the union succeeded in overcoming ethnic, occupational, and geographical divides to build a powerful industrial union. Still, the union's internal dynamics that permitted it to challenge the anthracite operators also drove repeated wedges between the union and its grassroots membership. And when the postwar crisis in anthracite mining emerged, the union had little interest in and was in no position to solve the problems of its constituency and the region. Economic development and decline are flip sides of the same coin—two centuries of the history of the Pennsylvania anthracite region.

The decline of the Pennsylvania anthracite region must be understood not only in broad historical perspective over the course of two centuries but also within an international framework. World coal production grew dramatically after World War II, increasing from 1.9 billion tons annually to more than 5.3 billion in 1990 (see table 8 in appendix 1). Yet, this aggregate growth occurred with precipitous declines of mining in historic coal-producing areas accompanied by vast reductions in employment, greater open-pit production, and the use of labor-saving technologies. In the United States, for example, annual coal output grew between 1920 and 2001 from 667 million to 1.1 billion tons as the number of mineworkers declined a full 90 percent, from 785,000 to 72,000 over the eighty-year period (see table 3 in appendix 1). Correspondingly, the locus of coal production shifted in the last decades of the twentieth century from western to nonwestern countries—to South Africa, China, India, and Australia. In 1990, China passed the United States as the world's leading coal producer, with Chinese miners laboring in dangerous and exploitative circumstances.[26]

Today coal is mined cheaply on new frontiers, including the American west, and oil, natural gas, and nuclear power have displaced coal as energy sources. In this changing context, coal mining has all but ceased in classic coal-mining regions of Western Europe. Newspaper headlines in April 2004 announced the closing of the last operating mine in France (in La Houve).[27] Coal has not been mined in Belgium and Portugal since the mid-1990s. Coal mining has declined in Germany and Spain and remarkably in Great Britain. In the United Kingdom, where coal was once the nation's leading industry, production declined from 220 million tons in 1950 to less than 33 million in 2001.[28] At the peak in production in 1957, more than 700,000 British coal miners were gainfully employed in 1,000 collieries; yet, currently, less than 10,000 men produce coal in eleven operating mines.[29]

Coal miners and their families in Western Europe have faced economic decline in significantly different ways than their counterparts in the Pennsylvania anthracite region, and the contrast is instructive. Great Britain offers a telling contrast. British mines closed while owned and administered by the national government; the closings generated intense national attention and concern; strong trade unions played buffer-ing roles; and the state provided expansive public assistance to miners and their communities.

The near half-century demise of the British coal industry occurred amid tumultuous events: the nationalization of the mines in 1947 with the National Coal Board (NCB) assuming administration of the mines, a state agency immediately saddled with huge debts due to the generous compensation paid to mine companies; the beginning of planned closings by the NCB in the 1960s; the rising militancy of the National Union of Mineworkers (NUM), the British counterpart of the UMWA; major national strikes in 1972 and 1974, the latter contributing to the toppling of a Conservative Party government; the coming to office of Prime Minister Margaret Thatcher in 1979, and her mounting determination to get the British government out of the business of mining coal and to defuse the NUM; a turbulent national strike in 1984–1985 aimed at curbing mine closings that splintered the NUM and left the union greatly weakened; privatization of electricity generation and the loss of British coal's captive market with subsequent conversions to gas-powered plants; protest and parliamentary debates in the early 1990s following disclosure of secret government plans to close collieries; and finally, the return of the surviving coal mines of the country to private hands in 1995.[30] Throughout the period, the fortunes of British coal gained headlines and public attention. Geography and demographics helped shape the national concern. Nine percent of the British population on the relatively small isle lives in close proximity to coal-producing areas.[31] Pennsylvania anthracite drew national concern in the first decades of the twentieth century, but from the 1930s on, the marginalized region remained largely out of sight and mind of the nation.

Yet, Socialist policies and forceful trade unionism in Britain mattered more than geography. While the mines were nationalized and the NUM exerted great pressure and influence, British coal miners and their communities remained the constant subjects of government-sponsored studies and recipients of substantial state assistance. Ongoing social science surveys found evidence of high unemployment in areas of mine closings, anger and frustration, family

tensions, migration of younger workers, resistance of older men to leaving their home communities, and environmental erosion—consequences similar to those in the Pennsylvania
anthracite region.[32] Social disruptions thus
marked the decline of British coal—a rosy portrait cannot be painted—but the welfare and
land reclamation programs implemented in
Great Britain to mitigate the impact of coal's decline are of a very different order of meaning
and magnitude than the much more modest
programs initiated in the United States.

When the National Coal Board began the
concerted closing of unproductive mines in the
1960s, the British government implemented
measures to employ and assist displaced miners.
Miners having to commute short distances received travel allowances. The government paid
entire relocation costs for those miners who
chose to move their families to distant mine
areas. The Redundancy Payments Bill, passed
by Parliament in 1965, further encouraged miners to leave the industry with generous lump-
sum awards. Additional legislation allowed miners
over the age of fifty-five to receive redundancy
payments for three years at 90 percent of their
last earnings and lowered the age at which they
would be eligible to draw full national pension
benefits. Redeployed and ex-miners also maintained health coverage under the British national health care system, which included programs serving their special medical needs.
Finally, the national government poured funds
into vocational retraining initiatives and job
placement services for former miners as well
into community redevelopment efforts.[33]

The measures enacted in the 1960s to cush-
ion the closing of the mines remain in place to
this day. In the 1980s, Prime Minister Thatcher,
to gain support for mine closings and her confrontations with the NUM, increased redundancy payments. Based on the number of their
years of employment in the trade, miners who
voluntarily agreed to leave mine work received
awards that rose to a maximum of £37,000 and
averaged £28,000. Between 1979 and 1993, during fourteen years of Conservative Party rule,
the national government paid an estimated £8
billion to miners affected by colliery closings.
Under legislation returning the mines to private
ownership, redundancy payments continue, but

the maximum has been reduced to £27,000
(with employment so reduced, there are now
few takers, resulting in a drastically reduced liability for private companies).[34]

The British government has attended to the
physical as well as the social environment of declining coal areas. Standards of care in land restoration are considered higher in Great Britain
than elsewhere in the world. In 1951, the Ministry of Agriculture began providing funds to restore depleted open pits to farmland; in subsequent closings of surface mines, new regulations
required complete rehabilitation for farming
within five years (at taxpayers' expense). In the
1960s and 1970s, after the NCB initiated extensive colliery closings, the government embarked
on ambitious reclamation projects that included
the building of a new planned town at a 520-acre
site that had two hundred mine shafts, and the
creation of a 1,000-acre country park.[35] Another
burst of coal field regeneration occurred in the
1990s and entailed the following: the formation
of a new government board, English Partner-
ships, to coordinate land restoration and economic redevelopment; the founding of several
nongovernmental organizations to raise private
funds for environmental cleanups; and subsidies
from the European Regional Development
Fund (and other agencies) of the European
Union. Spending approached £1 billion in the
last half of the decade. Work at fifty-six closed
mine sites created 20,000 new jobs.[36]

Conflict, pessimism, and hard times had their
place in the demise of British coal, but the responses to the decline of coal areas in Great
Britain contrast sharply with those in the Pennsylvania anthracite region. Different institutional frameworks made for different histories:
government ownership, national welfare, land
reclamation and redevelopment programs, and
strong trade union pressure in the United Kingdom; private financial interests, reliance on localism, and ineffectual unionism in the United
States.

With variations, responses to the closing of
mines in other Western European nations echo
the British case. The French government nationalized coal mines in 1948 and over the next
half century shaped the phasing out of coal production with heavily financed programs for job
retraining and redeployment of former miners.[37]

In Belgium, after violent strikes protesting the closing of state-owned mines in the late 1960s, the government reached an accord that permitted closings only after providing mineworkers alternative employment. In the 1980s, the Belgian government further earmarked substantial funds for redundancy payments, landscape revitalization, and redevelopment of coal lands for educational, cultural, and recreational purposes.[38] In Germany, the federal government consolidated coal companies in 1968 into a single, highly subsidized firm. The government responded to the need to close mines with vocational training and retirement programs that reduced the average age of the remaining mineworkers to thirty-two. The German government has also invested heavily in coalfield reclamation projects that cover 115 square miles of land and their conversion into parks, golf courses, shopping centers, and business enterprise zones. Among the 75,000 workers still employed in the industry, one-third are engaged in land restoration and waste recycling.[39] Economic decline is thus not an abstract, uniform, or unilinear process, but one shaped by national institutions. An international perspective—with an eye to Western Europe—highlights the failures of capital, labor, and government that contributed to the economic decline of the Pennsylvania anthracite region and the plight of its residents.[40]

The closing of the mines disrupted the lives of Pennsylvania anthracite coal miners and their counterparts in Western Europe. Yet residents of the anthracite region did not have the benefit of the institutional supports available to European mineworkers; they and their families faced economic decline under fundamentally different circumstances. Given the strong European contrast, looking closer to home raises a number of questions: Do the experiences of anthracite miners resemble those of the tens of millions of American men and women displaced during the massive loss of industrial jobs in recent decades?[41] What can be learned by comparing developments in the anthracite region to the varied cases of deindustrialization in the United States more generally?

The widespread closing of manufacturing plants in the last quarter of the twentieth century—particularly in the Midwest, once the nation's industrial heartland and now dubbed its Rust Belt—generated a spate of studies. Surveys have found that sustained employment loss has a personal impact on the lives of people similar to that found in the Pennsylvania anthracite region. Social psychology research initially highlighted the seeming traumas of deindustrialization, with reports of high incidence of physical and psychological health problems, alcohol and drug abuse, crime, and family disorder in communities besieged by plant closings—consequences broadcast by journalists and popular social commentators.[42] Subsequent long-term research found no clear patterns of distress, but instead evidence of successful coping and resilience. These studies echoed the anthracite experience: of women's increasing and vital contributions to family income and community life; gender role reversals; migration of educated, younger workers to areas with better employment prospects; the reluctance of older men to leave; deep connections to family and community; and sizable return migration to hometowns. In contrast to the anthracite region, some researchers found little nostalgia for lost industries or work.[43] An important commonality emerges with public policy implications: occupationally displaced men and women are active agents, conducting their lives not as purely economic actors and maintaining local bonds and institutions in ways that deserve respect and encouragement from robust government programs.

The findings of more extended social science research also confirm a lesson from the Pennsylvania anthracite region. Economic decline requires a long-term perspective. Newer studies of economic decline have moved the clock back to see the onset of dramatic recent losses of U.S. manufacturing in an earlier era. The beginnings of the eclipse of the New England textile industry and such venerable manufacturing centers as Philadelphia can be dated to the 1920s; similarly, the sunset of the steel and automobile industries has roots in the 1940s and 1950s.[44] In this respect, the Pennsylvania anthracite region provides one of the earliest cases of deindustrialization (as well as redevelopment). The luring of silk firms from Paterson, New Jersey, to anthracite mining communities by commercial interests as early as the 1880s proved a long-term disaster for Paterson and began the evolving history of efforts at economic stabilization and diversification in anthracite. Thus again, seen in a larger national context, development and decline are flip sides of the same coin. The long history of recovery initiatives in Pennsylvania anthracite resembles analogous campaigns in Rust Belt communities. The general record in America's industrial heartland shows limited success and mixed results, as in the anthracite region, although it is too early to predict long-term outcomes in areas battered recently by massive job loss. It does not bode well, however, that many of these heartland communities also rely on prisons and the treatment of wastes in their redevelopment efforts.[45]

In some ways anthracite economic decline differed from subsequent examples of decline in the United States. *Disinvestment*—the loss of productive capacities and prospects for future investment—marks all cases, yet the extractive nature of Pennsylvania anthracite makes it distinctive. Capital decomposition has occurred in various ways in the United States across the twentieth century: to undo ruinous competition, corporate takeovers have involved the liquidation of acquired companies, particularly those operating with obsolete technologies;[46] producers of small batches of specialty goods closed their doors in the face of competition from mass-production firms and consumer acceptance of standardized products;[47] and, most commonly, enterprises shut their operations in one community and relocated elsewhere to take advantage of low-cost labor or others factors of production.[48] The anthracite region differs from these scenarios. The anthracite companies could not move elsewhere to maintain their businesses. The mines closed following deliberate management decisions to abandon mining and to diversify assets. With the involvement of outside investors, who had no interest or experience in the region and who intended to profit through the buying and selling of firms, the possibilities for reinvestment in the area all but disappeared.

The economic decline of the anthracite region also differs from more recent cases of U.S. deindustrialization with the absence of racial conflict as a significant ingredient. An element in the closing of plants in the Northeast and Midwest begins with the massive migration of African Americans to northern industrial cities

during and after World War II. Racial tensions subsequently contributed to white flight to the suburbs, leaving the cities depopulated and with declining tax bases (allowing for limited municipal investment in redevelopment). Since the new migrants also secured industrial jobs at the moment of the beginnings of capital flight, their growing relief needs coupled with increased urban unemployment overwhelmed the capacities of local governments. Simultaneously, accompanying social problems also made northern cities unattractive to potential new businesses, undermining subsequent reindustrialization efforts.[49] The decline of anthracite was not shaped or complicated by the racial hostilities and dynamics that have characterized urban deindustrialization in recent times. African Americans have never represented more than 1 percent of the population of the anthracite region, and while ethnic conflict has been part of the area's history, racial friction historically has not.[50]

The different role of race in anthracite is evoked by Mike Knies, a former mineworker. As a young man, he responded to the urgings of a friend in Philadelphia to migrate there and train as a store manager. He arrived at the end of World War II, at a time of heightened racial tensions in the city. "Things started to get rough down there, with the minorities that came up during the war from the South and it didn't agree with me," Mike recalled. At one point, a group of black men assaulted him. He decided to return to Lansford, his hometown, and took work as a blacksmith in Lehigh Navigation Coal's Lansford shops.[51] At least for him, the anthracite region, even in decline, was a haven from the racial tensions of the urban north.

Finally, the role of trade unions in economic decline is another area for comparison between the anthracite region and communities in the Rust Belt. Generally trade union leaders remained quiescent in the face of plant closings, and challenges to corporate decisions came primarily from workers and their community supporters. However there were exceptions. In the 1940s and 1950s, officials in the Textile Workers Union of America determinedly proposed plans to textile firm executives to convince them to maintain operations in New England; they also led the way in redevelopment initiatives.[52] In Kenosha, Wisconsin, Local 72 of the United

Auto Workers led a valiant, though unsuccessful, campaign in the 1980s to gain a reprieve and concessions when the Chrysler Corporation threatened to close an assembly plant.[53] While there were other positive union responses, the overall record is bleak, with trade union leaders either unwilling or unable to counter the undoing of U.S. manufacturing.[54]

Notable opposition to plant closings surfaced but frequently without the support of trade unions. In Youngstown, Ohio, a vocal movement, led by religious leaders and community and labor activists, emerged in the late 1970s to block the closure of steel mills. Protesters also pressed for the seizure by the city through eminent domain of a mill slated for demolition and called for federal loans for modernization to permit public ownership and operation. The United Steel Workers of America (USWA) did not endorse or assist the community challenge.[55] In Pittsburgh, similar community-based movements protested downsizing decisions of executives of the U.S. Steel Corporation and lobbied for special relief services for displaced steelworkers. While local labor leaders joined the insurgency, national officers of the USWA not only refused support but also ultimately placed the local union in receivership.[56] The almost uniformly bleak record of U.S. trade unions in fighting deindustrialization is sharply at odds with labor movements in Western Europe and Canada.[57] There, trade union pressure placed brakes on corporate decision making and forced the enactment of well-funded national programs for assistance to displaced workers, economic redevelopment, and environmental recovery.

The role of the UMWA in Pennsylvania anthracite mirrors the responses of other unions to deindustrialization, but the story here is not just of weak response and inaction in the face of massive loss of jobs, but of damaging policies toward mine closings and, equally important, toward workers' health and pension protections. UMWA leaders put the union's institutional interests first and provided only token responses to the needs of anthracite members and retirees.

Two pertinent questions arise: Why, with or without union support, was there so little collective response to the final closings of the mines? Why were there no equalization of work campaigns in the late 1940s and the 1950s to match

the protests of the Great Depression? Individual protests did emerge, but they were short-lived and never gained widespread support. For example, a small minority of Panther Valley mineworkers did oppose the Lehigh Navigation Coal Company's plan in 1954 to consolidate the processing of coal in one breaker and their picket lines did block implementation of the company's back-to-work proposal. But their efforts to turn the protest into a general strike were singularly unsuccessful. Similarly, when a group of mineworkers laid off by Reading Anthracite in July 1957 tried to rally other miners to close a strip-mining operation in Wadesville and to demand the spreading of work, they met stiff resistance from mine operators, UMWA officials, and other mineworkers whose jobs were threatened by the protest. Both LNC and Reading Anthracite refused to budge and soon announced the permanent closing of all their mining operations.

Following the mine closings, in the southern field and elsewhere throughout the anthracite region, residents acted cooperatively, not solely as individuals, assisting each other in job searches and family and community survival. But mass movements against closings, such as those in Kenosha, Youngstown, and Pittsburgh, did not materialize. With a longer history of decline, perhaps reality had sunk in, that anthracite would no longer be mined, that capital or government would not reinvest in the area in meaningful ways. The hopes of the region's residents now centered on their children and the integrity of their communities.

Comparisons illuminate the distinctions and similarities between the long-term economic decline of the Pennsylvania anthracite region and the more recent grievous loss of manufacturing jobs in America's industrial heartland. A final comparison with other U.S. mining regions in decline further illuminates the case of anthracite. Particularly appropriate in this vein is Central Appalachia, comprising historic coal-producing counties in southern Ohio, West Virginia, western Virginia, Kentucky, Tennessee, and northern Alabama.[58]

Significant differences between the two regions are reflected in the fact that coal mining has been sustained in Central Appalachia. During a boom period in the 1970s, accompanying the oil embargo crises of the decade, the region was home to 74 percent of the coal-mining workforce of the United States and the source of 73 percent of the coal mined in the country.[59] The period saw increases of employment—mobilizations of women's groups even forced the unprecedented hiring of women to work in the mines—and gains in real income for Appalachian miners.[60] The UMWA maintained a presence in the region, rank-and-file militancy remained a feature as well of labor relations, and Central Appalachia experienced classic extended (and violent) coal strikes in the late 1970s and early 1980s.[61]

The history of coal mining continues in Appalachia, unlike in Pennsylvania anthracite, under less-than-ideal circumstances. The region holds vast reserves of bituminous coal; steam power for electricity generation is its major market. Lying relatively close to the surface, the soft coal of Central Appalachia is efficiently extracted mechanically through extensive open-pit strip mining. Similarly, underground veins of bituminous are wide and can be mined with longwall cutting machines. Since the 1950s, capital-intensive technologies have increasingly been applied in Central Appalachian mines, above and below ground.[62] Subsequent cost savings in production have sustained the industry, in sharp contrast to the case of Pennsylvania anthracite, but mechanization has greatly reduced the labor force. The boom in jobs in bituminous in the 1970s proved short-lived, and after 1982 the historical downturn in coal employment in the region resumed. In 1950, the number of coal miners in Central Appalachia stood at 237,600; with rapid technological innovation, the figure collapsed to 99,500 in 1960. Employment in the area reached a new low of 80,000 in 1968, rose with the 1970s surge to 188,000 by 1978, slid with the burst of the bubble to 129,000 by 1984, and fell with a second stage of mechanization to 48,000 by 2000. Throughout these decades of vast reductions of employment, annual production of bituminous coal in Central Appalachia increased steadily from 267 million tons in 1950 to 363 million tons in 2000.[63]

With the loss in employment, but not in production, many parallels can be drawn to Pennsylvania anthracite. Central Appalachian coal counties have witnessed massive out-migration,

high levels of unemployment, low per-capita earnings, rising labor force participation rates of married women, and heavy reliance on government assistance programs. Health problems and environmental decay mark the region as well.[64] The record is mixed to negative on successive initiatives at redevelopment, economic diversification, and job creation and training. As in Pennsylvania anthracite and elsewhere, community members have banded together to attract new businesses, with success usually entailing new light industrial positions for women, in garment-making especially (arriving apparel firms characteristically have had short tenures in the area).[65] Central Appalachia in the last half of the twentieth century has also seen the purchasing of coal lands and facilities by outside national (and international) corporate conglomerates, including major oil companies. The liquidation of assets has not been as severe as in Pennsylvania anthracite since bituminous mining has been sustained, but the buying and selling of properties to manipulate stock prices have similarly led to mine closings. Outside owners have also leased operations to renege on and avoid union contracts and lobbied for the deregulation of both environmental and safety standards. Recently Appalachian miners have been forced into battles to uphold their health care and retirement benefits, as coal companies declaring bankruptcy have been allowed by the courts to forgo benefit obligations to current and retired mineworkers. Coal mining is sustained in the area with a shrinking workforce and deteriorating conditions of employment.[66]

The coal-producing communities of Central Appalachia have been marked historically by a characteristic not associated with Pennsylvania anthracite: chronic poverty. More than one-fourth of the residents of the coal counties of Appalachia live below federal government–established poverty levels, a proportion nearly twice the national average.[67] The area has provided grist for generations of social commentators and investigators who have brought attention to the existence of poverty amid American affluence.[68] Appalachian coal communities are overwhelmingly rural, and subsistence agriculture until recent times has largely been the only alternative or supplement to coal employment; without competition for labor, the coal companies of the area have had a low-wage and seemingly isolated and acquiescent labor force at their command.[69] The residents of Pennsylvania anthracite have experienced dark economic times, but through the nineteenth and twentieth centuries they had the advantage of living in a region that underwent extensive economic development early in the region's history. Endemic poverty has not been part of their history.

◆

The lessons that emerge from both international and U.S. comparisons of regional economic and industrial decline are necessarily ambiguous. The emphasis may be on either the inevitability of anthracite's decline (a high-cost fuel yielding to cheaper fuels) or the inadequacy of institutional responses (with European examples especially in mind). Both global changes in energy markets and institutional failure contributed to the region's enduring slide—the ultimate conclusion of this study—but what is striking in interviewing elderly residents or migrants from the anthracite region is their resilience in the face of catastrophic economic decline. Residents made their individual choices within the constraints that the local economy presented, and those that remained have a strong sense of individual identity and a connection to the region's history that for them outweighs the material deprivation that has been their lot.

Two elderly interviewees expressed what kept people in the anthracite region and the sense of accomplishment in the face of adversity that is common for their generation. Theresa Pavlocak lost her first husband in a mining accident, remarried, moved to New Jersey for twenty years, but then returned to the Panther Valley in retirement. While she and her second husband made a life for themselves in New Jersey, the death of a son in Vietnam brought them home. They buried their son in a cemetery in Summit Hill and lived out their remaining years on the familiar streets of Lansford. She commented about many others in her generation:

A lot of the people will not leave the valley. No matter how hard [it is], they love it here and they stay. And they're managing. You look at the homes around here, they're all remodeled. They're all taken care of. They have a lot of beautiful cars. But

they worked for it. It wasn't given to them. They worked for it.[70]

Family, friends, churches, and ethnic organizations provided close networks that compensated for the lack of material success. Mike Sabron, a mineworker for forty years, said much the same thing when asked to reflect on his experiences after the mines closed.

[If I had it to do over again,] I'd do the same thing. When I think back, I could have stayed in Linden [in northern New Jersey], but I didn't like it. I just loved [it] right here. There's a mixture around here. You name it, we have it—the League of Nations. . . . this east end was all Polish . . . on the other end we had a lot of Italians. And the majority of people here in Summit Hill are Irish. . . . One guy . . . we started together in the bagging plant in 1931, and we also worked in No. 9 and were bowling buddies.

All my kids are good. They're great to me. Almost nine years since my wife passed away, and the kids never fail to call me every week. They come home often. My kids wanted me to go and live with them. No, I want to die here, baby. I worked hard for this.

Interviewed in 1993, Sabron lived another ten years. He spent his last year and a half in a nursing home in nearby Nesquehoning and died in St. Luke's Miners Memorial Hospital in Coaldale. He had "worked hard," and he got his wish—to live his remaining days in the Panther Valley.[71] Sabron shared with his friends and neighbors an abiding sense of his place in history. On the closing of the Lanscoal Company in June 1972 he founded an organization, the Last of the Panther Valley Deep Coal Miners, consisting of the eighteen surviving men who had been the valley's last underground miners, working in the old No. 9 mine for Lanscoal from 1960 until the mine's closing in June 1972. The purpose of the group was to keep alive the memory of underground mining in the Panther Valley and to that end they held dinners in August every year commemorating the anthracite era and their part in it. In 1977 the group erected a coal miners' monument in Lansford's downtown park, using stones taken from the old headquarters

Miners' monument, Lansford, 1978. Photograph by George Harvan.

building of LC&N when it was dismantled. The monument included sculpted images of miners at work and an inlaid central cross made of anthracite.[72]

The annual dinner coupled with a brief service at the miners' monument reflected the miners' sense of identity and history. The memorial services at the monument invariably included prayers, offered in the first years by Father Emil Sopoliga, pastor of St. John's Byzantine Catholic Church in Lansford. His prayer at the 1978 service reflected the way coal mining shaped the religious observances of anthracite miners:

My Dear Lord,
    I dig coal.
    I have tried to utilize to the best of my ability Your talent in serving You as a miner, and I now offer You this simple prayer of praise and thanksgiving. . . . Lord God, as You gave to Adam the admonition of working in the sweat of one's brow, so I offer You my sweat and my labor, asking that it be the cause and remission of my sins.[73]

Typically, after a memorial service, the group met for dinner, in the first years at the Edgemont Lodge, formerly the LC&N country club, and later at the AmVets building on Ridge Street. An account of the first meeting of the group called it the "night of long memories." Successive anniversary meetings offered tributes to men and women who had contributed to the miners' cause, speeches about mining's history in the region, acknowledgment of members who had passed away, and a renewal of the camaraderie that was so much a part of the miner's work and life. In 1977 former anthracite-region congressman, Daniel J. Flood, attended the dedication ceremony for the miners' monument, and the next year Flood, coauthor of the 1969 Coal Mine Health and Safety Act, that included the first federal black-lung compensation provision, received the group's kudos and sang the miners' praises in a brief speech.[74] At the group's twelfth anniversary, one speaker summarized the group's significance, calling the Deep Coal Miners "a precious tool in the preservation of the rich and rugged history of the deep coal mining days. You fellows are keeping the lamp lit." Several times the last superintendent of the Bethlehem Mines spoke to the group; on other occasions a local historian and schoolteacher, Bill Richards, spoke about the valley's anthracite history and reinforced the group's central purpose. As Richards put it, "We cannot let the anthracite miners' heritage die."[75] And although the industry was dying, and the Deep Coal Miners themselves did pass away, they maintained that sense of identity and purpose to the end. At a ceremony at the opening of a new emergency room at St. Luke's Miners Memorial Hospital in October 2002, the last two surviving members of the group participated, and Stanley Stanek said with pride, now more than thirty years after the closing of the last underground mine in the valley: "We're the last of the Panther Valley deep coal miners."[76]

The Deep Coal Miners were not alone in their efforts to preserve the Panther Valley's heritage. In July 1972, just two weeks after the closing of the valley's last deep mine, the first issue of the *Valley Gazette* appeared. A monthly newspaper, the *Valley Gazette* continued to publish for thirty-three years, covering the history of anthracite mining, events in the Panther Valley, such as the big band era and area sports. Local newspaperman, Ed Gildea, founded the paper and served as its only editor. At its beginning, he articulated a set of goals parallel to those of the Deep Coal Miners. "The Gazette," he wrote, "will be historically oriented, concerned with illuminating and preserving the history and heritage of the area, primarily the Panther Valley." The paper published original pieces, often in the form of interviews or reminiscences, and also reprinted news stories from earlier periods. In addition, under the direction of Panther Valley photographer, George Harvan, the paper reprinted a rich array of historical photographs. Gildea also addressed current issues facing the valley. The commitment to place, so important in others' individual responses to anthracite's economic decline, was evident as well. In his opening editorial, Gildea remarked, "Part of the Gazette's philosophy is that the Panther Valley is one of the best places in the world to be living. The newspaper is hereby dedicated to doing whatever it can to make it even better."[77]

While the newspaper did offer some coverage of current events, its focus has been much more consistently on the Panther Valley's history. The focus on the past had appeal. In the issue that marked the paper's first anniversary, the editor commented that monthly press runs had increased from two thousand to four thousand copies. Back issues were selling out at the local news agency in Lansford. And word was reaching out beyond the local area as well. "We now have a mailing list," Gildea wrote, "with over 400 names on it. We think it's the most complete list anybody has anywhere of former Panther Valley area folks." But, most important, in Gildea's view, the paper was contributing to the community's present. He noted:

There's still much to be done to keep the Panther Valley area from losing the vitality evident during the hey day of its deep coal mines. We feel the support given to the Gazette in its first year is an indication of the concern that present and former residents feel toward their home towns. Continuing to deserve this support and working for things that will benefit the Panther Valley area and its citizens will be our objective in the years ahead.[78]

The paper met this goal by bringing the past back to life for Panther Valley residents and former residents. Major stories on the great anthracite strike of 1902, the equalization march of August 1933, the sit-down strike of 1937, and the glory days of "Kiddie Kloes," Lansford's major garment factory, were interspersed with accounts of the Dorsey Brothers and Nick Nichols of big-band fame, a local muckraking reporter, Corry Breslin, the meteoric rise of local notable Evan Evans from water-boy to company president at Lehigh Navigation Coal, and a Korean War POW story.[79] In addition to original articles, the paper frequently reprinted whole pages from back issues of the *Coaldale Observer*, the Lansford *Evening Record*, and other local newspapers. Editor Gildea had no difficulty getting stories and photographs that spoke to the goals he had set, and the paper grew from its initial sixteen to thirty-two pages in its first year of publication. The paper's size and circulation grew in succeeding years as well. In 1975 the paper ex-

Lanscoal miners, 1966. These are the men who became the Last of the Panther Valley Deep Coal Miners. Photograph by George Harvan.

panded to thirty-six pages an issue, and by the paper's fifth anniversary, Gildea was sending out nine hundred copies monthly to subscribing former residents of the region.[80]

Gildea wrote many of the stories, but he also drew on an array of local authors who specialized in reminiscences of the Panther Valley. Jay D. Frantz was a regular contributor, whose "Memories to Share" provided repeated vistas onto the community's past. Emily Havrischak offered a monthly column, "I Remember the Patch," for a year beginning in November 1973, evoking the Panther Valley's patch-town past. Noted journalist, labor activist, and politician, James "Casey" Gildea (no relation to the editor), wrote five original historical articles between 1976 and 1983 that drew on his half century of involvement in labor and civic affairs.[81]

While there was a decidedly nostalgic tone to many of the stories appearing in the *Valley Gazette*, there was also a critical edge. Ed Gildea

had absorbed over the years the miner's taste for the underdog and a strong aversion to the powerful, particularly corporate power in the anthracite region. His coverage of labor struggles in the Panther Valley was an important aspect of this perspective, but it seeped into a much broader range of the articles that appeared in the paper. Consider "The Golden Miner," a story of a mid-nineteenth century anthracite miner and poet. The piece opens by summarizing the story of Welsh-born William Alexander Davies:

He was the spokesman of his people, a poet among the trodden and wanting anthracite coal miners of Schuylkill and Luzerne counties. Despised by the moneyed coal operators for his unremitting loyalty to his fellow workingmen, he walked the Schuylkill Valley and searched as far north as East Nanticoke . . . for work in the damned holes of darkness—all the while composing and reciting his versed sentiments which were the very cause of his being a blacklisted laborer in the mine shafts and collieries.[82]

A shared commitment to the common workingman linked Ed Gildea and fellow newspaperman, Casey Gildea, editor and publisher of the *Coaldale Observer* between 1910 and 1958. Son of a blacklisted coal miner who became a teacher, Casey Gildea rose to local prominence with the emergence of the equalization movement in the early 1930s, served two terms in Congress, and eventually led the miners in their final conflict with Lehigh Navigation Coal when the mines closed in 1954. Ed Gildea maintained the muckraking tradition of his predecessor, wrote about his struggles in the newspaper, and opened its pages to Casey's reminiscences. When Casey left the area in the 1980s, he left Gildea an extensive surviving run of the *Observer*. The *Valley Gazette* continued to reprint full pages of the *Observer* for another thirty years. The *Valley Gazette*, like the Last of the Panther Valley Deep Coal Miners, maintained the community's link to its mining and labor and immigrant past. As the last generation of under-

ground miners passed from the scene, Ed Gildea did all he could to keep their memory alive in the present.[83]

The *Valley Gazette* also provided a publication outlet for the work of George Harvan, a Lansford resident whose photography helped keep alive the memory of anthracite deep mining. The son of an immigrant miner, Harvan grew up in the Panther Valley and, except for five years in the Pacific during and just after World War II, spent his entire life in Lansford. He took up photography on his own after graduation from high school and gained his first professional experience as a photographer in the occupation army in Japan between 1945 and 1947. Upon his return from the Pacific, he began work in the area on a local newspaper, did freelance work for LC&N, and continued in the photography department of Bethlehem Steel after the mines closed. He had several opportunities to move with his craft to work for the Associated Press in Tokyo or the National Gallery in Washington, D.C., but he chose the strong family and community networks of the anthracite region. Passing up opportunities for career advance-

George Harvan, in front of No. 9 Mine, ca. 1969.

ment, he instead spent five decades documenting the heritage of the last generation of underground mining in the Panther Valley. His choice to remain in the anthracite region resembled that of many others of his generation after the mines closed.[84]

Place meant everything to George Harvan, and he gained his sense of identity in the course of photographing area miners during anthracite's decline. Beginning in 1960, Harvan began documenting a tiny mining operation, the Lanscoal Company, whose twenty employees were for the next twelve years the only underground miners in the Panther Valley, where LC&N had once provided work for more than eight thousand men in and around fourteen collieries. Taking time off from his regular photographic assignments with Bethlehem Steel, Harvan donned a hard hat and lamp and walked into the No. 9 mine to record the lone, ongoing remnant of once mammoth mining operations. He caught the men at work, resting during breaks, and sharing beer and song at the end of a hard work week. The miners came to trust Harvan—he was one of them—and he recorded the rhythms of their work, including the final car of coal that came from the No. 9 mine in June 1972 as the Lanscoal Company closed.

With the closing of the last underground mine in the Panther Valley, Harvan became photographic editor of the *Valley Gazette* and covered community events in the area for another three decades. He captured the annual dinners of the Last of the Panther Valley Deep Coal Miners, photographed the remaining abandoned, decaying breakers around the region, and began documenting Pennsylvania sites farther afield from his anthracite base.[85]

Harvan's longevity provided him a remarkable vantage point from which to record five decades of anthracite's decline and its impact on the region's residents. His classic 1955 photo shows the abandoned No. 8 breaker in Coaldale in the distance with row upon row of empty coal cars in the foreground. Another photograph, shot at about the same time, offers a contemporary comment on anthracite's decline in graffiti chalked on a coal car being used as a dumpster: "Go west young man. No future here" and "Valley shot to hell!" Harvan did not stage these scenes. He had the skill and prescience to capture what he saw around him and to record the desolation with moving clarity and understatement. As he recorded the region's decline, his focus always remained on the strength and resilience of the miners. A catalog of one of his

Washing the dust down, Lanscoal miners, 1967. Photograph by George Harvan.

View of Coaldale breaker and empty coal cars, 1955. Photograph by George Harvan.

"Valley Shot to Hell,"
Coaldale, 1953. Photograph
by George Harvan.

Joe "Gimbo" Geusic, Lanscoal miner, ca. 1967.
Photograph by George Harvan.

Locust Summit breaker with lone remaining worker, 1994. Photograph by George Harvan.

exhibitions began, appropriately enough, with individual head-and-shoulder portraits of nine Lanscoal miners, each attired in his work clothes with hard hat and headlamp. People, more than the technology of anthracite mining, mattered to Harvan.[86]

Harvan and Gildea together made the *Valley Gazette* a resource for their community's response to the closing of the mines. In the process they came to champion the establishment of an anthracite-mining museum in Lansford. In the paper's first issue, Gildea declared, "The first order of business is the Number 9 mine," and he argued that it would take local commitment and effort to turn it into a tourist mine that would "provide things to do for our young people and senior citizens," and even more important, "establish a new sense of pride in our home towns." This call came less than a month after the closing of the mine and the erection of a concrete barrier across its entrance to discourage vandalism and protect area residents.[87]

There would be repeated proposals for a No. 9 Mine Museum in succeeding years, but not until 1989 did people collaborate to make progress toward this goal. A group of Lansford residents, led by David and Lora Tokosh, incorporated the Panther Creek Valley Foundation, holding their first meeting in the basement of St. Ann's Church in Lansford on April 9.[88] Between their fundraising efforts and a $5,000 grant from an area foundation, these first activists established a nonprofit foundation and got the project off the ground. The group adopted by-laws in March 1990, secured a lease from LC&N, and over the next two years devoted their energies to cleaning up the No. 9 Wash Shanty on the site, reroofing the building, installing electrical service, and providing restrooms for potential visitors. When the wash shanty had been in use earlier, the water, steam heat, electricity, and sewer service had all been supplied by LNC. Now the group had to approach local utility companies to arrange for these services and pay for them. By 1992 the museum opened in the former wash shanty as the group focused on soliciting donations of mining artifacts and memorabilia. By January 1994 the *Valley Gazette* ran a story with the headline, "No. 9 Mine tours would draw tourists by the thousands." A year later the group removed the concrete barrier and under state supervision evaluated the prospects of restoring the mine and opening it up for tourists. By the winter of 1996 restoration had proceeded to the point that a volunteer team was working in the mine almost every weekend, first cleaning out the accumulated muck and debris of more than twenty years and then driving a second escape route that would have to be accessible before any mine tours would be permitted by the state. In 1997 the

museum attracted more than 1,500 visitors and the group hopefully projected that mine tours would begin in 1999.[89]

With the No. 9 Wash Shanty Museum attracting visitors, volunteers proceeded with the hard work of preparing the abandoned mine. Local construction companies donated equipment and modest foundation, state, and county grants supported the volunteers, now a younger generation than those who had kept alive the memory of anthracite mining since the mine's closing in 1972. One of the area's last mine operators donated a battery-powered mine motor to bring visitors into the mine. Finally, in June 2002, mine tours began Fridays to Sundays each week through November. The first summer, as many as a hundred took the mine tours on some weekends. State and federal funding permitted the museum to hire consultants to plan for possible renovation and the construction of a separate building for museum administration.[90] As the museum and mine tour neared the end of their third season in the fall of 2004, mine visitation surpassed five thousand. The operation remains an almost entirely voluntary affair, though with more than three hundred dues-paying members there is significant community support. The foundation's officers and museum volunteers remain a largely elderly group, men and women who experienced the closing of the mines in the Panther Valley in the 1950s and 1960s and who are committed to preserving that history for younger generations. They are the region's persisters who chose to stay and struggle rather than move to the more prosperous communities beyond the anthracite region.

One of those volunteers, Jack Yalch, a former newspaperman from Lansford, has participated in all the heritage activities that emerged in the Panther Valley after the mines closed. Not a mineworker himself, still Yalch served as toastmaster at the annual dinners sponsored by the Last of the Panther Valley Deep Coal Miners. He wrote a monthly column, "Backtalk," in the *Valley Gazette*, and he was active in the efforts that created the No. 9 mine museum and mine tour. In a piece written for the nation's bicentennial, Yalch expressed a perspective that was no doubt shared by many of his Panther Valley neighbors:

Exterior view, No. 9 Mine entrance, Lansford, 1974. Photograph by George Harvan.

Together we managed to weather many storms and the people who inhabit the grandest nine square miles on earth can look upon their accomplishments and sacrifices . . . with pride.

What the future holds remains to be seen. . . . But come what may, we shall endure.

And in the meantime, remember that God created just one other place a little bit better than the Panther Valley. He named it Heaven and he kept it for Himself.[91]

The contradiction between the material reality of the Panther Valley in decline, on the one hand, and Yalch's perception of it as "the grandest nine square miles on earth" speaks volumes about the power of place and history in the life of the anthracite region across the twentieth century. It reveals well the way that constructed historical memory serves a protective function for anthracite region residents. Mike Sabron, George Harvan, and Jack Yalch have created for themselves and shared with their neighbors a

positive vision of the history of anthracite coal mining and the region. They emphasize the strength and character of area residents in the face of hardships rather than the hardships themselves. The failures of the region's mining companies, of the UMWA, and of all levels of government to respond effectively to economic crisis as well as conflicting perspectives on the closing of the mines and redevelopment fade from view. Focusing on the resilience of local residents in crisis and lionizing their responses serve to diminish personal losses. Their construction of the region's history has permitted them to maintain a sense of purpose in the face of catastrophic loss.[92]

While the region's young people left in droves on graduation from high school, older residents remained; and of those who did migrate, many returned home to retire. The place and its popular narratives have a powerful grip on those who were born and have spent all their lives there. Those who remain have self-

New life for an old mine, No. 9 Mine on reopening, 1995. Notice the stalagmites growing from the floor of the mine gangway. Photograph by George Harvan.

consciously chosen to stay, and in doing so they maintain a profound continuity with the past. Though they have little of the affluence that so distinguishes the United States in the early years of the twenty-first century, their history and strong sense of place make their lives personally rich and meaningful. Mike Sabron worked in and around the mines of the Panther Valley for forty years and lived another three decades through the valley's decline after the mines closed. He had what he wanted in life and he would not leave. He wanted to live and die in the "grandest nine square miles on earth," and he made that wish come true.

Tables

Table 1. Demographic Statistics for the United States and Pennsylvania Anthracite Region, 1930–2000

| | Population | | | | Age Profile | | | |
| | United States | | Five Anthracite Counties* | | United States | | Five Anthracite Counties* | |
Year	Total	% Change	Total	% Change	% Age 20–39	% Age 65+	% Age 20–39	% Age 65+
1930	122,775,046		1,182,895		31.8	5.4	**	**
1940	131,669,275	+7	1,159,714	−2	32.2	5.0	29.0	5.5
1950	151,178,906	+15	1,024,887	−12	30.9	8.1	31.2	8.2
1960	183,212,093	+21	911,557	−11	25.9	9.1	23.1	12.4
1970	203,211,926	+11	886,260	−3	25.8	9.9	19.4	11.3
1980	226,545,805	+11	885,283	0	32.0	11.3	26.4	16.4
1990	248,709,873	+10	853,390	−4	33.0	12.6	27.8	19.6
2000	281,421,906	+13	836,239	−2	29.0	12.4	24.7	16.0
Percent Change 1930–2000:		+129		−29				

*Statistics are for Carbon, Lackawanna, Luzerne, Northumberland, and Schuylkill counties.
**Age data for 20–29 and 30–39 intervals are unavailable for the anthracite region in 1930.

Sources: 1930: U.S. Dept. of Commerce, Sixteenth Census of the United States: 1940, Population, Second Series, Characteristics of the Population, vol. 2, part 1 (Washington, D.C., 1943), p. 20 (for U.S.); U.S. Dept. of Commerce, Fifteenth Census of the United States: 1930, Population, vol. 3, part 2 (Washington, D.C., 1933), pp. 661–63, 665 (for Pa.).
1940: U.S. Dept. of Commerce, Sixteenth Census of the United States: 1940, Population, Second Series, Characteristics of the Population, vol. 2, part 1 (Washington, D.C., 1943), p. 20 (for U.S.); U.S. Dept. of Commerce, Seventeenth Census of the United States: 1950, Characteristics of the Population, vol. 2, part 38 (Washington, D.C., 1952), pp. 187, 191–94 (for Pa.).
1950: U.S. Dept. of Commerce, Seventeenth Census of the United States: 1950, Characteristics of the Population, vol. 2, part 1 (Washington, D.C., 1952), p. 89 (for U.S.); vol. 2, part 1 (Washington, D.C., 1952), p. 89 (for U.S.); vol. 2. part 38, pp. 187, 191–94 (for Pa.).
1960, 1970: U.S. Dept. of Commerce, Nineteenth Census of the United States: 1970, Characteristics of the Population, vol. 1, part 1 (Washington, D.C., 1972), p. 89 (for U.S.); vol. 1, part 40, pp. 247, 253, 254, 256, 258 (for Pa.).
1980: U.S. Dept. of Commerce, Census of the Population: 1980, Characteristics of the Population, vol. 1 (Washington, D.C., 1984), chapter D, Detailed Population Characteristics, part 1, p. 7 (for U.S.); vol. 1, part 40, pp. 908, 910–12 (for Pa.).
1990, 2000: Online. U.S. Census Bureau, American Fact Finder. http://factfinder.census.gov (U.S. and Pa.).

Table 2. Socioeconomic Statistics for the United States and Pennsylvania Anthracite Region, 1930–2000

	Employment/Unemployment Rates				Median Family Income ($)			Per Capita Income ($)		
	United States		Five Anthracite Counties*							
Year	Total Labor Force	% Unemployed	Total Labor Force	% Unemployed	United States	Five Anthracite Counties*	% Differ- ence	United States	Five Anthracite Counties*	% Differ- ence
1930	48,588,730	25.1	420,893	46.5	**	**	**	**	**	**
1940	52,789,499	9.6	453,904	24.6	**	**	**	**	**	**
1950	59,071,655	4.8	394,812	7.0	3,073	2,788	−9.3	**	**	**
1960	68,144,079	5.1	358,510	10.0	5,660	4,716	−16.7	1,810	1,460	−19.3
1970	80,943,299	4.4	362,651	4.2	9,590	7,910	−17.5	3,139	2,593	−17.4
1980	104,449,817	6.5	383,316	8.5	19,917	16,825	−15.5	7,298	5,982	−18.0
1990	123,473,450	6.3	394,566	6.0	35,225	29,752	−15.5	14,420	11,796	−18.2
2000	137,668,798	5.8	393,459	5.5	50,046	42,246	−15.6	21,587	17,886	−17.1

*Statistics are for Carbon, Lackawanna, Luzerne, Northumberland, and Schuylkill counties.
**Income figures unavailable

Sources: Statistics on employment and unemployment rates:

1930: Online. U.S. Historical Census Browser. http://fisher.lib.virginia.edu/collections/stats/histcensus(U.S. and Pa.).
1940: U.S. Dept. of Commerce, *Sixteenth Census of the United States: 1940, Population, Second Series, Characteristics of the Population*, vol. 2 (Washington, D.C., 1943), part 1, p. 6 (for U.S.); vol. 2, part 6, pp. 57, 60, 62 (for Pa.).
1950: U.S. Dept. of Commerce, *Seventeenth Census of the United States: 1950, Characteristics of the Population*, vol. 2 (Washington, D.C., 1952), part 1, p. 99 (for U.S.); vol. 2, part 38, pp. 204, 207, 209 (for Pa.).
1960: U.S. Dept. of Commerce, *Census of Population: 1960, Characteristics of the Population*, vol. 1 (Washington, D.C., 1961), part 1, p. 213 (for U.S.); vol. 1, part 40, pp. 481, 483–85 (for Pa.).
1970: U.S. Dept. of Commerce, *Nineteenth Census of the United States: 1970, Characteristics of the Population*, vol. 1 (Washington, D.C.: 1972), part 1, p. 371 (for U.S.); vol. 1, part 40, pp. 661, 663, 664 (for Pa.).
1980: U.S. Dept. of Commerce, *Census of the Population: 1980, Characteristics of the Population*, vol. 1 (Washington, D.C.: 1984), chapter D, Detailed Population Characteristics, part 1, p. 118 (for U.S.); vol. 1, part 40, pp. 943, 945–47 (for Pa.).
1990, 2000: U.S. Census Bureau, American Fact Finder, accessed online at http://factfinder.census.gov (U.S. and Pa.).

Statistics on median family income and per capita income:

1950: U.S. Dept. of Commerce, *Seventeenth Census of the United States: 1950, Characteristics of the Population*, vol. 2, part 1, p. 104 (for U.S.); U.S. Dept. of Commerce, Bureau of the Census, *County and City Data Book (A Statistical Abstract Supplement)* (Washington, D.C.: GPO, 1952), p. 339 (for Pa.).
1960: U.S. Dept. of Commerce, *Census of Population: 1960, Characteristics of the Population*, vol. 1 (Washington, D.C., 1961), part 1, pp. 232, 286 (for U.S.); vol. 1, part 40, pp. 502, 504, 505, 507 (for Pa.).
1970: U.S. Dept. of Commerce, *Nineteenth Census of the United States: 1970, Characteristics of the Population*, vol. 1 (Washington, D.C., 1972), part 1, p. 398 (for U.S.); vol. 1, part 40, pp. 679, 681, 682 (for Pa.).

Table 3. Coal Production and Employment: United States, Pennsylvania Anthracite Region, 1900–2000

Year	Coal Production (in tons) United States	Coal Production (in tons) Pa. Anthracite Region	Coal Employment United States	Coal Employment Pa. Anthracite Region
1900	269,679,000	57,363,396	348,581	143,824
1917	618,361,053	100,445,299	759,251	156,148
1920	667,430,197	89,636,036	784,621	149,117
1930	544,387,116	68,776,559	644,006	151,171
1940	519,389,554	51,526,454	533,267	90,790
1950	571,607,701	46,339255	483,239	75,231
1960	443,160,739	17,721,113	189,679	20,269
1970	627,236,943	9,248,001	144,480	6,286
1980	830,000,000	5,983,149	225,000	3,429
1990	1,029,000,000	3,414,967	131,000	2,258
2000	1,074,000,000	3,905,096	71,522	945

Sources: For U.S. coal production, 1917: http://energy.er.usgs.gov/products/openfile/OFR97-447. For U.S. coal production, 1920–1990: http://arch.rivm.nl/env/int/hyde/eisp_coal2.html, (with conversion from metric to U.S. tons) and http://www.eia.doe.gov/emeu/northamerica/engdata.htm. For U.S. coal employment, 1870–1970: U.S. Bureau of the Census, *Historical Statistics of the United States, Colonial Times to 1970, Part I* (Washington, D.C., 1975), pp. 592–93. For Pennsylvania anthracite coal production and employment, see table 1 at http://www.dep.state.pa.us/dep/deputate/minres/bmr/annualreport/2000. For U.S. coal production and employment for 1980 and 1990, see U.S. Bureau of the Census, *Statistical Abstract of the United States, 1999*, p. 715. For U.S. coal employment for 2000, see http://www.eia.doe.gov/cneaf/coal/cia/html/t40p01p1.html.

Table 4. Population Statistics for Five Countries of the Pennsylvania Anthracite Region, 1820–1900 (percentage changes in parentheses)

Country	1820	1830	1840	1850	1860	1870	1880	1890	1900
Carbon				15,686	21,024 (+35)	28,144 (+34)	31,923 (+13)	38,624 (+21)	44,510 (+15)
Lackawanna							89,269	142,088 (+59)	193,831 (+36)
Luzerne	20,027	27,379 (+37)	44,006 (+61)	56,072 (+27)	90,234 (+61)	160,915 (+78)	133,065 (−17)	201,203 (+51)	257,121 (+28)
Northumberland	15,424	18,133 (+18)	20,027 (+10)	23,272 (+16)	47,895 (+106)	41,444 (−13)	53,123 (+28)	74,698 (+40)	90,911 (+22)
Schuylkill	11,339	20,744 (+83)	29,053 (+40)	60,713 (+109)	89,501 (+47)	116,428 (+30)	129,974 (+12)	154,163 (+19)	172,927 (+12)
Total	46,790	66,256 (+42)	93,086 (+40)	155,743 (+67)	248,655 (+60)	346,931 (+40)	437,354 (+26)	610,776 (+40)	759,300 (+24)
Total Numbers of Anthracite Coal Mineworkers						35,600	73,373 (+106)	119,919 (+63)	143,824 (+20)

Note: The creation and redrawing of county boundaries are responsible for empty cells and certain shifts in the figures in the table. Reliable occupational statistics are unavailable before 1870.

Source: For population figures: U.S. Historical Census Browser, http://fisher.lib.virginia.edu/collections/stats/histcensus;for the numbers of anthracite mineworkers, http://www.dep.state.pa.us/dep/deputate/minres/bmr/annualreport/1999.

Table 5a. Foreign-Born Population in Anthracite Counties, 1850–1900

	1850		1860		1870		1880		1890		1900	
County	N	%	N	%	N	%	N	%	N	%	N	%
Carbon	4,514	29	5,324	25	6,964	25	5,636	18	6,580	17	7,275	16
Lackawanna							26,917	30	46,399	33	55,727	29
Luzerne	12,567	22	23,486	26	54,688	34	35,716	27	64,103	32	72,962	28
Northumberland	454	2	3,920	8	4,325	10	4,763	9	9,573	13	12,106	13
Schuylkill	18,377	30	26,267	29	30,856	27	26,146	20	31,533	20	32,668	19
Total	36,912	24	58,997	24	96,856	28	99,178	23	158,188	26	180,738	24

Table 5b. Birthplaces of Foreign-Born Population in Anthracite Counties, 1880–1900 (percentage of all foreign-born)

Year	England/Wales	Germany	Ireland	Poland/Russia	Other
1880	31	18	41	2	8
1890	29	16	24	8	23
1900	21	12	16	26	25

Note: Reliable data not available on foreign-born in population before 1850 and on places of birth of the foreign-born before 1880. Lackawanna County not formed until census year of 1880. N=number.

Source: U.S. Historical Census Browser, http://fisher.lib.virginia.edu/collections/stats/histcensus.

Table 6. Net Out-migration by Decade for Lackawanna County 1950–1990, by Age and Gender

Year	Males	Age Group*	Females
1950s			
	29.2%	15–24	29.3%
	14.4	25–34	15.1
	10.9	35–44	10.6
	6.8	45–54	7.7
	5.0	55–64	1.2
	14.0	Overall	13.9
1960s			
	9.4	15–24	19.9
	−1.9	25–34	−0.8
	−1.3	35–44	−1.0
	−1.0	45–54	−1.7
	0.5	55–64	−3.1
	1.0	Overall	2.4
1970s			
	10.4	15–24	14.6
	0.5	25–34	0.0
	−3.9	35–44	−0.9
	2.1	45–54	0.5
	−3.8	55–64	−5.7
	1.8	Overall	2.3
1980s			
	18.3	15–24	18.3
	0.2	25–34	1.7
	−1.6	35–44	0.2
	2.4	45–54	1.1
	2.4	55–64	−2.6
	4.7	Overall	4.6

*Age group intervals refer here to age intervals at the beginning of each decade.

Source: Based on published federal census population figures, 1950–1990. Calculations employ age- and gender-specific death rates from *Historical Statistics of the United States: Colonial Times to 1970* (Washington, D.C.: GPO, 1976), pp. 60–62; post-1970 death rates are found in published census volumes.

Table 7. Distribution of Younger and Older Residents, Anthracite Counties, 1950–1990

Age group	1950	1970	1980	1990
Under 30	54.8%	45.9%	45.2%	40.2%
Over 65	5.3%	11.8%	14.3%	17.0%
Sample Population	15,909	9,419	10,380	10,097

Sources: IPUMS samples for reporting districts including the major anthracite-producing counties of northeastern Pennsylvania, 1950, 1970, 1980, and 1990.

Table 8. World Coal Production and for Selected Countries, 1950–2001

	Annual Coal Production (in millions of tons)					
	1950	1960	1970	1980	1990	2001
World	1853	2658	3019	4182	5348	5227
United States	562	435	610	830	1029	1128
China	43	397	354	684	1190	1459
Germany*	328	468	475	532	514	231
Soviet Union	261	510	624	790	882	—
Russia	—	—	—	—	349	273
Ukraine	—	—	—	—	147	93
India	33	53	77	126	233	385
Australia	24	37	67	116	226	363
S. Africa	26	38	55	132	193	250
Poland	83	113	173	254	237	178
United Kingdom	220	197	145	144	104	33
France	53	58	40	25	13	2

*Statistics for Germany include both East and West Germany during the period when the nation was divided.

Sources: B. R. Mitchell, *International Historical Statistics*, 4th ed. (New York: Stockton Press, 1998), vol. for "The Americas, 1750–1993"; "Europe, 1750–1993"; and "Africa, Asia, and Oceania, 1750–1993"; *International Energy Annual 2002*, http://www.eia.doe.gov/emeu/iea/coal.html (for 1980–2001).

Employment Records of the Mining Operations
of the Lehigh Coal & Navigation Company

The employment records of the Lehigh Coal and Navigation Company (LC&N) in the archives of the National Canal Museum in Easton, Pennsylvania, consist of two file drawers of 5" × 8" cardboard file cards. All told there are more than 22,600 individual records. The file cards are printed on two sides with a grid and various labels that enabled clerical staff in a personnel office to record systematic information about all employees in the company's mining operations. These personnel records are similar to those adopted by other major U.S. firms in the first two decades of the twentieth century as a push toward more professional hiring and record keeping developed in large corporations.[1]

The LC&N records offer a single employment record for each worker in the company's mining operations between 1917 and May 1954, when the company closed its mines.[2] After the closing, the firm did lease its Panther Valley mines to a succession of local companies, but it no longer formally hired the mineworkers in those operations and its records provide no information about employment in the October 1954–February 1960 period, the last years of major underground mining in the Panther Valley.

The records provide a wealth of information about the composition of the mining workforce over this thirty-seven-year period and the patterns of workers' careers at the company. The records begin with an array of personal information about the employee at the time of his hire or at the time that this set of records began. A fair number of employees were obviously working for the company when this new set of record-keeping procedures was adopted, and these records note the date of an individual's first employment with the company and provide a reconstruction of the work history of the individual through 1917.

Personal data on each employee include name, address, date of record, date of birth, nationality, citizenship status, marital status, number of children, and children's ages. The value of this information is limited in part by the fact that this information was recorded only once, at the date the record was started, and changes in address, marital or citizenship status, and number of children were not subsequently recorded.

The main body of each record consists of a grid for recording successive work stints for each individual employed by the company. There are separate columns for occupation, operation, rate of pay, beginning date, ending date, and remarks. That final column held an array of often valuable information, including previous places of residence, previous employment, subsequent employment, reasons for terminating employment, and information concerning industrial accidents or deaths.

The selection, coding, and recording of cases into an employment database was an intricate and lengthy undertaking. Given the size of the workforce, we decided to draw a random sample from the original population, selecting one of every eight employees, providing a final sample of 2,833 workers. Because some employees worked a single stint over a very brief period of time, while others (43 in all) had twenty or more separate stints over what might be a fifty-year period, we developed a coding scheme that preserved as much of the original data as possible. We used the flexible data entry and coding features of the Statistical Analysis System (SAS) to create the database and a combination of SAS and the Statistical Package for the Social Sciences (SPSS) to analyze the data.[3]

We analyzed the resulting database in a variety of ways. Taking advantage of the date information in the records, we analyzed cross-sections of the company's workforce at specific moments in time and thus compared the composition of the workforce or career patterns for workers in 1920, 1930, 1940, and 1950. We could analyze workers' sequences of jobs and thus gain a sense of changing occupational ladders over time or occupational mobility over individual careers. We did only limited analysis of marital status and family information contained in the database, preferring to use more reliable information from decennial census samples between 1920 and 1990 to explore those dimensions of mineworkers' lives. Periodically we employed individual work records to summarize the career trajectories of individual mineworkers at the company.

Shortly after the publication of this book, we will deposit electronic versions of the original SAS database and the subsequently derived SPSS database at the Pennsylvania State Archives in Harrisburg for use by future scholars and students.

Children and Women Wage Earners in Anthracite Silk Mill Communities, 1907–1908

The national concern over child labor generated by the work of the National Child Labor Committee and other reform organizations in the early twentieth century led to a major, nineteen-volume study, *Report on Condition of Woman and Child Wage-Earners in the United States* published in 1911. The fourth volume of this *Report* focused on the silk industry and is of particular interest to our study of the decline of the anthracite region in that it presents raw survey data for woman and child wage earners employed in a sample of thirty-six silk mills in Luzerne and Lackawanna counties in the anthracite region of Pennsylvania. This volume also included extensive data on woman and child wage earners in Paterson silk mills, though we did not transcribe and analyze that portion of the study data.[4]

The data offer a rich view of the immigrant family economy as it existed in a number of communities in the anthracite region of Pennsylvania in the first decade of the twentieth century, what we have called the region's apogee. The data was collected between October 1907 and June 1908, and from the comments of the volume's authors, it seems likely that the Pennsylvania data refers to employment in 1908. This fact leads to one of the limitations of the Pennsylvania data, noted in the volume's introduction. Due to the "general industrial depression which began in the fall of 1907," the report probably considerably undercounts the number of silk mill workers in coal-mining families.[5] Still, the data provide the best surviving view of the family wage economy in communities in which anthracite mining and silk manufacturing dominated the local economy.

Rather than rely exclusively on the detailed analysis and tables that the report presented, we gravitated to the tables of raw survey data reprinted as tables xxvii–xxix in the volume's final chapter. We keyed and analyzed the individual-level data reported there for three classes of silk mill employees: children under sixteen, single women sixteen years of age and over, and married women. In addition, from the presentation of the individual data in the tables, it proved possible to aggregate the data on children within the same family and offer additional analysis for which the family served as the unit of analysis.

In our discussion of child labor and the family economy in chapter 2, we are able to draw on both the findings of the report and our reanalysis of the Pennsylvania portion of the original data compiled by the study's investigators. We are able to make connections between variables unexplored in the original report and aggregate or disaggregate the evidence presented as best suits our analytic needs.

Shortly after the publication of this book, we will deposit an electronic version of the SPSS system file based on the anthracite-region portion of this data set in the Pennsylvania State Archives in Harrisburg for use by future scholars and students.

APPENDIX 4

Anthracite Communities: The 1920 Federal Manuscript Census

The 1920 federal manuscript census for population became available in 1992, early enough to be included in the initial planning for this study. Given that the anthracite region was near the height of its prosperity at this date, it seemed appropriate to examine census enumerations for a number of communities in 1920. We might have relied on a 1-percent sample drawn from all anthracite communities, but a public-use sample invariably masks the identity of specific communities and therefore cannot provide community-specific data that are so valuable in analyzing family and employment patterns. We chose rather to generate a series of samples and populations from selected communities in 1920 that seemed representative of the range of cities and towns with significant anthracite-mining operations.

We wanted in this part of our study to cover the range of anthracite mining settings, in terms of geography and size of community. We selected four cities and towns, including Scranton, Lansford, and portions of Mount Carmel and Coal townships. Scranton provided the largest urban center in the anthracite region, with a population in 1920 exceeding 130,000. Located in the northern field of the region, Scranton offered extensive employment outside of coal mining to both men and women. Lansford, in contrast, was a midsized borough with 9,600 residents, which offered little employment for residents other than coal mining. The community was dominated by its major employer and landowner, the Lehigh Coal & Navigation Company (LC&N), known to residents as the Old Company, which had the headquarters of its mining division in Lansford. Located in the Panther Valley, Lansford was at the eastern end of the southern anthracite field. Lansford had another factor working toward its selection, the fact that we had access to the employment records of LC&N, and thus might be able to compare and contrast the perspectives offered of mineworkers and their families from the company and census sources. Finally, Mount Carmel and Coal townships provided clusters of patch towns located in the western half of the middle field of the region. The enumeration dis-

tricts selected from these townships for analysis consisted typically of a number of mining villages each with populations of around 200. While Scranton and Lansford had experienced considerable growth and recent immigration from southern, central, and eastern Europe, these two groups of patch towns were dominated by the native-born children of earlier Irish, English, and Welsh immigrants. These four communities reflected the range of settings in which mining was carried out in the anthracite region in 1920 and thus offered useful comparisons among large urban centers, midsized towns, and tiny patch towns, and across the three major fields within the region.

We employed varying sampling strategies given the different sizes of the four communities. For Scranton and Lansford, both of which had far more residents than we could reasonably key into a database and analyze, we drew random samples from the census enumerations for the two communities. Aiming for samples with 1,000 or more cases for each community, we employed a 1-percent random sample for Scranton and a 13-percent random sample for Lansford. These fractions resulted in sample sizes of 1,354 and 1,259 for Scranton and Lansford respectively. For Mount Carmel and Coal townships we selected individual enumeration districts with 1,326 and 944 residents respectively. We analyzed the entire populations of these two districts and did not sample as we had been compelled to do for the two larger communities.[6]

Recording of information about residents in the 1920 census enumerations was straightforward once we had developed a codebook to guide the work. We recorded first individual-level variables about each person coded, information that had been recorded by census enumerators following identical sets of instructions from the Census Bureau. After recording twenty-five variables about the individual, we coded another twelve variables focusing on the households in which these individuals resided. These included the relationship of the individual to the household head, the gender, nativity, and occupation of the head of household—sometimes our coded individual—as well as information concerning the number of residents in the household, the number of sons and daughters in the household and their school attendance and/or occupations. We noted the town of each individual, the enumeration district, and, for cities, the ward of residence.[7] This geographical information permitted the kinds of comparisons that we offer in chapter 2, where we explore the composition of mining communities in 1920 and the nature of the working-class family economy that developed across communities.

Shortly after the publication of this book, we will deposit an electronic version of the SPSS system file based on these four samples from the 1920 Federal Manuscript Census in the Pennsylvania State Archives in Harrisburg for use by future scholars and students.

Public Use Microdata Samples of Anthracite Region Censuses, 1930–1990

At the beginning of work on this book in the early 1990s, the latest federal population census available for examination was that of 1920. To trace the employment and family patterns of anthracite-region residents in the 1930 and later censuses, we had to rely on Public Use Microdata Samples (PUMS) prepared under the direction of the U.S. Census Bureau.[8] Originally, these census samples were available in electronic format on nine-track tapes from the Inter-university Consortium for Political and Social Research at the University of Michigan. Over the course of the study, the Minnesota Population Center reprocessed these census samples and made them available to researchers over the World Wide Web as the Integrated Public Use Microdata Samples (IPUMS).[9] We employed both the PUMS and IPUMS data sets as they provided access to the same bodies of 1-percent samples from the decennial federal population censuses that were relevant to our research. The IPUMS data sets had one major advantage over their PUMS predecessors in that researchers at the University of Minnesota had found ways to link individual and household variables that had been recorded in distinct PUMS census samples, thus making it possible to do multivariate analysis employing both individual- and household-level data.

The decennial censuses posed somewhat different questions from one decade to the next, and the U.S. Census Bureau coded geographical information contained in the censuses somewhat differently from one census to the next, so we were not always able to provide strictly comparable analysis across census years. For instance, the 1960 PUMS sample did not provide geographical identifiers that permitted one to extract sample members residing in the anthracite region from other Pennsylvania residents. Thus, the discussions of fathers and mothers in chapter 6 and of sons and daughters in chapter 7 make comparisons across the 1950, 1970, 1980, and 1990 census samples. Because the Census Bureau sought to protect the privacy of individuals whose census responses appear in these samples, geographical variables are provided only for aggregated areas with at least 100,000 in population. It thus proves impossible for any census sample to construct a sample population consisting precisely of sampled individuals residing in the anthracite region. One can select geographical areas that approximate reasonably closely the anthracite region, but some small anthracite communities may be excluded and some communities outside the anthracite region are definitely included in the resulting samples. Given the way the Census Bureau constructed its geographical variables, there is no way to avoid this difficulty entirely. We worked to minimize the difficulties associated with this problem by examining families of mineworkers or garment workers in our samples rather than simply all residents. By careful analysis across census years, the geographical uncertainty in the census samples need not prove of major concern.

Shortly after the publication of this book, we will deposit electronic versions of the SPSS system files based on the IPUMS samples for the anthracite region for the years between 1950 and 1990 in the Pennsylvania State Archives in Harrisburg for use by future scholars and students.

The PP&L Sample, 1940–1987

Annual reports of the major electrical utility in the anthracite region, the Pennsylvania Power & Light Company (PP&L), permitted the construction of the core database employed in the discussion of reindustrialization efforts in the anthracite region in chapter 5. In 1940 PP&L established an Industrial Development Department to keep tabs on industrial prospects that might be encouraged to relocate operations within the utility's service region. The name of this department changed over the years, but quite consistently it prepared an annual report (probably intended primarily for internal company use) of its work in the preceding year and developments in terms of plant relocations, expansions, and closings that had implications for the company's sales of electricity.[10]

In the company's Allentown offices, we found a lengthy series of annual reports for the Industrial Development Department that covered the period 1940–1987 with gaps only during World War II. Each year had a listing of firms that relocated or expanded within the service region, from which one could construct an extensive database for further record linkage and analysis. We recorded individual entries from the annual reports, noting the date, firm name, original location, location in the region, nature of product, numbers of male and female employees anticipated, value of payroll, value of products, reason for moving, and how recruited. Eventually we alphabetized this master list, culled out duplicate entries that cropped up for the same firm, and proceeded to link this list of relocating or expanding companies with records we had of financial assistance offered by local chambers of commerce, local industrial development agencies, and the Pennsylvania Industrial Development Authority (PIDA). We also traced firms through published industrial directories for Pennsylvania at three-year intervals, recording the continued operation of the firm and the number of employees where noted. Finally, we linked firms in local telephone books in major metropolitan areas (Scranton and Wilkes-Barre) as further evidence of their continuing in business.

In the process of building a database on relocating and expanding firms, we benefited from the assistance of individuals in Scranton, Wilkes-Barre, and Hazleton who had been active in reindustrialization campaigns. Frequently firm names changed upon mergers, and in order to distinguish between firm closings and name changes one had to know something of the business history of firms recorded in the database. We are grateful for the assistance of Austin Burke in Scranton, Joseph Yenchko in Hazleton, and Stephen Barrouk and Edward Schechter in Wilkes-Barre for making records available for our use and for sharing their extensive knowledge of these reindustrialization efforts.

The ultimate result of the record-linkage process was a database with 1,252 firms that either relocated in or expanded within the service area of PP&L over almost five decades. Most of the analysis presented in chapter 5 is based on firms that entered the service area in the period 1940–1970, dates selected to permit tracing of firms for at least twenty-five years after they first entered PP&L records. At times, though, we present analyses of cross-sections of the records, comparing the experiences of firms that relocated in particular decades or examining the changing nature of private and public assistance to firms over time.

Shortly after the publication of this book, we will deposit an electronic version of the SPSS database based on the PP&L records and subsequent record linkage at the Pennsylvania State Archives in Harrisburg for use by future scholars and students.

Panther Valley and Migrant
Oral History Interviews

Initially our collection of systematic data came from the kinds of written records that traditionally provide the primary records that historians employ in their research—the employment records of the Lehigh Coal & Navigation Company, published government reports, and the federal decennial population censuses of communities in the anthracite region of northeastern Pennsylvania. To gain fuller access to the attitudes and behavior of individuals during the period of industrial decline, we decided to interview men and women who had experienced economic decline in the post–World War II years.

We focused our oral history interviewing by concentrating our efforts with residents of the Panther Valley at the eastern end of the southern anthracite field. This focus built on two sources we had already uncovered for the communities in the Panther Valley, particularly the LC&N employment records and the *Valley Gazette*, a local monthly newspaper that had been published in Lansford since 1972.

At the outset we used occupational criteria, seeking out former miners and garment workers as the two leading occupational groups in the anthracite region. We made a conscious effort to interview equal numbers of men and women, thinking in this way to address the gendered character of responses to economic decline. We began with individuals who were referred to us as having stories to tell about their experience of economic decline, and we approached additional individuals who were referred by our initial interviewees. We tried to balance men and women interviewed and also sought out a number of interviewees who had just been coming of age as the mines closed and had not themselves been employed prior to the onset of industrial decline in the region. Occasionally, other occupational criteria influenced our selections as we sought out a number of LC&N managers to provide that additional perspective. We were also pleased to find a number of former mineworkers who had been employed in strip-mine operations and former employees at LC&N who had not been miners but had worked in the Lansford shops or the forestry department and brought the perspective of "outside" mineworkers to their remarks.

From critiques of earlier oral history projects, we were aware of the importance of not limiting our interviews entirely to lifetime residents of the Panther Valley.[11] Out-migration was an important response to the closing of the mines, and we needed to find a way to include permanent migrants from the region in our interview group. From early interviews and a variety of contemporary written sources, we were aware that anthracite-region migrants had often gone to the Fairless Hills area northeast of Philadelphia and to a number of communities in northern New Jersey. With this knowledge, we reached out to institutions and individuals in these two areas to provide us with referrals for possible interviewing. With help from Charlie Marshall and Walter Sobieski, we met with members of the United Automobile Workers retirees' group in Linden, New Jersey. James Coon, the president of the Steelworkers' Organization of Active Retirees (SOAR) in Fairless Hills, Pennsylvania, was helpful in reaching out to retired steelworkers. With their assistance, we found two large pools of migrants who were eager to talk about their earlier lives in the anthracite region and the nature of their experience as out-migrants.

In the end we interviewed about one hundred individuals, equally divided between men and women, with migrants comprising somewhat more than a quarter of the group. We prepared transcripts of seventy-eight of these interviews and indexed seventy of them, the diminishing numbers representing judgments we had to make about the relative value of different interviews given limitations of time and resources. We have made extensive use of the indexes and transcriptions in the course of our analysis and writing of chapters 6 and 7.[12]

Shortly after the publication of this book, we will deposit original audio cassettes, transcripts, and indexes of these oral history interviews at the Pennsylvania State Archives in Harrisburg.

Panther Valley High School Graduates' Survey

To provide family, residence, and career information on the generation of sons and daughters who came of age during the most intense period of anthracite's decline, 1946–1960, we prepared a two-page questionnaire for distribution to members of the alumni associations of Panther Valley public and parochial high schools. We gained the support of leaders of the alumni associations for Lansford, Coaldale, Nesquehoning, Summit Hill, and Marian High Schools.[13] The Marian association also provided names and addresses of graduates of its predecessors, St. Ann's and St. Mary's High Schools, both of which had graduates in the period of interest.

We sent out questionnaires to some 2,335 graduates with known addresses with a return rate of about 6 percent of questionnaires as undeliverable. Of the 2,182 likely to have been delivered, we received 572 completed, usable responses, just over 26 percent of the questionnaires. Of the responses returned, a few came from graduating classes outside our original time frame. Six came from graduates in 1940 or 1945, and another six from graduates between 1961 and 1963. We included these responses in the database we created. What was particularly useful about the survey was the high proportion of questionnaires returned by Panther

Valley alumni who had migrated out of the anthracite region since their graduations, more than 65 percent of respondents. Analysis of their responses permitted us to explore systematic differences between graduates based on their subsequent migration experience, providing a useful complement to the oral histories we compiled with persisters and out-migrants from the Panther Valley.

Because area high school graduates resided over such a large area, we made no systematic effort to interview respondents. From twenty respondents we did receive reasonably detailed letters that accompanied returned questionnaires, and we have used respondents' comments in these letters where they spoke to issues highlighted by analysis of the survey. Occasionally, we followed up on remarks in letters with a brief phone interview to clarify issues raised in the correspondence. With these few exceptions, our evidence for the high school survey came exclusively from the questionnaires themselves.

Shortly after the publication of this book, we will deposit the original returned questionnaires and an electronic version of the SPSS database constructed from the responses at the Pennsylvania State Archives in Harrisburg.

Notes

Introduction

1. Interview with Mike Sabron, June 28 and 29, 1993. For first-person narratives based on interviews with Sabron and other current and former residents of the Pennsylvania anthracite region, see Thomas Dublin, *When The Mines Closed: Stories of Struggles in Hard Times* (Ithaca: Cornell University Press, 1998).

2. U.S. Bureau of the Census, *Historical Statistics of the United States: Colonial Times to 1957* (Washington, D.C.: Government Printing Office, 1960), p. 355. See also Louis C. Hunter, *A History of Industrial Power in the United States, 1780–1930* (Charlottesville: University Press of Virginia, 1985), 2:431.

3. Priscilla Long, *Where the Sun Never Shines: A History of America's Bloody Coal Industry* (New York: Paragon House, 1989), pp. 4, 118; Barbara Freese, *Coal: A Human History* (Cambridge, Mass.: Perseus, 2003), pp. 137–38.

4. Pennsylvania population statistics for counties in 1800 and 1900 are accessible at the Historical Census Browser at http://fisher.lib.virginia.edu/collections/stats/histcensus/. We have used Luzerne County in 1800 and Luzerne, Lackawanna, and Schuylkill counties in 1900 to best approximate the region at those two dates. For the changing boundaries of anthracite counties over time, see *The Atlas of Pennsylvania* (Philadelphia: Temple University Press, 1989), p. 83. The land areas of counties have been calculated from Pennsylvania county data available from the U.S. Census Bureau, "State and County *Quick-Facts*," at http://quickfacts.census.gov/qfd/states/42000.html.

5. The most useful accounts of the changing character of immigrant working-class life in the anthracite region across the nineteenth century are Kevin Kenny, *Making Sense of the Molly Maguires* (New York: Oxford University Press, 1998); William D. Jones, *Wales in America: Scranton and the Welsh, 1860–1920* (Cardiff: University of Wales Press, 1993); Victor Greene, *The Slavic Community on Strike: Immigrant Labor in Pennsylvania Anthracite* (Notre Dame: University of Notre Dame Press, 1968).

6. The most important contemporary studies of the development of anthracite are Peter Roberts, *The Anthracite Coal Industry: A Study of the Economic Conditions and Relations of the Cooperative Forces in the Development of the Anthracite Coal Industry of Pennsylvania* (New York: Macmillan, 1901); Peter Roberts, *Anthracite Coal Communities: A Study of the Demography, the Social, Educational and Moral Life of*

the Anthracite Regions (New York: Macmillan, 1904). The most accessible recent history of the anthracite region is Donald L. Miller and Richard E. Sharpless, *Kingdom of Coal: Work, Enterprise, and Ethnic Communities in the Mine Fields* (Philadelphia: University of Pennsylvania Press, 1985).

7. The most widely read and discussed book in the 1980s predicting the enduring economic and international decline of the United States is Paul M. Kennedy, *Rise and Fall of the Great Powers: Economic Change and Military Conflict from 1500 to 2000* (New York: Random House, 1987).

8. Barry Bluestone and Bennett Harrison, *The Deindustrialization of America: Plant Closings, Community Abandonment, and the Dismantling of Basic Industry* (New York: Basic Books, 1982); *The Downsizing of America* (New York: Times Books, 1996).

9. Joseph E. Stiglitz, *The Roaring Nineties: A New History of the World's Most Prosperous Decade* (New York: W.W. Norton, 2003).

10. For British commentary on Great Britain's lost economic glory, see Keith Hutchinson, *The Decline and Fall of British Capitalism* (New York: Scribner, 1950); M. W. Kirby, *The Decline of British Economic Power Since 1870* (London: Allen & Unwin, 1981); John Eatwell, *Whatever Happened to Britain: The Economics of Decline* (London: British Broadcasting System, 1982); and Sidney Pollard, *Britain's Prime and Britain's Decline: The British Economy, 1870–1914* (London: E. Arnold, 1989). For works by American scholars analyzing factors leading to British economic decline, see William Lazonick, *Competitive Advantage on the Shopfloor* (Cambridge, Mass.: Harvard University Press, 1990); Alfred D. Chandler, *Scale and Scope: The Dynamics of Industrial Capitalism* (Cambridge, Mass.: Harvard University Press, 1990), pt. 3.

11. For diagnoses of the failing U.S. economy of the mid-1970s to mid-1990s, see Michael J. Piore and Charles F. Sabel, *The Second Industrial Divide: Possibilities for Prosperity* (New York: Basic Books, 1984); Michael A. Bernstein and David E. Adler, eds., *Understanding American Economic Decline* (New York: Cambridge University Press, 1994). In 1992, comments of a leading Japanese political leader inferring that America's fall from economic grace was due to the laziness of U.S. workers sparked animated discussion. See David E. Sanger, "A Top Japanese Politician Calls U.S. Work Force Lazy," *New York Times,* January 21, 1992, pp. D1, D11. For a discussion of the industrial policy debate of the early

1980s, see Otis L. Graham, *Losing Time: The Industrial Policy Debate* (Cambridge, Mass: Harvard University Press, 1992).

12. Pennsylvania in 1990 had one of the highest proportions of retirees among all states, clear evidence of the aging of its population in recent decades. More is at work here than simply economic decline in the anthracite region alone; the region evidently reflects broader statewide trends. See David Glassberg, "Sense of History in Pennsylvania: Work, Craft, Ethnicity, and Place," *Pennsylvania History* 60 (October 1993): 520.

13. The most extreme example of intertwined environmental and social decay is the ghost town of Centralia in the southern anthracite field, victim of a long-burning underground coal seam that cannot be extinguished. The federal government has condemned the town and bought out most residents, evicting those who would not accept the buyouts. See Renée Jacobs, *SLOW BURN: A Photodocument of Centralia, Pennsylvania* (Philadelphia: University of Pennsylvania Press, 1986); J. Stephen Kroll-Smith and Stephen Robert Couch, *The Real Disaster Is Above Ground: A Mine Fire and Social Conflict* (Lexington, Ky.: University Press of Kentucky, 1990); David DeKok, *Unseen Danger: A Tragedy of People, Government, and the Centralia Mine Fire* (Philadelphia: University of Pennsylvania Press, 1986). We discuss Centralia more extensively in chapter 8.

14. We discuss the use of the Integrated Public Use Microdata Samples (IPUMS) of the United States Manuscript Census of Population in appendix 5. The statistics presented here are based on IPUMS samples for reporting districts including the major anthracite-producing counties of northeastern Pennsylvania, 1950 and 1990, extracted from the Web site of the Minnesota Population Center at http://www.ipums.umn.edu/. Table 1 in appendix 1 provides comparable data for somewhat different age groups.

15. For the 1914 and 2000 employment figures for anthracite, see the Web site of the Pennsylvania Department of Environmental Protection at http://www.dep .state.pa.us/dep/deputate/minres/bmr/annualreport/2000 /table_01.htm (accessed March 8, 2005).

16. Phenomenal recent demand for Colorado coal is the subject of Kirk Johnson, "Coal Market Zooms in on Mines in Colorado," *New York Times,* June 16, 2004, p. A16.

17. The literature here is vast, but useful overviews include Mike Parker, *The Politics of Coal's Decline: The Industry in Western Europe* (London: Earthscan Publications, 1994); David Waddington, Chas Critcher, Bella Dicks, and David Parry, *Out of the Ashes: The Social Impact of Industrial Contraction and Regeneration on Britain's Mining Communities* (London: The Stationery Office, 2001); William Ashworth, *The History of the British Coal Industry,* vol. 5, *1946–1982: The Nationalized Industry* (Oxford: Clarendon Press, 1986); D. Ian Scargill, "French energy: The end of an era for coal," *Geography* 76, no. 2 (1991): 172–75; Erik Swyngedouw, "Reconstructing Citizenship, the Re-Scaling of the State and the New Authoritarianism: Closing the Belgian Mines," *Urban Studies* 33 (1996), 1499–1521; René Leboutte,

"Les bassins industriels en Europe: Production et mutation d'un espace, 1750–1992," *EUI Working Papers in History* (San Domenico, Italy: European University Institute, 1993); Lewis Siegelbaum and Daniel J. Walkowitz, *Workers of the Donbass Speak: Survival and Identity in the New Ukraine, 1989–1992* (Albany: State University of New York Press, 1995).

18. On the unevenness of U.S. industrialization, see Philip Scranton, *Proprietary Capitalism: The Textile Manufacture at Philadelphia, 1800–1885* (New York: Cambridge University Press, 1983); Walter Licht, *Industrializing America: The Nineteenth Century* (Baltimore: Johns Hopkins University Press, 1995), chap. 2.

19. For an anthology of studies on the impact of recent plant closings, see Paul D. Staudoher and Holly E. Brown, eds., *Deindustrialization and Plant Closure* (Lexington, Mass.: D.C. Heath, 1987). Book-length case studies include: Jean Gordus, Paul Jarky, and Louis Ferman, *Plant Closing and Economic Dislocation* (Kalamazoo, Mich.: W. E. Upjohn Institute for Employment Research, 1981); Terry F. Buss and F. Stevens Redburn with Joseph Waldron, *Mass Unemployment: Plant Closing and Community Mental Health* (Beverly Hills, Cal.: Sage, 1983); Carolyn C. Perucci, Robert Perucci, Dena B. Targ, and Harry R. Targ, *Plant Closings: International Context and Social Costs* (New York: A. de Gruyter, 1988); Marie Howland, *Plant Closings and Worker Displacement: The Regional Issue* (Kalamazoo, Mich.: W. E. Upjohn Institute for Employment Research, 1988); Adam Seitchik and Jeffrey Zumitsky, *From One Job to the Next: Worker Adjustment in a Changing Labor Market* (Kalamazoo, Mich.: W. E. Upjohn Institute for Employment Research, 1989); Carrie R. Leana and Daniel Feldman, *Coping with Job Loss: How Individuals, Organizations, and Communities Respond to Layoffs* (New York: Lexington Books, 1992); Scott Camp, *Worker Response to Plant Closings: Steelworkers in Johnstown and Youngstown* (New York: Garland, 1995). For review essays on plant closings that both catalog recent studies and point to varying interpretations, see Angelo Kinicki, "Personal Consequences of Plant Closings: A Model and Preliminary Test," *Human Relations* 33 (1985): 197–212; Peter J. Leahy and Xiannuan Lin, "Plant Closings: A Comparison to Natural Disasters," *American Journal of Economics and Sociology* 51 (1992): 333–48; Carrie R. Leana and Daniel D. Feldman, "The Psychology of Job Loss," *Research in Personnel and Human Resources Management* 12 (1994): 271–302; Nancy R. Vosler, "Displaced Manufacturing Workers and Their Families: A Research-Based Practice Model," *Families in Society: The Journal of Contemporary Human Services* 39 (1994): 105–17.

20. The classic work on immigrant chain migration is John S. Macdonald and Leatrice D. Macdonald, "Chain Migration, Ethnic Neighborhood Formation, and Social Networks," *Milbank Memorial Fund Quarterly* 42 (January 1964): 82–97.

1. Creating the Anthracite Region

1. For the geological history of the Pennsylvania anthracite region, see William E. Edmunds and Edwin F.

Koppe, *Coal in Pennsylvania* (n.p.: Commonwealth of Pennsylvania, Topographic and Geologic Survey, 1968), pp. 3–10, also accessible online at http://www.dcnr.state.pa .us/topogeo/education/coal/es7.pdf; Donald L. Miller and Richard E. Sharpless, *The Kingdom of Coal: Work, Enterprise, and Ethnic Communities in the Mine Fields* (Philadelphia: University of Pennsylvania Press, 1985), pp. 3–7.

2. Edmunds and Koppe, *Coal in Pennsylvania,* p. 15. See also Charles H. Shultz, ed. *The Geology of Pennsylvania* (Harrisburg: Pennsylvania Geological Survey, 1999), chap. 36.

3. Surveys of the topography and early development of the region are found in Hudson Coal Company, *The Story of Anthracite* (New York: Hudson Coal Co., 1932), chap. 2; Barbara Freese, *Coal: A Human History* (Cambridge, Mass.: Perseus, 2003), chap. 5. Useful contemporary nineteenth-century accounts of the three anthracite fields are found in "The Anthracite Coal Trade of Pennsylvania," *North American Review,* no. 90 (January 1835): 241–56; James Macfarlane, *The Coal-Regions of America: Their Topography, Geology, and Development* (New York: D. Appleton, 1873), pp. 7–62; H. M. Chance, *Second Geological Survey of Pennsylvania: 1883: Report of the Mining Methods and Appliances Used in the Anthracite Coal Fields,* 2 vols. (Harrisburg: Board of Commissioners for the Second Geological Survey, 1883), 1:6–8. See also an 1840 map of the three fields in Michael Knies, *Coal on the Lehigh, 1790–1827: Beginnings and Growth of the Anthracite Industry in Carbon County, Pennsylvania* (Easton, Pa.: Canal History and Technology Press, 2001), facing p. 1.

4. For an excellent discussion of how mining methods in the anthracite region varied across the northern, middle, and southern fields, see Eli T. Conner, "Anthracite and Bituminous Mining," *Coal Age* 1 (Oct. 14, 1911): 2–6; (Oct. 21, 1911): 42–45; and (Oct. 28, 1911): 76–79.

5. For the human history of the Pennsylvania anthracite region before and directly after European settlement, see *Historic Resources Study: Delaware and Lehigh Canal National Heritage Corridor and State Heritage Park* (Easton, Pa.: Hugh Moore Historical Park and Museums, 1992), pp. 16–21, 99–112; Paul A. W. Wallace, *Indians in Pennsylvania,* 2nd ed. (Harrisburg, Pa.: Pennsylvania Historical and Museum Commission, 1991), pp. 103–4, 108–17, 459–62, 549–56. See also Stephen A. Flanders, *Atlas of American Migration* (New York: Facts on File, 1998), pp. 9–10.

6. Wallace, *Indians in Pennsylvania,* chaps. 16 and 18.

7. Wallace, *Indians in Pennsylvania,* pp. 155–66; C. Hale Sipe, *The Indian Wars of Pennsylvania,* 2nd ed. (Lewisburg, Pa.: Wennawoods, 1995; originally published in 1931), pp. 459–62, and 549–56. For the best concise treatment of both Indian-settler and Connecticut-Pennsylvania conflicts, see Charles F. Petrillo, "Insulting a King: The Naming of Wilkes-Barre," http://www1.wilkes .edu/history/naming (accessed March 14, 2004).

8. Miller and Sharpless, *Kingdom of Coal,* p. 2.

9. Miller and Sharpless, *Kingdom of Coal,* p. 9.

10. Oscar Jewell Harvey and Ernest Gray Smith, *A History of Wilkes-Barre, Luzerne County, Pennsylvania,* 6 vols. (Wilkes-Barre, 1909–1930), 4:1810–11; Stewart Pearce, *Annals of Luzerne County,* 2nd ed. (Philadelphia: J. B. Lippincott, 1866), pp. 365–66; *Historic Resources Study,* p. 207. Our thanks for Robert Janosov for alerting us to these sources.

11. Miller and Sharpless, *Kingdom of Coal,* p. 7.

12. Knies, *Coal on the Lehigh,* chap. 1. A competing version of the Ginder story, based on oral tradition, "has it that a despondent Ginter, possibly cheated and certainly forlorn, wandered off to disappear forever in the great coal lands to the west." Miller and Sharpless, *Kingdom of Coal,* p. 9. This poetic version has not convinced Michael Knies, the most recent historian to examine the account.

13. Hudson Coal Company, *Story of Anthracite,* pp. 26–28.

14. Hudson Coal Company, *Story of Anthracite,* pp. 38–40.

15. H. Benjamin Powell, *Philadelphia's First Fuel Crisis: Jacob Cist and the Developing Market for Pennsylvania Anthracite* (University Park: Pennsylvania State University Press, 1978).

16. Knies, *Coal on the Lehigh,* pp. 35–36.

17. Vincent Hydro Jr., *The Mauch Chunk Switchback: America's Pioneer Railroad* (Easton, Pa.: Canal History and Technology Press, 2002), pp. 9–12; Knies, *Coal on the Lehigh,* p. 67.

18. Hydro Jr., *Mauch Chunk Switchback,* pp. 15–31; Miller and Sharpless, *Kingdom of Coal,* pp. 26–27; *Historic Resources Study,* pp. 143–49, 209–14; *A History of the Lehigh Coal and Navigation Company* (Philadelphia: William S. Young, 1840), p. 47.

19. Miller and Sharpless, *Kingdom of Coal,* pp. 34–36; *Story of Anthracite,* pp. 47–51; Knies, *Coal on the Lehigh,* p. 66. The "wildcat" nature of enterprise in the Schuylkill coal basin before the 1860s is told and analyzed well in: C. K. Yearley Jr., *Enterprise and Anthracite: Economics and Democracy in Schuylkill County, 1820–1875* (Baltimore: Johns Hopkins University Press, 1961), chaps. 1–4; Anthony F. C. Wallace, *St. Clair: A Nineteenth-Century Coal Town's Experience with a Disaster-Prone Industry* (Ithaca: Cornell University Press, 1981), chaps. 1 and 5.

20. Hudson Coal Company, *Story of Anthracite,* pp. 78–80; William H. Shank, *The Amazing Pennsylvania Canals* (York, Pa.: American Canal and Transportation Center, 1981), pp. 49–52.

21. Miller and Sharpless, *Kingdom of Coal,* pp. 36–37; *Story of Anthracite,* pp. 61–67. See also Robert M. Vogel, *Roebling's Delaware and Hudson Canal Aqueducts* (Washington, D.C.: Smithsonian Institution Press, 1971), pp. 1–3, 25–26, 40.

22. Miller and Sharpless, *Kingdom of Coal,* p. 53; Yearley, *Enterprise and Anthracite,* p. 27.

23. Jules Bogen, *The Anthracite Railroads: A Study in American Railroad Enterprise* (New York: Ronald Press, 1927), chaps. 2–5.

24. Miller and Sharpless, *Kingdom of Coal,* pp. 68–74; W. Ross Yates, *Lehigh University: A History of Education in Engineering, Business, and the Human Condition*

(Bethlehem: Lehigh University Press, 1992), pp. 17–22; Burton W. Folsom, *Urban Capitalists: Entrepreneurs and City Growth in Pennsylvania's Lackawanna and Lehigh Regions, 1800–1920* (Scranton: University of Scranton Press, 2001; originally published in 1981), chap. 8; Bogen, *Anthracite Railroads,* pp. 126–34; *Historic Resources Study,* pp. 164–69.

25. Miller and Sharpless, *Kingdom of Coal,* pp. 55–58; Yearley, *Enterprise and Anthracite,* chap. 6; Bogen, *Anthracite Railroads,* pp. 28–31.

26. Bogen, *Anthracite Railroads,* pp. 22–23, 35–36, 49–51; Marvin W. Schlegel, *Ruler of the Reading: The Life of Franklin B. Gowen, 1836–1889* (Harrisburg, Pa.: Archives Publishing Company of Pennsylvania, 1947), chaps. 2 and 5.

27. Grace Palladino, *Another Civil War: Labor, Capital, and the State in the Anthracite Regions of Pennsylvania, 1840–68* (Urbana: University of Illinois Press, 1990), pp. 25–27, 132; Eliot Jones, *The Anthracite Coal Combination in the United States* (Cambridge, Mass: Harvard University Press, 1914), pp. 28–33, 50–53; Bogen, *Anthracite Railroads,* chap. 3; Wallace, *St. Clair,* pp. 417–18; *Historic Resources Study,* pp. 176–77; Chester Lloyd Jones, *The Economic History of the Anthracite-Tidewater Canals* (Philadelphia: University of Pennsylvania, 1908), pp. 144, 145–46. See also Jean Strouse, *Morgan: American Financier* (New York: Random House, 1999), pp. 251–54, 322–24.

28. Folsom, *Urban Capitalists,* chaps. 2 and 3; W. David Lewis, "The Early History of the Lackawanna Iron and Coal Company: A Study in Technological Adaptation," *Pennsylvania Magazine of History and Biography* 96 (1972): 424–68.

29. Phoebe E. Gibbons, "The Miners of Scranton," *Harper's New Monthly Magazine* 55 (Nov. 1877): 916–27.

30. Bogen, *Anthracite Railroads,* chap. 4; Daniel Hodas, *The Business Career of Moses Taylor: Merchant, Finance Capitalist, and Industrialist* (New York: New York University Press, 1976), chaps. 9 and 10; Rowland Berthoff, "The Social Order of the Anthracite Region, 1825–1902," *Pennsylvania Magazine of History and Biography* 89 (1965): 276–77.

31. Craig L. Bartholomew and Lance E. Metz, *The Anthracite Iron Industry of the Lehigh Valley* (Easton, Pa.: Center for Canal History and Technology, 1988), pp. 41–42. For anthracite's central role in the rise of the United States to industrial supremacy, see Alfred D. Chandler Jr., "Anthracite Coal and the Beginnings of the Industrial Revolution in the United States," *Business History Review* 46 (Summer 1972): 141–81.

32. Jones, *Anthracite Coal Combination,* chaps. 3 and 4.

33. Bogen, *Anthracite Railroads,* pp. 94–95, 119–21.

34. Miller and Sharpless, *Kingdom of Coal,* p. 56.

35. Miller and Sharpless, *Kingdom of Coal,* p. 47; Kevin Kenny, *Making Sense of the Molly Maguires* (New York: Oxford University Press, 1998), p. 48.

36. Edward J. Davies II, *The Anthracite Aristocracy: Leadership and Social Change in the Hard Coal Regions of Northeast Pennsylvania, 1800–1930* (DeKalb: Northern Illinois University Press, 1985), pp. 112–15.

37. The ability of the Reading to purchase coal lands was not just a function of its wealth but also its ability to manipulate shipping rates for coal. By sharply raising rates during strikes in the early 1870s, the Reading had been able to deny independent operators access to markets. Operators began to see the handwriting on the wall and became more amenable to selling out their holdings given the overwhelming power of the Reading. See Kenny, *Making Sense of the Molly Maguires,* pp. 143–48.

38. Sheldon Spear, *Chapters in Northeastern Pennsylvania History: Luzerne, Lackawanna, and Wyoming Counties* (Shavertown, Pa.: Jemags, 1999), pp. 3–4; Davies, *Anthracite Aristocracy,* pp. 19–30; Miller and Sharpless, *Kingdom of Coal,* pp. 77–79; Jones, *Anthracite Coal Combination,* p. 102.

39. Jones, *Anthracite Coal Combination,* pp. 86, 104. See also the historical sketch of the "Lehigh and Wilkes-Barre Coal Company," at http://www.gingerb.com/history.htm (accessed March 14, 2004).

40. Jones, *Anthracite Coal Combination,* pp. 40–46, 54–57.

41. Bogen, *Anthracite Railroads,* pp. 69–72.

42. Jones, *Anthracite Coal Combination,* p. 150.

43. Ibid., pp. 67–73, 78–79, 113–31, 151–55.

44. Ibid., pp. 147–51, 160–73.

45. Ibid., pp. 73–97.

46. Ibid., pp. 104–110. The anthracite combination included the Reading, Lehigh Valley, Delaware & Hudson, the Pennsylvania, the Erie, the Delaware, Lackawanna & Western, and the New York, Ontario, and Western.

47. For a detailed and scathing contemporary expose of the anthracite cartel, see Scott Nearing, *Anthracite: An Instance of Natural Resource Monopoly* (Philadelphia: John C. Winston, 1915). Nearing was a leading social critic and activist of the day; he would later become famous as a pioneer environmentalist and advocate of living the simple life on the land, organic farming, and vegetarianism.

48. The contrast between the transformation of work processes in iron and steel making across the nineteenth century with those of anthracite mining could hardly be starker. At the outset of the nineteenth century, charcoal iron furnaces dotted the countryside, to give way over the course of the century to the massive Bessemer steel furnaces of Pittsburgh and Bethlehem. Anthracite mining experienced no such dramatic technological transformation in the same period. See Edward C. Kirkland, *Industry Comes of Age: Business, Labor, and Public Policy, 1860–1897* (New York: Holt, Rinehart and Winston, 1961), pp. 165–66, 172, 174–75; Michael Nash, *Conflict and Accommodation: Coal Miners, Steel Workers, and Socialism, 1890–1920* (Westport, Conn.: Greenwood, 1982), pp. 16–17.

49. For a description of mining techniques, see Wallace, *St. Clair,* pp. 8–14; Yearley, *Anthracite and Enterprise,* pp. 108–11; Miller and Sharpless, *Kingdom of Coal,* pp. 85–87, 96; *Historic Resources Study,* pp. 215–16. For

contemporary accounts of mining methods, see Chance, *Second Geological Survey of Pennsylvania: 1883*, 1:55–59; Peter Roberts, *The Anthracite Coal Industry: A Study of the Economic Conditions and Relations of the Cooperative Forces in the Development of the Anthracite Coal Industry of Pennsylvania* (New York: MacMillan, 1901), chap. 2.

50. These two main methods of mining coal in the anthracite region are well described in *Story of Anthracite*, pp. 120–23, 134–35.

51. Miller and Sharpless, *Kingdom of Coal*, chap. 4; M. A. Walker, "Preparation of Anthracite Coal," *Coal Age* 1 (Oct. 28, 1911): 71–73; S. V. Tench, "Preparation of Anthracite Coal," *Coal Age* 2 (Nov. 16, 1912): 681–82.

52. "Anthracite Coal Statistical Summaries, 1870–1999," table 1, accessed at http://www.dep.state.pa.us/dep/deputate/minres/bmr/annualreport/1999 (accessed March 8, 2005).

53. Kenny, *Making Sense of the Molly Maguires*, chaps. 3–6.

54. Population figures from the Historical Census Browser at http://fisher.lib.virginia.edu/collections/stats/histcensus/ (accessed March 14, 2004). John M. Laslett, "British Immigrant Colliers, and the Origins and Early Development of the UMWA, 1870–1912," in John M. Laslett, ed., *The United Mine Workers of America: A Model of Industrial Solidarity?* (University Park: Pennsylvania State University Press, 1996), pp. 31–34, 46; Miller and Sharpless, *Kingdom of Coal*, p. 181.

55. For the development of Scranton, see Folsom, *Urban Capitalists*, chaps. 2–4, 7; for the Woolworth story, see Burton Folsom Jr., "The Scranton Story," http://www.fee.org/vnews.php?nid=1948 (accessed September 3, 2004).

56. Population statistics for Scranton and Wilkes-Barre in 1900 are online at http://www.census.gov/population/documentation/twps0027/tab13.txt (accessed September 3, 2004).

57. For Wilkes-Barre's development, see Folsom, *Urban Capitalists*, chap. 5; Davies, *Anthracite Aristocracy*, chaps. 3–4.

58. Davies, *Anthracite Aristocracy*, chaps. 5–6; Miller and Sharpless, *Kingdom of Coal*, pp. 70–75; Kenny, *Making Sense of the Molly Maguires*, p. 51.

59. Harold W. Aurand, *Population Change and Social Continuity: Ten Years in a Coal Town* (Selinsgrove, Pa.: Susquehanna University Press, 1986), pp. 20–24.

60. Thomas Dublin, "Two Hundred Years in the Panther Valley," in George Harvan, *The Coal Mines of Panther Valley* (Bethlehem: Lehigh University Art Galleries, 1995), pp. 5–10.

61. Berthoff, "Social Order of the Anthracite Region," 272.

62. In its proliferation of ethnic churches, Coaldale was representative of anthracite communities more generally. Shamokin, forty-eight miles west of Coaldale, had some thirty different churches. See *The Shamokin Area Centennial: 1864–1964* (Shamokin, Pa., 1964), pp. 88–108. For the Welsh churches in the Wilkes-Barre area, see Ellis W. Roberts, *Journey Through Welsh Hills and Amer-ican Valley* (Wilkes-Barre, Pa.: Wyoming Historical and Geological Society, 1986), p. 91.

63. Joseph H Zerbey, *History of Coaldale* (Pottsville: J.H. Zerbey, 1934); *Coaldale: One Hundred Twenty-Five Years of Progress* (1952), commemorative history in possession of authors.

64. Tony Wesolowsky, "A Jewel in the Crown of Old King Coal: Eckley Miners' Village," *Pennsylvania Heritage Magazine* 22, no. 1 (1976): 30–37, also online at http://www.phmc.state.pa.us/ppet/eckley/page1.asp?secid=31 (accessed March 8, 2005).

65. Stephen G. Warfel, *A Patch of Land Owned by the Company* (Harrisburg: Pennsylvania Historical and Museum Commission, 1993), pp. 6–7. For the company store, see Miller and Sharpless, *Kingdom of Coal*, p. 142; Harold W. Aurand, *Coalcracker Culture: Work and Values in Pennsylvania Anthracite, 1835–1935* (Selinsgrove, Pa.: Susquehanna University Press, 2003), pp. 104–5.

66. Stephen Crane, "In the Depths of a Coal Mine," *McClure's Magazine* 3 (August 1894): 209; Alan Derickson, *Black Lung: Anatomy of a Public Health Disaster* (Ithaca: Cornell University Press, 1998), p. 14, quotes the geologist responsible for Pennsylvania's Second Geological Survey (1883), who wrote: "Miners working in dry dusty workings, and breaker hands working in breakers where coal is prepared dry, undoubtedly become the victims, sooner or later, of the disease known as 'miners' consumption,' 'miners' asthma,' 'miners' anemia,' and by various other names."

67. "Anthracite Coal Statistical Summaries, 1870–1999" in table 1 at http://www.dep.state.pa.us/dep/deputate/minres/bmr/annualreport/1999 (accessed March 8, 2005). For a thorough analysis of these figures, see Mark Aldrich, *Safety First: Technology, Labor, and Business in the Building of American Work Safety, 1870–1939* (Baltimore, Md.: Johns Hopkins University Press, 1997), pp. 300–303.

68. Aldrich, *Safety First*, p. 15.

69. Perry K. Blatz, *Democratic Miners: Work and Labor Relations in the Anthracite Coal Industry, 1875–1925* (Albany: State University of New York Press, 1994), pp. 28–29.

70. Miller and Sharpless, *Kingdom of Coal*, pp. 110–13; Roberts, *Journey Through Welsh Hills*, pp. 73–78.

71. Wallace, *St. Clair*, pp. 296–302.

72. Alexander Trachtenberg, *The History of Legislation for the Protection of Coal Miners in Pennsylvania, 1824–1915* (New York: International Publishers, 1942), pp. 41–46. The provisions of the 1870 law went well beyond those of 1869, but they still failed to provide adequate ventilation to remove hazardous methane gas from the mines' workings. Anthony F. C. Wallace provides a trenchant critique of the law's inadequacies. See *St. Clair*, pp. 305–9.

73. Trachtenberg, *History of Legislation*, pp. 107, 115, 118, 128–29; "Anthracite Coal Statistical Summaries, 1870–1999" in table 1 at http://www.dep.state.pa.us/dep/deputate/ minres/bmr/annualreport/1999 (accessed March 8, 2005).

74. Trachtenberg, *History of Legislation*, pp. 135–41.

75. Aldrich, *Safety First*, fig. 2.1, p. 42; quote on p. 75. While fatality rates in anthracite declined in the last decades of the nineteenth century, absolute numbers of deaths shot up from 211 in 1870 to 644 in 1905. See Aldrich, *Safety First*, pp. 300–303.

76. Aurand, *Coalcracker Culture*, pp. 100–102, 114, 118; Peter Roberts, *Anthracite Coal Communities: A Study of the Demography, the Social, Educational and Moral Life of the Anthracite Regions* (New York: MacMillan, 1904), pp. 259–73; Berthoff, "Social Order of the Anthracite Region," 284–85.

77. Blatz, *Democratic Miners*, p. 33.

78. Ivana Krajcinovic, *From Company Doctors to Managed Care: The United Mine Workers' Noble Experiment* (Ithaca: ILR Press of Cornell University Press, 1997), pp. 18–19.

79. *History of Schuylkill County, PA* (New York: W. W. Munsell, 1881), p. 96; "Ashland Regional Medical Center: History," http://www.ashlandregional.com/history.html (accessed August 18, 2004); *History of Luzerne County, Pennsylvania* (Chicago: S.B. Nelson, 1893), p. 530; Miller and Sharpless, *Kingdom of Coal*, pp. 113–16; *Coaldale: One Hundred Twenty-five Years of Progress*, pp. 42–45. Private Catholic hospitals also played important roles in treating miners who were accident victims; see Roberts, *Journey Through Welsh Hills*, pp. 99–105.

80. The classic statement of this independence of the contract miner is found in Carter Goodrich, *The Miner's Freedom* (Boston: Marshall Jones, 1925). Although Goodrich wrote about bituminous miners, his generalizations hold equally for contract miners in anthracite.

81. Miller and Sharpless, *Kingdom of Coal*, pp. 125, 149; Aurand, *Coalcracker Culture*, pp. 82–84.

82. *Annual Report of the Secretary of Internal Affairs of the Commonwealth of Pennsylvania. Part III. Industrial Statistics. Volume VII, 1878–79* (Harrisburg, Pa.: Lane S. Hart, 1880), pp. 322–52.

83. These findings are based on analysis of the fifty anthracite miners in this survey. Lest one be concerned about the small number of cases in the survey, or think that 1879 might have been an unusually bad year for miners, this finding is corroborated by Perry Blatz's considerably larger study of the income of anthracite mineworkers in 1890. Blatz found that only 24 percent of the miners in that year earned sufficient income to support their families. Perry K. Blatz, "Ever-Shifting Ground: Work and Labor Relations in the Anthracite Coal Industry, 1868–1903" (Ph.D. diss., Princeton University, 1987), pp. 208–11. Other studies suggest that 1879 was not a particularly poor year as far as miners' earnings were concerned. Between 1890 and 1896, for instance, mineworkers' annual wages averaged $330. See Michael Nash, *Conflict and Accommodation*, p. 35.

84. Aurand, *Coalcracker Culture*, p. 22.

85. Kenneth C. Wolensky and Judith Rich, *Child Labor in Pennsylvania* (Harrisburg: Pennsylvania Historical and Museum Commission, 1998), p. 3.

86. *Report on Condition of Woman and Child Wage-Earners in the United States in 19 Volumes. IV: The Silk Industry* (Washington, D.C.: Government Printing Office,

1911), pp. 19–20; Richard Dobson Margrave, *The Emigration of Silk Workers from England to the United States in the Nineteenth Century: With Special Reference to Coventry, Macclesfield, Paterson, New Jersey and South Manchester* (New York: Garland, 1986), pp. 330–33; Alfred Charles Krause, "The Silk Industry of Pennsylvania" (unpublished M.A. thesis, University of Pennsylvania, 1933), p. 33; Philip Scranton, "Introduction" and "An Exceedingly Irregular Business: Structure and Process in the Paterson Silk Industry, 1885–1910," in Philip Scranton, ed., *Silk City: Studies on the Paterson Silk Industry, 1860–1940* (Newark: New Jersey Historical Society, 1985), pp. 4, 40.

87. Bonnie Stepenoff, *Their Fathers' Daughters: Silk Mill Workers in Northeastern Pennsylvania, 1880–1960* (Selinsgrove: Susquehanna University Press, 1999), p. 27.

88. Spear, *Chapters*, p. 2; *Valley Gazette*, no. 37 (July 1975): 22; quote is from Stepenoff, *Their Fathers' Daughters*, p. 26.

89. Palladino, *Another Civil War*, pp. 48–51; Harold W. Aurand, *From the Molly Maguires to the United Mine Workers* (Philadelphia: Temple University Press, 1971), p. 66.

90. Palladino, *Another Civil War*, pp. 101–4, 146–47, 147–62.

91. Aurand, *From the Molly Maguires to the United Mine Workers*, pp. 66–77, 81–93.

92. Aurand, *From the Molly Maguires to the United Mine Workers*, pp. 75, 81; Kenny, *Making Sense of the Molly Maguires*, pp. 141–42. See also Schlegel, *Franklin B. Gowen*, chaps. 2 and 5.

93. *Laws of Pennsylvania, 1866. Pamphlet Law 99* (Harrisburg, 1866).

94. Aurand, *From the Molly Maguires to the United Mine Workers*, pp. 88–93; Maier B. Fox, *United We Stand: The United Mine Workers of America, 1890–1990* (n.p.: United Mine Workers of America, 1990), pp. 4–7; Kenny, *Making Sense of the Molly Maguires*, pp. 168–81.

95. Wallace, *St. Clair*, p. 403.

96. Before the formation of the Pennsylvania State Police in 1905, the private Coal & Iron Police, operating under state commissions, provided the main police power within Pennsylvania. State legislation in 1865 and 1866 authorized railroads, mines, and iron works to organize private police forces, whose authority was reinforced by the state. They were a major force putting down strikes in the state's bituminous and anthracite regions until 1931 when Governor Gifford Pinchot revoked the state commissions for these private police forces. See "The People Versus the Private Army," at http://historymatters.gmu.edu/d/5661/, which provides the testimony of James Maurer, president of the Pennsylvania Federation of Labor before the U.S. Commission of Industrial Relations in 1915. See also "Coal and Iron Police," at http://www.mcintyrepa.com/coalandironpolice.htm.

97. The Molly Maguires have been thoroughly rehabilitated in the popular culture and historical memory of the anthracite region today. Family reunions and mock trials in recent years have commemorated the events with considerable nostalgia and none of the moralizing that

characterized contemporary treatments. For a sampling of the recent revival of interest, see Jim Haldeman, "Some Descendants met for the first time at Molly Maguire Weekend," *Valley Gazette,* no. 288 (Aug. 1996): 7–9. For additional stories and a letter to the editor, see the same issue, pp. 2, 16–17, 28–30. See also related material in *Valley Gazette,* no. 345 (May 2001): 12–13, and (July 2001): 3–4; Virginia S. Wiegand, "Mining the Past for Delayed Justice a Century Later: A 'Retrial' for 'The King of the Mollies,'" *Philadelphia Inquirer,* December 1, 1994. In 1979, Governor Milton Shapp responded to this pressure and granted John Kehoe a posthumous pardon. See Valerie Anne Lutz, "The Old Country in the New World," an online exhibit of the American Philosophical Society, http://www.amphilsoc.org/library/exhibits/wallace/technology.htm (accessed August 18, 2004).

98. Kenny, *Making Sense of the Molly Maguires,* chaps. 6–8.

99. For strike statistics for the anthracite region, 1881–1894, see Blatz, "Ever-Shifting Ground," pp. 105, 323.

100. Fox, *United We Stand,* pp. 18–19, 22–29; Aurand, *From the Molly Maguires to the United Mine Workers,* pp. 131–36.

101. Blatz, "Ever-Shifting Ground," p. 409.

102. Blatz, *Democratic Miners,* pp. 56–59; Fox, *United We Stand,* pp. 84–86; Victor Greene, *The Slavic Community on Strike: Immigrant Labor in Pennsylvania Anthracite* (Notre Dame: University of Notre Dame Press, 1968), chap. 7; Aurand, *From the Molly Maguires to the United Mine Workers,* pp. 137–42; Spear, *Chapters,* pp. 15–16. For a book-length treatment of these events, see Michael Novak, *The Guns of Lattimer: The True Story of a Massacre and a Trial, August 1897–March 1898* (New York: Basic Books, 1978).

103. Blatz, "Ever-Shifting Ground," p. 375.

104. Nash, *Conflict and Accommodation,* pp. 62–63; Blatz, *Democratic Miners,* p. 85.

105. Blatz, *Democratic Miners,* pp. 70–78; see also Greene, *Slavic Community on Strike,* pp. 156–57.

106. For Mitchell's early life, see Craig Phelan, *Divided Loyalties: The Public and Private Life of Labor Leader John Mitchell* (Albany: State University of New York Press, 1994), pp. 2–20.

107. For Mitchell's appealing character and his beliefs, see Phelan, *Divided Loyalties,* pp. 11–13, 48–9, 119, 202, 359–60; Greene, *Slavic Community on Strike,* pp. 163–64, 199–203.

108. Christopher J. Cyphers, *The National Civic Federation and the Making of a New Liberalism, 1900–1915* (Westport, Conn.: Praeger, 2002), p. 43. For a classic early view of the NCF, see James Weinstein, *The Corporate Ideal in the Liberal State, 1900–1918* (Boston: Beacon Press, 1968), esp. pp. 37–39.

109. Blatz, *Democratic Miners,* pp. 88–89. The issue of the distribution of coal cars to miners had two dimensions that concerned the strikers. One entailed the number of cars received by contract miners; the other concerned the emergence of a subcontracting system in which a small number of miners were allotted such a large number of cars that they became subcontractors employing other miners in exploitative ways. See Robert P. Wolensky, "The Subcontracting System and Industrial Conflict in the Northern Anthracite Coal Field," in *The "Great Strike": Perspectives on the 1902 Anthracite Coal Strike* (Easton, Pa.: Canal History and Technology Press, 2002), pp. 74–75.

110. For the anthracite strike of 1900, see Phelan, *Divided Loyalties,* chap. 3; Blatz, *Democratic Miners,* chap. 4, especially pp. 94–98; Fox, *United We Stand,* pp. 86–89. For a view of the strike from the perspective of rank-and-file Slavic mine workers, see Greene, *Slavic Community on Strike,* pp. 157–76. An excellent contemporary summary of mineworkers' grievances is found in "The Strike of the Pennsylvania Coal-Miners," *Harper's Weekly* XLIV (Sept. 29, 1900): 912.

2. Apogee and Descent

1. For a description of the meeting at the White House, see George E. Mowry, *The Era of Theodore Roosevelt and the Birth of Modern America, 1900–1912* (New York: Harper & Row, 1962), pp. 136–37.

2. For the full text of Baer's declaration on July 17, see Craig Phelan, *Divided Loyalties: The Public and Private Life of Labor Leader John Mitchell* (Albany: State University of New York Press, 1994), p. 180.

3. Mowry, *Era of Theodore Roosevelt,* p. 139.

4. These final, cautious moves toward arbitration are well treated in Robert Wiebe, "The Anthracite Strike of 1902: A Record of Confusion," *Mississippi Valley Historical Review* 48 (1961): 229–51, see especially 246–48. See also Ron Chernow, *The House of Morgan: An American Banking Dynasty and the Rise of Modern Finance* (New York: Atlantic Monthly Press, 1990), pp. 106–8; Jean Strouse, *Morgan: American Financier* (New York: Random House, 1999), pp. 448–53.

5. Joseph P. McKerns, "The 'Faces' of John Mitchell: News Coverage of the Great Anthracite Strike of 1902 in the Regional and National Press," in *The "Great Strike": Perspectives on the 1902 Anthracite Coal Strike* (Easton, Pa.: Canal History and Technology Press, 2002), pp. 29–42.

6. The classic study of the strike is Robert J. Cornell, "The Anthracite Coal Strike of 1902" (Ph.D. diss., Catholic University of America, 1957). For more recent treatments, see Phelan, *Divided Loyalties,* chap. 5; Perry K. Blatz, *Democratic Miners: Work and Labor Relations in the Anthracite Coal Industry, 1875–1925* (Albany: State University of New York Press, 1994), chap. 6. On the fears of national political leaders of social disorder, see Phelan, *Divided Loyalties,* p. 183.

7. Blatz, *Democratic Miners,* pp. 103–9.

8. Ibid., pp. 125–31. For a full treatment of Mitchell's mediating role in the National Civic Federation, see James Weinstein, *The Corporate Ideal in the Liberal State, 1900–1918* (Boston: Beacon Press, 1968), passim.

9. Blatz, *Democratic Miners,* pp. 134–36; Phelan, *Divided Loyalties,* pp. 170–75, 181–82.

10. Quoted in, Victor Greene, *The Slavic Community on Strike: Immigrant Labor in Pennsylvania Anthracite*

(Notre Dame: University of Notre Dame Press, 1968), p. 196.

11. Ron Mihalko, "The Strike of 1902," *Valley Gazette*, no. 15 (Sept. 1973): 27, and no. 16 (Oct. 1973): 14, 19, 22; Ed Gildea, "Valley Views," *Valley Gazette*, no. 42 (Dec. 1975): 4; *Report to the President on the Anthracite Coal Strike of May-October 1902, by the Anthracite Coal Strike Commission* (Washington, D.C.: Government Printing Office [GPO], 1903), p. 137.

12. From the *Report of the Anthracite Coal Strike Commission,* as quoted in Donald L. Miller and Richard E. Sharpless, *The Kingdom of Coal: Work, Enterprise, and Ethnic Communities in the Mine Fields* (Philadelphia: University of Pennsylvania Press, 1985), p. 268.

13. Blatz, *Democratic Miners,* p. 135; Phelan, *Divided Loyalties,* pp. 179–80. For the extent of violence during the strike of 1902, see Lance E. Metz, "The Role of Intimidation and Violence in the Great Anthracite Coal Strike of 1902," in *The "Great Strike,"* pp. 43–66.

14. James F. J. Archibald, "The Striking Miners and Their Families," *Collier's Weekly* 30, no. 2 (October 11, 1902): 6.

15. Archibald, "Striking Miners," 6–7.

16. Phelan, *Divided Loyalties,* pp. 191–97. Commission members included Judge George Bray, Brigadier General John M. Wilson, union leader E. E. Clark, Bishop John Lancaster Spaulding, editor Edward Parker, and former mine owner Thomas H. Watkins. See Miller and Sharpless, *Kingdom of Coal,* pp. 279–80.

17. Phelan, *Divided Loyalties,* pp. 198–99; Blatz, *Democratic Miners,* pp. 166–69.

18. Blatz, *Democratic Miners,* pp. 178–79, 181, 190–91, 208–9; quotation on 190–91.

19. Ibid., pp. 192–202.

20. Ibid., pp. 209–11. See also Patrick M. Lynch, "Pennsylvania Anthracite: A Forgotten IWW Venture, 1906–1916" (M.A. thesis, Bloomsburg State College, 1974). The IWW was also active among anthracite-region silk workers; see Bonnie Stepenoff, *Their Fathers' Daughters: Silk Mill Workers in Northeastern Pennsylvania, 1880–1960* (Selinsgrove: Susquehanna University Press, 1999), pp. 81–83.

21. Blatz, *Democratic Miners,* pp. 215–17.

22. Ibid., pp. 217–18, 222–23.

23. Ibid., pp. 229–39.

24. Eliot Jones, *The Anthracite Coal Combination in the United States* (Cambridge, Mass.: Harvard University Press, 1914), chap. 4; Jules Bogen, *The Anthracite Railroads: A Study in American Railroad Enterprise* (New York: Ronald Press, 1927), pp. 242–51.

25. "Anthracite Coal Statistical Summaries, 1870–1999," at http://www.dep.state.pa.us/dep/deputate/minres/bmr/annualreport/1999/table_01_anthracite_coal_statistical_summaries.htm (accessed March 8, 2005).

26. Peter Roberts, *The Anthracite Coal Industry: A Study of the Economic Conditions and Relations of the Cooperative Forces in the Development of the Anthracite Coal Industry of Pennsylvania* (New York: MacMillan, 1901), p. 121; Commonwealth of Pennsylvania, *Annual Report of the Secretary of Internal Affairs,* pt. 3 (1910),

p. 285; Anne Bezanson, "Earnings of Coal Miners," *The Annals* CXI (January 1924): 10.

27. The wage figure for 1901 is based on data presented in *Report to the President on the Anthracite Coal Strike of May-October 1902,* pp. 181–83. Since adult wages seemed of most relevance to any comparison of earnings over time, we have deleted boys' occupations from the tables and then added into the resulting mean figures the earnings of contract miners in 1901 weighted according to the share that contract miners constituted of the anthracite labor force. For contract miners' annual earnings, see *Report,* pp. 135, 177; for the proportion of contract miners in the anthracite work force, see *Reports of the Immigration Commission: Immigrants in Industries, Part 19, Anthracite Coal Mining* (Washington, D.C.: GPO, 1911), pp. 600–7. For 1921 earnings, see Bezanson, "Earnings of Coal Miners," 10. For cost-of-living changes over the period, see Bureau of the Census, *Historical Statistics of the United States: Colonial Times to 1957* (Washington, D.C.: GPO, 1960), p. 127—see the Burgess and Douglass columns for relevant statistics. This argument is confirmed by the independent analysis of Paul H. Douglas, *Real Wages in the United States, 1890–1926* (Boston: Houghton Mifflin, 1930), p. 353.

Converting these figures into their equivalents in 2004 dollars indicates that mineworkers' average annual earnings rose from $11,204 in 1901 to $15,652 in 1921, an overall increase in real income of 39.7 percent. See "What is your dollar worth?" on the Web site of the Federal Reserve Bank of Minneapolis at http://woodrow.mpls.frb.fed.us/research/data/us/calc/ (accessed March 8, 2005).

28. Douglas, *Real Wages in the United States,* pp. 154, 353, 653.

29. Harold A. Aurand, *Population Change and Social Continuity: Ten Years in a Coal Town* (London: Associated University Presses, 1986), p. 116.

30. John E. Bodnar, "Socialization and Adaptation: Immigrant Families in Scranton, 1880–1890," *Pennsylvania History* 43 (1976): 147–62; the proportion here is based on combining tables VI and VII in the article.

31. *Report on Condition of Woman and Child Wage-Earners in the United States in 19 Volumes. IV: The Silk Industry* (Washington, D.C.: GPO, 1911), pp. 456–583. These findings are based on a statistical reanalysis of the raw survey data reprinted in this report. For a recent treatment of the importance of working daughters in the anthracite-region family economy, see Stepenoff, *Their Fathers' Daughters.*

32. *Report on Condition of Woman and Child Wage-Earners,* p. 119.

33. Ibid., p. 310. For additional contemporary coverage of the employment of women and girls in the silk industry in the Pennsylvania anthracite region, see Florence Lucas Sanville, "A Woman in the Pennsylvania Silk-Mills," *Harper's Monthly Magazine* 120 (April 1910): 651–62; "Home Life of the Silk-Mill Workers," *Harper's Monthly Magazine* 121 (June 1910): 22–31.

34. Child labor in the anthracite region received the attention of the Anthracite Coal Strike Commission; see Stepenoff, *Their Fathers' Daughters,* pp. 60–69. For the

campaign to regulate the employment of children in an-
thracite mining, see Owen Lovejoy, "The Extent of Child
Labor in the Anthracite Coal Industry," *The Annals of the
American Academy of Political and Social Science* 20
(1907): 35–49; Francis H. Nichols, "Children of the Coal
Shadow," *McClure's* 20 (February 1903): 435–44; Walter
N. Trattner, *Crusade for the Children: A History of the
National Labor Committee and Child Labor Reform in
America* (Chicago: Quadrangle, 1970), pp. 70–75; Wolen-
sky and Rich, *Child Labor in Pennsylvania,* p. 4.

35. The numbers for each time period are 26, 43, and
109, respectively, and refer to LC&N employees on the
company payroll in January 1920.

36. The population here consists of a 1-percent Scranton sample, a one-in-eight Lansford sample, and the entire enumerations for Coal and Mount Carmel townships
from 1920 federal manuscript census of population, totaling 4,883 anthracite region residents. See appendix 4 for a
discussion of the sampling methods, the numbers of cases,
and microfilm citations for each of the four communities.

37. *Report on Condition of Woman and Child
Wage-Earners,* pp. 250–51. The study reports on all children in anthracite silk communities, while our statistics
are limited to families in which the male head of household was employed in anthracite. For the 1920 census
samples, school attendance for seven- to thirteen-year-
olds in anthracite families exceeded 97 percent; for
fourteen- to fifteen-year-olds, the comparable figure was
73 percent.

38. Bodnar, "Socialization and Adaptation," 159, 161;
see comment in note 30 above. For the 1920 Scranton
census sample, this proportion is based on analysis of 121
children between six and twenty years of age residing with
parents born in southern, central, and eastern Europe.
More detailed analysis of the 1920 census reveals that
Scranton offered considerably more employment for miners' children than was the case in Lansford, or Mount
Carmel or Coal townships. The more complex economy of
the anthracite region's largest city made for a greater
range of employment opportunities outside of the mines
and thus more work for miners' daughters. Sheldon Spear
attributes the evident decline in child labor in anthracite
in the Wilkes-Barre area to a combination of mechanization and legal change. He places child labor in the mines
within a broader urban context in *Chapters in Northeastern Pennsylvania History: Luzerne, Lackawanna, and
Wyoming Counties.* (Shavertown, Pa.: Jemags, 1999),
pp. 165–72.

39. For a less sanguine view of child labor in anthracite, see U.S. Children's Bureau, *Child Labor and the
Welfare of Children in an Anthracite Coal-Mining District,* pub. no. 106 (Washington, D.C.: GPO, 1922), based
on a Children's Bureau study of Shenandoah and surrounding towns in 1919. There is also evidence that child
labor actually increased during the Depression decade.
See Pennsylvania Bureau of Women and Children, *"Children Preferred": A Study of Child Labor in Pennsylvania*
(Harrisburg: Department of Labor and Industry, 1937),
pp. 25–26.

40. Rounded population figures include Carbon,

Lackawanna, Luzerne, Northumberland, and Schuylkill
counties and are available at the "Historical Census
Browser," http://fisher.lib.virginia.edu/collections/
stats/histcensus (accessed March 8, 2005).

41. John Oliver La Gorce, "The Industrial Titan of
America," *The National Geographic Magazine* 35, no. 5
(May 1919): 377, as quoted in John Beck, *Never Before in
History: The Story of Scranton* (Northridge, Calif.: Wind-
sor Publications, 1986), p. 88. For Scranton in its heyday,
see Beck, *Never Before in History,* chaps. 6–7.

42. Spear, *Chapters in Northeastern Pennsylvania
History,* p. 36.

43. *Fourteenth Census of the United States Taken in
the Year 1920: Population, 1920* (Washington, D.C.: GPO,
1921), 1:594, 2:58, 2:729–31.

44. These last statistics are based on the analysis of a
1-percent random sample of Scranton drawn from the
1920 Federal Manuscript Census of Population, T625,
reels 1578–81.

45. Beck, *Never Before in History,* pp. 89–90; *Greater
Scranton Chamber of Commerce: 100 Years of Service,
1867–1967,* [Scranton, Pa.: Harper & Row, honorary publisher, 1967], pp. 21–22.

46. *Fourteenth Census of the United States Taken in
the Year 1920: Population, 1920* (Washington, D.C.: GPO,
1921), 1:590; except for the population figure, statistics in
this paragraph come from analysis of a sample from the
enumeration of Lansford in the 1920 Federal Manuscript
Census of Population, T625, reel 1543.

47. "A Look Inside the Churches of Lansford," *Valley
Gazette,* no. 49 (July 1976): 15, 16–19.

48. Marie L. Obenauer, "Living Conditions Among
Coal Mine Workers of the United States," *The Annals*
CXI (January 1924): 22.

49. *Fourteenth Census,* 1:597.

50. The statistics here are based on the analysis of all
residents recorded in 1920 for these particular portions of
Mount Carmel and Coal townships: 1920 Federal Manuscript Census of Population, T625, reel 1612, enumeration district 105; T625, reel 1610, enumeration district 64.

51. *New York Times,* June 8, 1902, p. 2, as quoted in
Aurand, *Coalcracker Culture,* p. 33.

52. Thirty-eight out of sixty delegates to the first annual convention of the GCU came from anthracite lodges
of the Union. See *Opportunity Realized: The Greek Catholic Union's First One Hundred Years, 1892–1992*
(Beaver, Pa.: Greek Catholic Union of the U.S.A., 1994),
pp. 12–13, 95. Similar proportions of the first lodges in the
PSCU and the Russian Brotherhood Organization came
from the anthracite region of Pennsylvania. See Anthony
X. Sutherland, *The Pennsylvania Slovak Catholic Union:
A Century of Brotherhood, 1893–1993* [Wilkes-Barre, Pa.:
PSCU, n.d.], pp. 7–8, 74–75; *The Truth* (the newsletter of
the Russian Brotherhood), 93, no. 3 (July 1995): 3. Our
thanks to Zenon Wasyliw who translated a good number
of stories from Russian and eastern European foreign-
language ethnic newspapers.

53. *Opportunity Realized,* pp. 20, 44.

54. Antanas Kucas, *Shenandoah St. George Lithuanian Parish in Commemoration of the Diamond Anniver-*

sary of St. George Parish, 1891–1966 (Shenandoah: St. George Parish, 1968), pp. 169–70, 179–80.

55. Thomas F. Sable, "Lay Initiative in Greek Catholic Parishes in Connecticut, New York, New Jersey and Pennsylvania, 1884–1909" (Ph.D. diss., Graduate Theological Union, Berkeley, 1984), pp. 62–63, as quoted in *Opportunity Realized,* p. 7.

56. Bureau of the Census, *Religious Bodies: 1916, Part I* (Washington, D.C.: GPO, 1919), table 63, pp. 304–5.

57. *1891–1991: A Century of Faith and Heritage* (Lansford, Pa.: St. Michael the Archangel Parish, 1995), p. 15; James B. Earley, *Envisioning Faith: The Pictorial History of the Diocese of Scranton* (Devon, Pa.: W.T. Cooke, 1994), pp. 85–103.

58. Kucas, *Shenandoah St. George Lithuanian Parish,* pp. 169–80, 208–19; Earley, *Envisioning Faith,* pp. 102, 109–25. On the broader conflicts that Greek Catholics and Russian Orthodox experienced within the Roman Catholic Church in the United States, see Bureau of the Census, *Religious Bodies: 1916, Part II* (Washington, D.C.: GPO, 1919), p. 259. For an overview of the emergence and early character of the Polish National Catholic Church, see *Religious Bodies: 1916, Part II,* pp. 546–48.

59. Sutherland, *The Pennsylvania Slovak Catholic Union,* p. 42; *Opportunity Realized,* pp. 32–33.

60. See appendix 2 for a description of this source and a discussion of its analytic possibilities.

61. "Brief History and Description of the Lehigh Coal and Navigation Company and of the Properties of that Company . . ." guide prepared for the annual meeting of the American Institute of Mining and Metallurgical Engineering, September 12–15, 1921, Wilkes-Barre; courtesy of Ed Gildea of Lansford.

62. Commonwealth of Pennsylvania, *Report of the Department of Mines of Pennsylvania, Part I, 1919–1920* (Harrisburg: J.L.L. Kuhn, n.d.), pp. 16–17. See also "Coal Preparation at Lehigh Navigation Coal Co.," *Coal Age* 40 (Dec. 1935): 502–6.

63. These findings and those that follow are based on analysis of individuals from the one-in-eight random sample drawn from the LC&N employment records. See appendix 2 for further discussion of the methods employed in working these records into electronic format.

64. "Anthracite Coal Statistical Summaries, 1870–1999," http://www.dep.state.pa.us/dep/deputate/minres/bmr/annualreport/1999/table_01_anthracite_coal_statistical_summaries.htm (accessed March 8, 2005).

65. Mark Aldrich, "The Perils of Mining Anthracite: Regulation, Technology and Safety, 1870–1945," *Pennsylvania History* 64 (Summer 1997): 363, 366–77. Fatalities per million-man-days worked in anthracite mining dropped from 14 in 1930 to below 10 by 1945 and below 5 by 1965. The reduction in fatal accidents no doubt reflected declining output and increased mechanization—strip mining precisely—but improved safety regulations and practice contributed as well. See http://www.dep.state.pa.us/dep/deputate/minres/bmr/annualreport/1999/table_10_fatalities_per_million_man_days.htm (accessed March 8, 2005).

66. *Report of the Department of Mines of Pennsylvania, Part I, 1919–1920,* pp. 293, 295.

67. The changing life chances of miners reported here reflect the fact that LC&N in 1919 was a particularly dangerous place to work. For LC&N, fatalities in 1919 averaged 4.42 per thousand employees; for the anthracite region as a whole, the comparable figure was 4.11. Moreover, for the entire period of time during which the typical LC&N miner was employed, 1906–1933, mine safety in the anthracite region as a whole improved significantly. Over the period, annual fatalities in anthracite mines averaged 3.23 per thousand workers. See Mark Aldrich, *Safety First: Technology, Labor, and Business in the Building of American Work Safety, 1870–1939* (Baltimore: Johns Hopkins University Press, 1997), p. 301. However, before taking comfort in the fact that "only" about 9 percent of anthracite miners in the early twentieth century died on the job, keep in mind that fatalities in U.S. mines occurred at roughly twice the rates common in European mines during the same period. See Aldrich, *Safety First,* p. 42; Crystal Eastman, *Work-Accidents and the Law* (New York: Charities Publication Committee, 1910), p. 47. Anthracite mines were considerably more dangerous than bituminous mines in the early twentieth century; see William Graebner, *Coal-Mining Safety in the Progressive Period: The Political Economy of Reform* (Lexington: University of Kentucky Press, 1976), p. 7.

68. Daniel J. Curran, *Dead Laws for Dead Men: The Politics of Federal Coal Mine Health and Safety Legislation* (Pittsburgh: University of Pittsburgh Press, 1993), p. 71.

69. Alan Derickson, "Occupational Disease and Career Trajectory in Hard Coal, 1870–1930," *Industrial Relations* 32 (Winter 1993): 98.

70. Alan Derickson, *Black Lung: Anatomy of a Public Health Disaster* (Ithaca: Cornell University Press, 1998), chap. 2.

71. Ibid., pp. 72, 74, 80, 95.

72. This analysis is based on final comments noted on the records of mineworkers employed by LC&N on January 1, 1920. Overall 218 were said to have died or suffered a fatal accident; 43 left due to illness; 11 had reference to an accident or workmen's compensation. Ninety received pensions or had applied for company pensions according to these comments; another 65 received UMWA pensions; 3 retired. The majority of final comments on individual records referred to layoffs, resignations, discharges, returns to Europe, and moves to other communities or other jobs.

73. These findings are based on annual lists of pensioners reported each March in the Minutes of the Board of Managers, Lehigh Coal & Navigation, 1918–1922. Microfilm in the Lehigh Coal & Navigation Collection, National Canal Museum, Easton, Pa.

74. Derickson, "Occupational Disease," 105–6.

75. *Scranton Times,* May 2, 1917, p. 10; *The Evening Record* (Lansford), May 23, 1918, p. 1. The employment of women in the breakers would have violated state protective legislation and thus never moved beyond the hypo-

thetical. On the taboo of women entering the mines, see interview with Nellie Valinski, Nov. 6, 1995, p. 3.

76. *Scranton Times,* May 9, 1917, p. 1.

77. *Evening Record,* July 15, 1918, p. 1; October 3, 1918, p. 4; October 10, 1918, p. 1.

78. *Scranton Times,* May 11, 1917, p. 19; July 18, 1918, p. 1; August 22, 1918, p. 5; September 5, 1918, p. 1; Blatz, *Democratic Miners,* pp. 231–39.

79. Blatz, *Democratic Miners,* pp. 231–32; *Evening Record,* February 7, 1918, p. 1.

80. Burton W. Folsom Jr., *Urban Capitalists: Entrepreneurs and City Growth in Pennsylvania's Lackawanna and Lehigh Regions, 1800–1990* (Baltimore: Johns Hopkins University Press, 1981), p. 117. For the use of coke in blast furnaces as early as the 1870s, see Frank J. Evan, "An Analysis of the Decline of the Anthracite Industry with Emphasis on Wyoming Valley from 1930 to Date with the View of Stimulating Industrial Diversity (M.B.A. thesis, University of Pennsylvania, 1950), p. 27; Richard R. Mead, "An Analysis of the Decline of the Anthracite Industry Since 1921" (Ph.D. diss., University of Pennsylvania, 1933), p. 14.

81. Bureau of Mines, *Minerals Yearbook, 1937* (Washington, D.C.: GPO, 1937), pp. 808–10. The earliest available figures we have for energy contributions of various fuels are for 1889, at which date anthracite provided 29 percent of energy consumption, and bituminous provided 58 percent.

82. "Anthracite Coal Statistical Summaries, 1870–1999," http://www.dep.state.pa.us/dep/deputate/minres/bmr/annualreport/1999/table_01_anthracite_coal_statistical_summaries.htm (accessed March 8, 2005).

83. Dever C. Ashmead, *Anthracite Losses and Reserves in Pennsylvania* (n.p.: Pa. Department of Forests and Waters, 1926), p. 9.

84. Ashmead, *Anthracite Losses,* pp. 35, 39, 41, 46. For similar but slightly different estimates on anthracite depletion, see Theodore Bakerman, *Anthracite Coal: A Study in Advanced Industrial Decline* (New York: Arno Press, 1979; originally completed as a University of Pennsylvania Ph.D. dissertation in Economics in 1956), pp. 18–31. On the higher cost of production in the southern field, see Mead, "An Analysis," 32, 72–73; Perry K. Blatz, "Ever-Shifting Ground: Work and Labor Relations in the Anthracite Coal Industry, 1868–1903" (Ph.D. diss., Princeton University, 1987), p. 40.

Ashmead's estimates proved reliable, and the declining production of anthracite coal in succeeding decades meant that anthracite reserves hardly declined between 1922 and 1943. Ashmead estimated that there were 16.4 billion tons of reserves in 1922; production over the next twenty-one years totaled 1.1 billion tons, but a 1943 estimate placed remaining reserves at 16.2 billion tons. At then-current yields and levels of production, anthracite production was likely to continue in Pennsylvania for another 160 years. Commonwealth of Pennsylvania, Department of Internal Affairs, Bureau of Statistics, Topographic and Geologic Survey, *Pennsylvania's Mineral Heritage: The Commonwealth at the Economic Crossroads of Her Industrial Development* (Harrisburg, 1944), p. 85.

85. Mead, "An Analysis," 42.

86. Evan, "An Analysis," 14–16.

87. Blatz, "Ever-Shifting Ground," 39.

88. Oil's share in energy use grew from the 10-percent range in 1917 to the 26- to 28-percent range by the mid-1930s; natural gas use grew in tandem, from about 3 percent at the turn of the century to the 9-percent range by the mid-1930s. See *Minerals Yearbook, 1937,* p. 810.

89. Evan, "An Analysis," 84, 90–92; Mead, "An Analysis," 27; Bakerman, *Anthracite Coal,* pp. 36–38, 57. Waterpower's share in energy use grew from about 2 percent in the late nineteenth century to 10 percent by the 1930s. See *Minerals Yearbook, 1937,* p. 810.

90. Bakerman, *Anthracite Coal,* pp. 45–46.

91. Evan, "An Analysis," 73; Mead, "An Analysis," 18.

92. Evan, "An Analysis," 46.

93. For regional sales of anthracite, see Mead, "An Analysis," 48, 158–59, 172.

94. For a helpful treatment of the labor crisis of 1920 and the United States Anthracite Coal Commission established by President Woodrow Wilson, see Harold Kenneth Kanarek, "Progressivism in Crisis: The United Mine Workers and the Anthracite Coal Industry during the 1920's" (Ph.D. diss., University of Virginia, 1972), chaps. 2 and 3.

95. The three-member commission included William L. Connell for the anthracite operators, Neal J. Ferry for the UMWA, and Ohio State University president, William O. Thompson, for the public. Blatz, *Democratic Miners,* p. 240.

96. Kanarek, "Progressivism in Crisis," 50–55; W. Jett Lauck, "Combination in the Anthracite Industry," testimony before the United States Anthracite Coal Commission (Washington, D.C., 1920), pp. 3–5. This published pamphlet is found in the Papers of W. Jett Lauck, Special Collections, University of Virginia Library. The most useful discussion of Lauck's relationship with the UMWA is found in Leon Fink, *Progressive Intellectuals and the Dilemmas of Democratic Commitment* (Cambridge, Mass.: Harvard University Press, 1997), chap. 7.

97. Kanarek, "Progressivism in Crisis," 67–70.

98. Maier B. Fox, *United We Stand: The United Mine Workers of America, 1890–1990* ([Washington, D.C.]: UMWA, 1990), pp. 193–95.

99. Melvyn Dubofsky and Warren Van Tine, *John L. Lewis: A Biography* (New York: Quadrangle/New York Times, 1977), part II; see pp. 57–66 for Lewis's accommodation to Secretary of Labor Wilson in settling the 1919 bituminous coal strike and for his ascendance to the UMWA presidency.

100. Kanarek, "Progressivism in Crisis," 72–89.

101. Ibid., 90–94.

102. Bogen, *Anthracite Railroads,* p. 217.

103. For evolving Supreme Court decisions leading to the segregation of production from transportation in the anthracite industry in the first decades of the twentieth century, see Bogen, *Anthracite Railroads,* pp. 216–27.

104. Ibid., pp. 227–36.

105. Ibid., pp. 228, 230–32.

106. Ibid., p. 52.

107. Ibid., pp. 237–41; Bakerman, *Anthracite Coal*, p. 53.

108. Bogen, *Anthracite Railroads*, pp. 261–65.

109. Harold K. Kanarek, "The Pennsylvania Anthracite Strike of 1922," *Pennsylvania Magazine of History and Biography* 99 (1975): 207–25. For the national dimensions of the strike movement, see Dubofsky and Van Tine, *John L. Lewis*, pp. 82–91.

110. Kanarek, "Pennsylvania Anthracite Strike of 1922," 211, 214–15.

111. Colin J. Davis, *Power at Odds: The 1922 National Railroad Shopmen's Strike* (Urbana: University of Illinois Press, 1997).

112. Kanarek, "Pennsylvania Anthracite Strike of 1922," 212–13, 216–18.

113. Ibid., 219–20.

114. For the anthracite coal strike of 1923, see Robert H. Zieger, "Pinchot and Coolidge: The Politics of the 1923 Anthracite Crisis," *Journal of American History* 52 (1965): 566–81.

115. Ibid., 573–74.

116. *Minerals Yearbook, 1945* (Washington, D.C.: GPO, 1945), p. 948.

117. For the anthracite strike of fall and winter 1925–26, see Robert Zieger, "Pennsylvania Coal and Politics: The Anthracite Strike of 1925–1926," *Pennsylvania Magazine of History and Biography* 92 (April 1969): 244–62; Harold K. Kanarek, "Disaster for Hard Coal: The Anthracite Strike of 1925–1926," *Labor History* 15 (Winter 1974): 44–62.

118. For Communist challenges within the UMWA in the 1920s, see Alan Singer, "Communists and Coal Miners: Rank-and-File Organizing in the United Mine Workers of America during the 1920s," *Science and Society* 55 (Summer 1991): 132–57; Walter T. Howard, "Communist Activism in the Pennsylvania Anthracite during the Great Depression," paper delivered at the Pennsylvania Historical Association, October 1995; Walter T. Howard, "The Communist Party in the Anthracite Coal Fields, 1919–1955," paper delivered at the Fall Lecture Series of the Eckley Miners' Village, October 2003. Professor Howard has published three books that speak to the issues explored in chapter 3 since we completed this book. Walter T. Howard, ed. *Anthracite Reds: A Documentary History of Communists in Northeastern Pennsylvania during the 1920s* (New York: iUniverse, 2004), and *Anthracite Reds*, vol. 2, *A Documentary History of Communists in Northeastern Pennsylvania during the Great Depression* (New York: iUniverse, 2004); Walter T. Howard, *Forgotten Radicals: Communists in the Pennsylvania Anthracite, 1919–1950* (Lanham, Md.: University Press of America, 2005). See also Steve Nelson, James R. Barrett, and Rob Ruck, *Steve Nelson: American Radical* (Pittsburgh: University of Pittsburgh Press, 1981), chap 6.

119. Kanarek, "Disaster for Hard Coal," 45–46.

120. Mead, "An Analysis," 169.

121. Kanarek, "Disaster for Hard Coal," 59; Mead, "An Analysis," 233.

3. The Anthracite Miners' New Deal

1. Maury Klein, *Rainbow's End: The Crash of 1929* (New York: Oxford University Press, 2001), p. 229. The classic study of the stock market crash of October 1929 is John Kenneth Galbraith, *The Great Crash, 1929* (Boston: Houghton Mifflin, 1954).

2. Diane Lindstrom, "Stock Market Crash of 1929," in Paul S. Boyer, ed., *The Oxford Companion to United States History* (New York: Oxford University Press, 2001), p. 748.

3. Gary M. Walton and Hugh Rockoff, *History of the American Economy* (Fort Worth, Tex.: Dryden Press, 1998), pp. 516–18.

4. William E. Leuchtenberg, *Franklin D. Roosevelt and the New Deal, 1932–1940* (New York: Harper & Row, 1963), pp. 17–62.

5. For a comprehensive study of the CIO, see Robert H. Zieger, *The CIO, 1935–1955* (Chapel Hill, University of North Carolina Press, 1995). For a view of labor during the New Deal period from the grassroots, see Staughton Lynd, ed., *"We Are All Leaders": The Alternative Unionism of the Early 1930s* (Urbana: University of Illinois Press, 1996). For the Depression in Pennsylvania, see Thomas H. Coode and John F. Bauman, *People, Poverty, and Politics: Pennsylvanians during the Great Depression* (Lewisburg, Pa..: Bucknell University Press, 1981).

6. "Annual Report of the Mining Department, 1923," p. 1, Lehigh Coal & Navigation Company records, Archives Center, National Museum of American History, Smithsonian Institution, Washington, D.C. (hereafter cited as LC&N-Smithsonian).

7. Ibid., p. 43, LC&N-Smithsonian.

8. "Annual Report of the Mining Department, 1927," pp. 1, 42; "Annual Report of the Mining Department, 1928," pp. 1, 9, LC&N-Smithsonian.

9. "Annual Report of the Mining Department, 1928," p. 10, LC&N-Smithsonian. Spelling as in the original typescript report.

10. The three accounts that follow are based on individual records taken from a collection of about 25,000 employment records for the coal operations of LC&N between 1917 and 1954. The collection is held by the National Canal Museum in Easton, Pa. Early in this project we microfilmed the records as part of the process of creating a 1-in-8 computerized random sample of the employment records. See appendix 2 for a fuller discussion of the work in preparing the LC&N employment sample.

11. These examples confirm that the UMWA contract with coal operators had no seniority clause and mining companies were free to layoff or rehire workers without regard to seniority. Although Kissner had been employed at LC&N for more than forty years, his layoff lasted almost six years; in contrast, 39-year-old Dibero was rehired after only four months of unemployment. In 1952, the UMWA contract added a seniority clause, but it came only two years before all of LC&N's mines closed for good. Alan Derickson, "Occupational Disease and Career Trajectory in Hard Coal, 1870–1930," *Industrial Relations* 32 (1993): 107.

12. "Annual Reports of the Mining Department, 1929," pp. 16, 29; "1930," p. 43, LC&N-Smithsonian.

13. "Annual Reports of the Mining Department, 1930," p. 15; "1931," pp. 9, 15; "1932," pp. 10, 16, LC&N-Smithsonian. On statistics for layoffs at Rahn Colliery, see "Anthracite Code: Presentation of 'The Equalization of Working Time,' by the Reverend John Pounder of Lansford, PA representing the Panther Valley Equalization Committee," p. 4, RG 9, box 6061, National Archives.

14. These figures and proportions are based on Lehigh Coal & Navigation annual reports and the company's annual Mining Department reports. In employing company reports, 1930 marks a transition point as LC&N organized a wholly owned subsidiary, Lehigh Navigation Coal Company (LNC) and appointed the head of its mining department, Jessy B. Warriner, president of LNC. This development involved no substantive change in the way the company operated, but it was part of the legal evolution of LC&N into a holding company and reflected management's desire to maintain separate accounting for the company's divisions. See W. Julian Parton, *The Death of a Great Company: Reflections on the Decline and Fall of the Lehigh Coal and Navigation Company* (Easton, Pa.: Canal History and Technology Press, 1998; originally published in 1986), pp. 24, 27; *The Story of the Old Company* (Lansford, Pa.: Lehigh Navigation Coal Company, 1941), pp. 48–49. We will continue using both terms, LC&N and LNC, with LC&N referring to the parent company and LNC to its mining operations in the Panther Valley. Contemporaries, however, used the terms inconsistently and interchangeably and in quotations we will make no effort to "correct" their usage.

15. *Evening Record* (Lansford), Dec. 8, 1932, p. 1; Dec. 10, 1932, p. 1; Dec. 19, 1932, p. 2; Dec. 21, 1932, p. 1; Dec. 24, 1932, p. 1; Dec. 27, 1932, p. 1; Jan. 13, 1933, p. 1. See also T. D. Lewis, general superintendent, LNC, to W. J. McCann, subdistrict president, UMWA, March 12, 1932, in RG 9, box 6064, National Archives.

16. *Evening Record,* Aug. 17, 1933, pp. 1, 4, 6, also reprinted in *Valley Gazette,* no. 6 (Dec. 1972): 1–2, 4–6; there is a picture of the demonstrators' delegation to LNC on p. 1. See also *Philadelphia Inquirer,* Aug. 18, 1933, p. 3. For an oral account of the mass demonstration in Lansford as recalled sixty years later by the son of the General Manager of the LNC, see Thomas Dublin, "Daniel Helms Remembers the Equalization March of 1933," *Valley Gazette,* no. 285 (May 1996): 28–29.

17. *Evening Record,* Aug. 17 and 18, 1933—see p. 1 stories and their continuations. See also *Philadelphia Inquirer,* Aug. 19, 1933, p. 6, and Aug. 20, 1933, p. 1. These events struck company officials as "mob" action, the term used repeatedly in internal management memos. See W. J. Reese to Mr. Warriner, 8-18-33 and a second unsigned memo to Warriner that date, both in "Labor Troubles" folder, Administrative File, carton 32, LC&N Records, MG-311, Pennsylvania State Archives, Harrisburg, Pa. (hereafter cited as LC&N, PSA).

18. *Evening Record,* Aug. 21, 1933, pp. 1, 6; quotation, p. 6; *Shenandoah Evening Herald,* Aug. 21, 1933, as reprinted in *Valley Gazette,* no. 138 (Dec. 1983): 26; *Valley Gazette,* no. 6 (Dec. 1972): 2. This popular uprising had repercussions well beyond the Panther Valley and the anthracite region more generally and provided the UMWA useful leverage in the negotiations for the NRA Code for bituminous coal. David Brody called the equalization campaign "an indispensable weapon" for Lewis in Washington. See David Brody, "Labour Relations in American Coal Mining: An Industry Perspective," in Gerald D. Feldman and Klaus Tenfelde, eds. *Workers, Owners and Politics in Coal Mining: An International Comparison of Industrial Relations.* (New York: Berg, 1990), p. 101.

19. *Evening Record,* Aug. 22, 1933, p. 1.

20. *Coal Age* 38 (Nov. 1938): 386–87. For a discussion of the preceding code-writing process for bituminous miners, see Coode and Bauman, *People, Poverty, and Politics,* pp. 142–43.

21. "Presentation of the Code of Fair Competition of the Anthracite Industry by Charles F. Huber, Chairman of the Code Committee," p. 8, RG 9, box 6061, National Archives.

22. NRA, Release no. 1455, Oct. 31, 1933, RG 9, box 6060; "Code of Fair Competition for the Anthracite Industry Submitted by the Anthracite Operators," and National Recovery Administration, *Proposed Code of Fair Competition for the Anthracite Industry* (Washington, D.C.: GPO, 1933), both in RG 9, box 6065, National Archives. The operators' typescript code and the administration's published code proposal are identical. For further coverage see *Coal Age* 38 (Nov. 1938): 386–87; *Evening Record,* Oct. 31, 1933, p. 1.

In 1907 the leading ten anthracite companies—all linked to major anthracite railroads—mined 78 percent of the region's output. In 1936, the leading ten firms mined or marketed fully 74 percent of total anthracite output. Jones, *Anthracite Coal Combination,* p. 104; Commonwealth of Pennsylvania, *Report of the Anthracite Coal Industry Commission* (Harrisburg: Murrelle, 1938), p. 331. Despite efforts on the part of the federal government to break up the anthracite cartel, all evidence indicates that the dominance of the leading firms remained undiminished over these decades.

23. *Coal Age* 38 (Dec. 1933): 426.

24. See the biographical sketch of Gildea by Tom Dublin in *Valley Gazette,* no. 276 (June 1995): 11–13.

25. "Speech of Hon. J. H. Gildea," 74th Cong., 1st sess., *Congressional Record* 79 (April 1, 1935), pt. 5: H 4788–90; James H. Gildea to Mr. McIntyre, Jan. 19, 1934, RG 9, box 6064, National Archives.

26. James H. Gildea to William H. Davis, Jan. 16, 1934, RG 1, box 6061, National Archives.

27. Resolutions of the Lithuanian Catholic Convention, May 27, 1934, and telegram to General Hugh Johnson from the Tamaqua NRA Committee, Nov. 14, 1933, both in RG 9, box 6060, National Archives.

28. "Statement of Thomas Kennedy," pp. 2, 3, 12, RG 9, box 6061, National Archives. See also *Coal Age* 38 (Dec. 1933): 427. For earlier evidence of UMWA support for equalization, see *Evening Record,* Sept. 20, 1933, p. 1; Nov. 18, 1933, p. 1. For a useful treatment of how John L.

Lewis came to support equalization as part of his effort to control rank-and-file activism in the anthracite region, see Douglas K. Monroe, "A Decade of Turmoil: John L. Lewis and the Anthracite Miners, 1926–1936" (Ph.D. diss., Georgetown University, 1976), pp. 219–31.

29. David Sherman, Tamaqua, to Franklin D. Roosevelt, Aug. 3, 1933, RG 9, box 6064, National Archives. Capitalization and punctuation as in original. For similar arguments, raised in the context of Shamokin and Coal Township (forty miles west of Tamaqua), see "Summary of Unemployment Situation, Shamokin-Coal Township Area—Shamokin, Penna., June 10th 1935," in Records of the Federal Mediation and Conciliation Service, RG 280, case 165–1812, National Archives.

30. "Statement of Thomas Kennedy," p. 15, RG 9, box 6061, National Archives. Although the national UMWA came around to support equalization, the issue remained controversial, and miners were appropriately skeptical concerning the conversion of their leaders.

31. "Anthracite Code: Presentation of 'The Equalization of Working Time,'" pp. 4, 5.

32. Ibid.; quotes, pp. 2, 6, 10.

33. J. B. Warriner to Deputy Administrator in charge of Anthracite Code, Dec. 5, 1933, RG 9, box 6064, National Archives, quotes on p. 4. See also *Evening Record,* Nov. 23, 1933, p. 1. Warriner was right in arguing that strip mining offered increases in labor productivity, but he was disingenuous in suggesting that stripping operations were a conservation measure. One only has to tour closed-up strippings to be struck by the environmental degradation they bring.

34. *Evening Record,* Oct. 4, 1933, p. 1; Oct. 10, 1933, pp. 1, 5; Oct. 11, 1933, p. 1; Oct. 24, 1933, p. 1; Oct. 25, 1933, p. 1; Nov. 3, 1933, p. 1.

35. J. B. Warriner to S. D. Warriner, 13 November 1933, LC&N, PSA. For the best description of this initial meeting and an indication of the community character of subsequent events, see Dr. John Pounder to John L. Lewis, December 12, 1933, District 7–President correspondence, Papers of the United Mine Workers of America, Historical Collections and Labor Archives, Pattee Library, Pennsylvania State University (hereafter cited as UMWA Papers).

36. Dr. John Pounder to John L. Lewis, December 12, 1933, District 7–President correspondence, UMWA Papers; see also a series of memos from W. J. Reese on "the strike situation," 6–13 December 1933, Labor Troubles folder, LC&N, PSA.

37. W. J. Reese to S. D. Warriner, 8 December 1933, Labor Troubles folder, LC&N, PSA.

38. *Evening Record,* Nov. 2, 1933, pp. 1, 6; John L. Lewis to Dr. John Pounder, December 15, 1933, District 7–President correspondence, UMWA Papers.

39. Dr. John Pounder to John L. Lewis, December 12, 1933, District 7–President correspondence, UMWA Papers. Pounder's correspondence proved important in convincing Lewis not to intervene in the conflict by revoking the charters of the Panther Valley locals pressing for equalization. See Monroe, "Decade of Turmoil," pp. 237–

39, for more on Lewis's exchanges with local leaders of the equalization movement.

40. *Evening Record,* Dec. 4, 5, 7, 11–16, 18–20, 26, 1933, all p. 1.

41. James H. Gildea to General Hugh S. Johnson, Jan. 16, 1934, RG 9, box 6061, National Archives. For a sample of one such biweekly equalization table, see *Coaldale Observer,* June 16, 1934, p. 1.

42. Once again, the internal memos of LC&N managers provide the richest view of the labor struggle for equalization in the winter and spring of 1934. See memos of 5–10 March 1934, and union commission proposal and statement, 13 March 1934, Labor Troubles folder, LC&N, PSA.

43. Eric Foner and John A. Garraty, eds., *Reader's Companion to American History* (Boston: Houghton Mifflin, 1991), p. 778.

44. *Evening Record,* Oct. 1, 1934, p. 1.

45. *Evening Record,* Oct. 13, 1934, p. 1.

46. Daily diary, Feb. 28, 1934, box 173, Papers of W. Jett Lauck (#4742), Special Collections, University of Virginia Library.

47. A similar story circulates in relation to resistance that miners in the territory of Reading Anthracite offered in opposition to efforts to wipe out coal bootlegging. As the story was told, "At the first appearance of the police, a signal has been given, a truck with a bell—the old tocsin of the insurrectionist and minute man—went speeding through the highways and side roads summoning the bootleggers from their holes." See John T. Flynn, "Bootleg Coal," *Collier's* (Dec. 3, 1936): 32.

48. Annual Report of the Mining Department, 1934, p. 18, LC&N-Smithsonian.

49. Annual Report of the Mining Department, 1933, p. 18, LC&N-Smithsonian.

50. *Report of Proceedings of The Twenty-sixth Successive Constitutional and Ninth Biennial Convention of DISTRICT No. 9, United Mine Workers of America, Held at Lykens, Pennsylvania, October 16th to 20th, 1934,* p. 109; *Evening Record,* Oct. 31, 1933, p. 1.

51. That even their district president, Michael Hartneady, remained unconvinced became apparent when he lashed out against the Panther Valley miners' achievement of equalization at the 1934 district convention. See *Evening Record,* Sept. 21, 1934, pp. 1, 5.

52. *Evening Record,* Aug. 14, 1937, p. 3, "Comparative Equalization Table," by Steve Gaydos.

53. "Tentative Agreement Between the Anthracite Operators and the United Mine workers of America, Districts 1, 7, and 9," [May 7, 1936], also reprinted in *Coal Age* 41 (June 1936): 267–68.

54. J. B. Warriner to S. D. Warriner, 3 September 1936, and unsigned LNC memo, 27 July 1937, LC&N, PSA.

55. *Evening Record,* Aug. 14, 1937, p. 3.

56. W. C. Trapnell, "The Employment Situation in the Pennsylvania Anthracite Region," pp. 11, 14, typescript document, Works Progress Administration, October 1935.

57. The economic history of DL&W in the years be-

fore the forced segregation of its coal lands is nicely summarized in Robert A. Janosov, "The Rise, Demise, and Resurrection of 'Blue Coal,'" in Sheldon Spear, *Chapters in Northeastern Pennsylvania History: Luzerne, Lackawanna, and Wyoming Counties* (Shavertown, Pa.: Jemags, 1999), pp. 135–37.

58. "Summary Review of Segregation and Divesting of Coal Properties and Distribution of 100% Stock Dividend, August 20, 1921," mimeographed document prepared by the law department, Lackawanna Railroad, January 1950, William W. Scranton Papers, box 78, folder 2, Historical Collections and Labor Archives, Pennsylvania State University (hereafter cited as HCLA).

59. These statistics are based on Glen Alden Company annual reports and tabulated annual returns made by the Company. See Commonwealth of Pennsylvania, Department of Mines and Mineral Industries, *Anthracite Division: Annual Reports, 1931–1945* (Harrisburg, Pa., 1960 reprint ed.), reports for 1931–1933, *passim.* Similar grim statistics on idled mines and unemployed miners, but dating from January 1934, are available in a report filed from the Scranton–Wilkes-Barre district to H. L. Kerwin, Director of Conciliation in the Department of Labor. See Records of the Federal Mediation and Conciliation Service, RG 280, box 276, case 170–9326, National Archives.

60. Monroe, "Decade of Turmoil," 146–52; quote on p. 152. For a critical reminiscence of the Lewis strong-arm tactics that gained approval of the anthracite tridistrict convention for this contract, see Steve Nelson, James R. Barrett, and Rob Ruck, *Steve Nelson: American Radical* (Pittsburgh: University of Pittsburgh Press, 1981), pp. 167–69.

61. Monroe, "Decade of Turmoil," 158.

62. John S. Drake, letter to the editor of the *Wilkes-Barre Independent,* 27 August 1933, in District 1–President Correspondence, UMWA Papers.

63. Monroe, "Decade of Turmoil," 189–94.

64. Joe Marianacch to John L. Lewis, March 17, 1932, District 1–President Correspondence, UMWA Papers. The original in Italian and an English translation survive together in the union files.

65. Monroe, "Decade of Turmoil," 201–09.

66. Monroe, "Decade of Turmoil," 240–65, and 285–322; quote is on p. 242; John Bodnar, *Anthracite People: Families, Unions, and Work, 1900–1940* (Harrisburg: Pennsylvania Historical and Museum Commission, 1983), pp. 52–53.

67. *Washington (D.C.) News,* Nov. 2, 1934, clipping in Records of the Federal Mediation and Conciliation Service, RG 280, case file 170–9326, National Archives; Bodnar, *Anthracite People,* pp. 1, 38, 47–54 and 94–95; Nelson *et al., Steve Nelson: American Radical,* pp. 169–72. The parallels between Maloney's assassination in 1936 and that of UMWA insurgent Jock Yablonsky in 1969 are so strong that it is fair to say that the authoritarian culture and resort to strong-arm tactics that characterized the UMWA's response to internal challenge in the 1930s had changed little as late as the 1960s. For a journalist's ac-

count of the later conflict within the union, see Brit Hume, *Death and the Mines: Rebellion and Murder in the United Mine Workers* (New York: Grossman, 1971).

68. Spear, *Chapters,* pp. 76–77, 84–90; Bodnar, *Anthracite People,* pp. 49, 52; Monroe, "Decade of Turmoil," 328; Nelson et al., *Steve Nelson,* pp. 171–72.

69. Monroe, "Decade of Turmoil," 318–19.

70. The following account relies heavily on annual reports of the Philadelphia & Reading Coal & Iron Company as well as a 1941 report of a bankruptcy examiner. See District Court for the Eastern District of Pennsylvania, "In the Matter of the Philadelphia and Reading Coal and Iron Company, Debtor. In Proceedings for the Reorganization of a Corporation, No. 19711: Summary of the Report of the Examiner, Nicholas G. Roosevelt, Examiner" [Dec. 1940] (hereafter cited as "Summary of the Report of the Examiner"). For a useful secondary account of the company's economic history, see James L. Holton, *The Reading Railroad: History of a Coal Age Empire,* vol. 2 (Laury's Station, Pa.: Garrigues House, 1992), especially chaps. 12–14.

71. The best explication of the intricacies of the federal antitrust case against the Reading Railroad that eventually required the ineffectual separation of producing and transporting functions appears in Thomas LeDuc, "Carriers, Courts, and the Commodities Clause," *Business History Review* 39 (1965): 70–72.

72. Annual Report of the P&RC&I Company, 1923, p. 15; Annual Report of the P&RC&I Company, 1928, p. 19; "Summary of the Report of the Examiner," 1941, p. 9.

73. Annual Report of the P&RC&I Company, 1928, pp. 8–9.

74. Annual Reports of the P&RC&I, 1935–1938; "Summary of the Report of the Examiner," 1941, pp. 1, 18.

75. Carter Goodrich, Hugh S. Hanna, and David J. Price, "Final Report of Anthracite Committee of Secretary of Labor, April 6, 1934," pp. 4–6, Department of Labor Library, Washington, D.C. See also "Summary of Unemployment Situation, Shamokin-Coal Township Area—Shamokin, Penna., June 10th 1935," in Records of the Federal Mediation and Conciliation Service, RG 280, case 165–1812, National Archives. For analysis of costs, see Richard R. Mead, "An Analysis of the Decline of the Anthracite Industry Since 1921" (Ph.D. diss., University of Pennsylvania, 1933), pp. 72–73. Unemployment in the anthracite region consistently exceeded that in Pennsylvania as a whole. In May 1935, for instance, WPA statistics placed the unemployment rate in the anthracite region at 27 percent compared to 19.5 percent for the state as a whole. Trapnell, "The Employment Situation in the Pennsylvania Anthracite Region," table A-19.

76. Rodney S. Lechleitner, "Bootlegging in Schuylkill County, 1930–1936" (M.A. thesis, Bloomsburg University, 1986), pp. 58, 61–62, 71.

77. Louis Adamic, *My America, 1928–1938* (New York: Harper & Brothers, 1938), p. 317; Flynn, "Bootleg Coal," 13. See also John C. Brennan, "Chapter II. Highlights of the Development of Bootleg or Illegal Mining,"

unpublished manuscript in Pennsylvania Writers Project, American Guide series, Job 54, Pennsylvania State Archives (hereafter cited as Pennsylvania Writers Project, Job 54).

78. *Report of the Anthracite Coal Industry Commission,* p. 44. For higher estimates on the number of coal bootleggers, upward of 20,000, including 13,000 direct miners, see Flynn, "Bootleg Coal," 12; Nelson et al., *Steve Nelson,* pp. 164–65.

79. Other estimates placed the number of bootleg operations at close to 5,000. See William H. Saye, "The Development and Present Status of the Bootleg Anthracite Industry and its Influence on Legitimate Producers" (M.A. thesis, Temple University, 1941), p. 40.

80. *Report of the Anthracite Coal Industry Commission,* p. 90; "Information Concerning Stolen Coal," Fall 1935, attachment to a letter from C. F. Huber, President, Anthracite Institute to W. Jett Lauck, March 29, 1937, box 225, Papers of W. Jett Lauck. Huber shared with Jett Lauck, chairman of the Anthracite Coal Industry Commission, a copy of the findings of a fall 1935 survey of coal bootlegging conducted by the Anthracite Institute, a company-sponsored coordinating and lobbying organization.

81. Interview with Henry and Dorothy Blum, June 8, 1995, pp. 10–12, 19. This account describes the bootleg operation of Henry's father on the property of the Reading Company in the Gilberton area. See also interview with Helen Mordock, May 15, 1995, pp. 27–29; Weir L. Shradley, "Bootleg Coal," 11, Field Note 150 (1936), Pennsylvania Writers Project, Job 54.

82. William F. Gustafson, "Bootleg Coal Mining, 1925–1953," *Proceedings of the Ninth Annual Conference on the History of Northeastern Pennsylvania: The Last 100 Years,* October 10, 1997, p. 4.

83. "Bootlegging," p. 1, July 7, 1939, Field Note 694, Pennsylvania Writers Project, Job 54; *Report of the Anthracite Coal Industry Commission,* p. 5.

84. Flynn, "Bootleg Coal," 30.

85. Saye, "The Development and Present State of the Bootleg Anthracite Industry," 70.

86. Quoted in Oliver Carlson, "Bootlegging Coal: A New Industry Appears on the Horizon," *Harper's Magazine,* April 1935, 616.

87. Quoted in Flynn, "Bootleg Coal," 34.

88. Ibid., 34.

89. Ibid., 30, 32. Another account described an even broader system of alarms: "Brewery whistles blow, school bells ring (and sometimes church bells too, with the clergyman on the rope), fire sirens whoop. Miners come on the run. Men, women, and children gather by the hundreds, by the thousands. They press in close about the police—and stare." Shradley, "Bootleg Coal," 4.

90. Carlson, "Bootlegging Coal," 620–21.

91. Gustafson, "Bootleg Coal Mining, 1925–1953," 10.

92. Adamic, *My America,* p. 322.

93. Saye, "The Development and Present State of the Bootleg Anthracite Industry," 82, 94–95.

94. Flynn, "Bootleg Coal," 32.

95. For the corporate perspective on coal bootleg-

ging, see "Memorandum of Opinion Expressed by George H. Jones, Superintendent of Stevens Coal Company, before the Pennsylvania Legislative Coal Commission at Shamokin, Northumberland County, PA., March 26, 1937," box 224, Papers of W. Jett Lauck; Flynn, "Bootleg Coal," 32, 34.

96. Flynn, "Bootleg Coal," 30.

97. Michael Kozura, "We Stood Our Ground: Anthracite Miners and the Challenge to Corporate Property, 1930–1941," in Staughton Lynd, ed., *"We Are All Leaders": The Alternative Unionism of the Early 1930s* (Urbana: University of Illinois Press, 1996), p. 213. The results of these police raids were not always favorable to the coal bootleggers. Malcolm Boyer recalled that as a ten- or eleven-year-old he witnessed his father being hauled off by Lehigh Valley Coal & Iron Police to Bloomsburg where he spent two days in jail before being released. See *Valley Gazette,* no. 300 (Aug. 1997): 14–17.

98. Flynn, "Bootleg Coal," 30.

99. Shradley, "Bootleg Coal," 4.

100. Randall M. Miller and William Pencak, eds., *Pennsylvania: A History of the Commonwealth* (University Park: Pennsylvania State University Press, 2002), pp. 296, 298; Coode and Bauman, *People, Poverty, and Politics,* 235–39.

101. See also Richard C. Keller, "Pennsylvania's Little New Deal," pp. 45–76 in John Braeman, Robert H. Bremner, and David Brody, eds., *The New Deal: The State and Local Levels* (Columbus: Ohio State University Press, 1975).

102. "Summary of Meeting Held by Governor George H. Earle at Elks Club, Mahanoy City, Tuesday, December 22, [1936,] with Leading Citizens of Mauch Chunk and Vicinity, in Connection with Tour of 'Bootleg' Coal Fields," pp. 1–2, box 222, Papers of W. Jett Lauck.

103. "Summary of Meeting Held by Governor George H. Earle at American Legion Building, Shamokin, Wednesday, December 23, [1936,] with Leading Citizens of Shamokin and Vicinity, in Connection with Tour of 'Bootleg' Coal Fields," p. 1, box 222, Papers of W. Jett Lauck.

104. *Coaldale Observer,* August 4, 1934, p. 1.

105. Adamic, *My America,* pp. 320–21; also quoted in Louis Adamic, "The Great 'Bootleg' Coal Industry," *The Nation* 140 (Jan. 9, 1935): 48.

106. Shradley, "Bootleg Coal," 16.

107. Shradley, "Bootleg Coal," 6.

108. United Polish Societies to Franklin Delano Roosevelt, 11 January 1934, Records of the Federal Mediation and Conciliation Service, RG 280, case 170–9360, National Archives.

109. T. B. Martin to Frances Perkins, 22 January 1934, Records of the Federal Mediation and Conciliation Service, RG 280, case 170–9360, National Archives.

110. Shradley, "Bootleg Coal," 15–16.

111. "Interview with Benjamin Riccotelli, Aug. 22, 1939, p. 2, Field Note 790, Pennsylvania Writers Project, Job 54.

112. Commonwealth of Pennsylvania, Anthracite Coal Industry Commission, *Report of Morris L. Ernst* (Harrisburg: n.p., 1937), p. 16.

113. Gustafson, "Bootleg Coal Mining, 1925–1953," 14.

114. *Evening Record,* Dec. 9, 1940, p. 1.

115. *Evening Record,* Feb. 6, 1941, p. 6; Saye, "The Development and Present State of the Bootleg Anthracite Industry," p. 93.

116. *Evening Record,* Feb. 6, 1941, p. 6.

117. *Evening Record,* Mar. 21, 1941, p. 1; July 31, 1941, p. 1; July 31, 1941, p. 1.

118. *The Shamokin Area Centennial: 1864–1964* [Shamokin, Pa., 1964], p. 67.

119. *Evening Record,* Mar. 17, 1942, p. 4; "Interview with Frank Colitz," Jan. 16, 1940, Field Note 970, Pennsylvania Writers Project, Job 54. Gustafson dates a formal end to bootlegging in 1953 when remaining small-scale lessees were brought under the jurisdiction of the Pennsylvania Department of Mines. See Gustafson, "Bootleg Coal Mining, 1925–1953," 14.

120. Local 1719 (Coaldale), Resolution, December 18, 1933, District 7–President correspondence, UMWA Papers.

4. Reprieve and Final Collapse, 1940–1970

1. J. A. Corgan and Marian I. Cooke, "Pennsylvania Anthracite," in U.S. Bureau of Mines, *Minerals Yearbook, 1945* (Washington, D.C.: Government Printing Office [GPO], 1945), table 25, p. 948.

2. U.S. Bureau of the Census, *Manufactures 1939: State series, Pennsylvania* (Washington, D.C.: GPO, 1942), pp. 4–5; U.S. Bureau of the Census, *Census of Manufactures: 1947,* vol. III, *Statistics by States* (Washington, D.C.: GPO, 1950), p. 47.

3. U.S. Bureau of the Census, *Census of Manufactures: 1947,* Vol. III, *Statistics by States,* pp. 533–34, provides county-level statistics for the garment industry in Lackawanna and Luzerne counties; adding in a reasonable estimate for Schuylkill County would suggest more than 25,000 garment workers in the anthracite region just after the close of World War II.

4. Arthur H. Reade, "The Impact of War on Pennsylvania Workers," Bulletin of the Pennsylvania Sate College, Bureau of Business Research, Bulletin no. 6 (1943), p. 6.

5. *Economic Conditions in the Anthracite Coal Regions: Letter from the Federal Anthracite Coal Commission,* 77th Cong., 2d sess., House Document 709 (Washington, D.C.: GPO, 1942), pp. 5–6.

6. Kathryn Sudol interview, Passaic, N.J., June 16, 1993, pp. 1, 4, 6.

7. Selina Woodring interview, Winfield, N.J., June 17, 1995, pp. 1, 4, 6–8.

8. For the original dues increase, see "37th Convention of UMWA Adjourns," *Anthracite Tri-District News,* Oct. 16, 1942, p. 2.

9. The origins of the January 1943 wildcat strikes can be dated to early September 1941 when mineworkers in Nesquehoning walked off their jobs to demonstrate against approved increases in union dues. Their protest spread through the Panther Valley and north to Hazleton and Wilkes-Barre, but it gained only limited support as various UMWA locals in the anthracite region had voted in favor of the new assessments. Disaffected miners returned to work in early October with the promise that Lewis would address the grievance. The issue continued to fester—with district 7 disciplined by the International—until a larger strike erupted in January 1943. See (Lansford) *Evening Record,* Sept. 10, 11, 12, 13, 22, 1941, all p. 1.

10. Records of the Federal Mediation and Conciliation Service, RG 280, see case 199–9070, box 663. The fullest contemporary account of the strike is offered in a Progress Report filed by Thomas Lambert, Commissioner of Conciliation, dated January 23, 1943, submitted to the U.S. Conciliation Service of the U.S. Department of Labor.

11. *Evening Record,* June 1, 10, 26, July 1, 6, 1943, all p. 1.

12. J. R. Sperry, "Rebellion Within the Ranks: Pennsylvania Anthracite, John L. Lewis, and the Coal Strikes of 1943," *Pennsylvania History* 40 (1973): 312; Makoto Takamiya, *Union Organization and Militancy: Conclusions from a Study of the United Mine Workers of America, 1940–1974* (Masenheim am Glan: Verlag Anton Hain, 1978), pp. 47–51; Melvyn Dubofsky and Warren Van Tine, *John L. Lewis: A Biography* (New York: Quadrangle/The NY Times, 1977), p. 439; Nelson Lichtenstein, *Labor's War at Home: The CIO in World War II* (Cambridge: Cambridge University Press, 1982), pp. 157–71.

13. Joseph McCluskey, president, and others of Local 1376, UMWA (Hazleton) to J. R. Steelman, director U.S. Conciliation Service, Dec. 19, 1942, in records of the Federal Mediation and Conciliation Service, RG 280, case 199–9070, box 663; Sperry, "Rebellion Within the Ranks," 297; undated, unsigned memo on 1942 anthracite dues strike, carton 32, Labor Troubles folder, LC&N Records, Pennsylvania State Archives.

14. Robert H. Zieger, *John L. Lewis: Labor Leader* (Boston: Twayne, 1988), p. 137. See also Walter T. Howard, "The National Miners Union: Communists and Miners in the Pennsylvania Anthracite, 1928–1931," *Pennsylvania Magazine of History and Biography* 125 (Jan./April 2001): 121. Zieger put the date of Lewis's last visit to the region in 1926; Howard placed it in September 1931, when UMWA officials came to Wilkes-Barre to resolve the Glen Alden equalization strike. Whatever the exact date, Lewis had alienated many anthracite miners by his standoffishness, and it is symptomatic of the divisions within the UMWA that anthracite miners alone appear to have protested the dues increase, while bituminous miners acceded to the International officers' leadership on this issue.

15. *Wilkes-Barre Times-Leader,* June 23, 1943, as quoted in Sperry, "Rebellion Within the Ranks," p. 311.

16. W. Julian Parton, *The Death of a Great Company: Reflections on the Decline and Fall of the Lehigh Coal and Navigation Company* (Easton, Pa.: Canal History and Technology Press, 1998; originally published in 1986), p. 41. There were signs of impending difficulties even with the strong performance. See "Lehigh's 123rd," *Business Week,* March 18, 1944, pp. 63–64.

17. Parton, *Death of a Great Company,* pp. 42, 44.

18. Ibid., pp. 41, 46, and 55.

19. *Evening Record,* Jan. 16, 20, Feb. 13, March 3, 1947; July 21, 1948, March 4, 1949, June 1, 1950, all p. 1.

20. Parton, *Death of a Great Company,* pp. 46, 47.

21. Panther Valley figures were slightly worse than those for the region as a whole. In 1944, anthracite mines operated 282 days on average—just over 5½ days per week. In 1953, the comparable figures for Coaldale No. 8 (one LNC mine) were 144 days, while the industry average that year was 163 days, barely 3 days per week on average. Frank Thomas DeLauritus, "Anthracite Coal: A Case Study of the Social Problems of a Declining Industry" (M.A. thesis in Labor and Industrial Relations, University of Illinois-Urbana, 1956), pp. 100, 184, and 206.

22. Parton, *Death of a Great Company,* p. 74.

23. The fullest renderings of these events are found in two sources that offer consistent accounts: DeLauritus, "Anthracite Coal," pp. 207–10; Parton, *Death of a Great Company,* pp. 71–81. See also Gregory S. Wilson, "Before the Great Society: Liberalism, Deindustrialization and Area Redevelopment in the United States, 1933–1965" (Ph.D. diss., Ohio State University, 2001), pp. 80–89. Contemporary accounts include *Coaldale Observer,* June 11, 1954, p. 1; *Philadelphia Inquirer,* June 14, 1954; *Time,* July 5, 1954, pp. 69–70. Longtime labor activist and retiree Ezra Cox addressed a mass meeting at Lansford Field, June 8, 1954. He supported the Tamaqua local and his presentation has been preserved on an audiotape from a local radio station's broadcast of the event. Tape and transcript in possession of the authors, courtesy of George Harvan. See also Wilson, "Before the Great Society," p. 85.

24. Frank Weir, "Mines Shut Till 4800 Raise Output," *Philadelphia Inquirer,* June 14, 1954, unpaginated clipping courtesy of George Harvan.

25. Parton, *Death of a Great Company,* pp. 74–81.

26. Only LC&N shares sold on the open market since the Lehigh Navigation Coal Company (LNC) was a wholly owned subsidiary of the parent company, hence our shift in terms here as we move from consideration of the mining operations of the firm to the broader economic fate of the parent holding company.

27. Parton, *Death of a Great Company,* pp. 63–69.

28. Ibid., p. 86.

29. Panther Valley Coal Company, Inc., Nesquehoning Outside Hourly Rates, September 7, 1954, District 25 UMWA Records, Collection 109, series 3, subseries A, box 7; Memorandum of Understanding between Coaldale Mining Company, Inc., and Local Unions Nos. 1536 and 1572 and Representatives of the District and International Organization, United Mine Workers of America, Nov. 16, 1954, and Supplementary Memorandum, May 20, 1957, District 25 UMWA Records, Collection 109, series 3, subseries A, box 2, Indiana University of Pennsylvania.

30. For annual production figures, see Commonwealth of Pennsylvania, Department of Mines, Anthracite Division, *Annual Reports,* 1952, 1957, and 1959; for employment figures for Coaldale Mining Company, see DeLauritus, "Anthracite Coal," pp. 207, 243, and 245; *Tamaqua Evening Courier,* Feb. 20, 1960, clipping in T. R.

Berger Collection, National Canal Museum, Easton, Pa.; *The Morning Call* (Allentown), Feb. 19, 1960, pp. 1, 7.

31. Dubofsky and Van Tine, *John L. Lewis,* p. 508.

32. Production figures are taken from Pa. Department of Mines annual reports, 1961–1970.

33. Parton, *Death of a Great Company,* pp. 91–97.

34. Ibid., pp. 108, 111.

35. Glen Alden production figures for 1930 and after come from the company's annual reports; for 1950, they come from the annual report of the Pa. Department of Mines. Industrywide statistics are found in DeLauritis, "Anthracite Coal," pp. 100 and 184. The Reading Anthracite company rivaled Glen Alden in much of this period, though with its filing for bankruptcy in 1937 and the closing of many of its collieries and the sale of much of its coal lands, the company emerged from World War II distinctly smaller than Glen Alden. See the annual reports for the Philadelphia & Reading Coal & Iron Corporation, 1923–1945. Good runs of both of these sets of annual reports are available at the Mudd Social Science Library at Yale University. For a useful history of the Philadelphia & Reading's coal operations and the impact of the company's bankruptcy filing, see "Summary of the Report of the Examiner," in Bankruptcy Reorganization Proceedings of P&RC&I Company, 1941. We have accessed a copy of this material in the Mudd Library, Yale University.

36. See the extensive correspondence between Lewis and Inglis in President-District 1 Correspondence file of the UMWA Papers, Historical Collections and Labor Archives (hereafter cited as HCLA).

37. Pa. Dept. of Mines, Annual Reports for 1917 and 1950.

38. Commonwealth of Pennsylvania, Department of Mines and Mineral Industries, Anthracite Division annual report, 1917 and *Anthracite Division: Annual Reports, 1931–1945* (Harrisburg, Pa., 1960 reprint ed.). This compilation of Pennsylvania annual reports permits one to reconstruct the operations of all major anthracite firms on an annual basis, recording employment and output at the level of individual collieries. Even as Glen Alden was cutting back on its production, it purchased a competitor, the Lehigh and Wilkes-Barre Coal Company, for $86 million in January 1930. See Robert A. Janosov, "The Rise, Demise, and Resurrection of 'Blue Coal,'" in Sheldon Spear, *Chapters in Northeastern Pennsylvania History: Luzerne, Lackawanna, and Wyoming Counties* (Shavertown, Pa.: Jemags, 1999), pp. 138–39.

39. 1954 Annual Report of the Glen Alden Coal Company, pp. 4, 10.

40. 1958 Annual Report of the Glen Alden Corporation, pp. 2, 9.

41. Tom Bigler, "The Glen Alden Story," p. 3, paper presented to the Luzerne County United Community Development Conference, Wilkes College, Wilkes-Barre, Pa., April 22, 1958.

42. Glen Alden annual reports: 1959, p. 2; 1960, pp. 1–2; 1961, p. 5; 1963, pp. 3–4.

43. Glen Alden annual reports: 1965, p. 2; 1966, p. 9; 1967, pp. 1–2, 27; 1969, pp. 1–2.

44. W. W. Everett to F. O. Case, interoffice memo,

May 1, 1954, in UMWA District 1 Correspondence, 1954–1960, HCLA. For an interesting analysis of what in November 1954 appeared to be the changing fortunes of Glen Alden, see a memo prepared by Donald Wales of Chicago, for the Glen Alden Board of Directors, box 78, folder 3, W. W. Scranton Papers, HCLA.

45. "Office Memorandum Re. Glen Alden," Jan. 14, 1954, W. W. Scranton Papers, box 78, folder 2, HCLA; *Wilkes-Barre Times-Leader,* May 9, 1956, Feb. 19, 1966, both stories in Glen Alden clippings file, Osterhout Library, Wilkes-Barre, Pa. For a parallel secondary account of the decline of Glen Alden's coal operations, see Janosov, "The Rise, Demise, and Resurrection of 'Blue Coal,'" 140–45.

46. Janosov, "The Rise, Demise, and Resurrection of 'Blue Coal,'" pp. 145–48.

47. For the early history of the St. Clair Coal Company, see Anthony F. C. Wallace, *St. Clair: A Nineteenth Century Coal Town's Experience with A Disaster-Prone Industry* (Ithaca: Cornell University Press, 1981), pp. 105–12, 441.

48. James L. Holton, *The Reading Railroad: History of a Coal Age Empire. Volume II: The Twentieth Century* (Laury's Station, Pa.: Garrigues House, 1992), p. 99.

49. Jules I. Bogen, *The Anthracite Railroads* (New York: Ronald Press, 1927), pp. 230–32.

50. The records of the bankruptcy proceedings for the P&RC&I, Bankruptcy Case 19711, are held in the National Archives and Records Administration, Philadelphia Branch (hereafter cited as NARA).

51. *Report of Mart F. Brennan, President of District No. 9, U. M. W. of A., To Twentieth-Eighth Consecutive and Eleventh Biennial Convention, at Williamstown, October 18th, 1938,* pp. 28–32. In the George J. Curilla Collection, Anthracite Heritage Museum, Scranton, Pa. (hereafter cited as Curilla Collection).

52. Affidavit, Aug. 2, 1939, box 10, NARA; Memorandum Opinion, March 27, 1939, box 5, NARA.

53. Petition to Increase Salary, June 15, 1940, box 10, NARA; Petition to Increase Salary, June 30, 1941, box 11, NARA; *New York Times,* July 13, 1938, p. 33. See chapter 3 for a discussion of the impact of these developments on the emergence of coal bootlegging on P&RC&I lands.

54. Report of Special Examiner, Nov. 14, 1940, box 1, p. 200, NARA.

55. Judge William Kirkpatrick replaced Judge Dickinson on Dickinson's death. Dickinson had appointed Howard Benton Lewis as his chief deputy, with the title of Special Master, and Lewis was the active voice in the court throughout the bankruptcy proceedings.

56. Sur Applications for Allowances, July 27, 1945, box 16, NARA.

57. Petition to Dispose of Lands, June 22, 1937, box 2, NARA; Petition to Dispose of Lands, Oct. 19, 1937, box 2, NARA; Special Master's Report on Petition to Sell Property, Oct. 19, 1938, box 3, NARA.

58. Continued Hearing on Petition for Amendment of Order Continuing Debtor in Possession, Dec. 27, 1937, box 2, NARA; Memorandum Opinion in Re: Motion to Vacate Order Leaving Debtor in Possession and to Appoint a Trustee, March 27, 1939, box 5, NARA; Hearing, Aug. 21, 1939, box 11, app., pp. 5–7, NARA; Report of Special Examiner, Nov. 14, 1940, box 1, app., p. 79, NARA.

59. Petition for the Appointment of a Trustee, Nov. 20, 1937, box 2, NARA.

60. Petition, Aug. 17, 1938, box 3, NARA; Petition, Oct. 5, 1938, box 3, NARA.

61. Petition of Frank Farsky, March 18, 1943, box 13, NARA; Answer of Debtor to Petition of Frank Farsky, May 11, 1943, box 13, NARA.

62. Petition for Order on Debtor in Possession to Pay Taxes Due Petitioners, April 10, 1939, box 7, NARA; Answer of Debtor to Petition to Direct Payment of Taxes, April 14, 1939, box 7, NARA; Sur Petition to Direct Payment of Taxes, May 10, 1939, box 7, NARA; Decree on Tax Issue, Dec. 29, 1943, box 17, NARA.

63. UMW Answer to Petition to Dispose of Lands and Borrow Money, June 25, 1938, box 3, NARA; Special Master's Report on Petition to Sell Property, Oct. 19, 1938, box 3, NARA; UMW Exceptions to Report of Special Master, Oct. 26, 1938, box 3, NARA.

64. Plan of Reorganization, March 14, 1944, box 16, NARA; Order Confirming Plan, July 10, 1944, unboxed envelope, NARA; Statement of Identity and Qualifications of Officers, unboxed envelope, NARA.

65. "Coal Company Gets Off Hook," *Business Week,* April 20, 1957, p. 100.

66. Ibid., pp. 99–100, 105.

67. *New York Times,* Jan. 5, 1955, p. 31.

68. *Pottsville Republican,* Dec. 30, 1955, p. 1; *New York Times,* Feb. 3, 1961, p. 31.

69. *New York Times,* Feb. 18, 1956, p. 23; April 1, 1959, p. 27; March 30, 1961, p. 37.

70. *New York Times,* March 21, 1958, p. 31; Dec. 15, 1959, p. 58; June 1, 1960, p. 53.

71. *New York Times,* Dec. 15, 1959, p. 58.

72. Stockholders' Report, 1936, March 22, 1937, St. Clair Coal Company Papers, Hagley Library (hereafter cited as SCCCP), Accession 1232, File 6, p. 2; Stockholders' Report, SCCCP, Accession 1232, File 6, March 26, 1938, p. 2; Stockholders' Report, SCCCP, Accession 1232, File 6, March 27, 1939, p. 2; *Pottsville Republican,* Nov. 16, 1957, p. 3.

73. Stockholders' Report, 1939, SCCCP, accession 1232, file 6, p. 1.

74. Ibid., p. 1; early leases signed between the Philadelphia and Reading Coal and Iron Company and the St. Clair Coal Company with stipulations for coal to be shipped on the Philadelphia and Reading Railroad can be found in SCCCP, accession 1663, box 3; *Pottsville Republican,* Sept. 28, 1957, p. 11.

75. Letter of President of St. Clair Coal Company to Mr. Edward Fox, Oct. 12, 1955, SCCCP, accession 1240, folder 18.

76. Letter of Jessie E. Smyth, Sept. 18, 1957, SCCCP, accession 1240, folder 18.

77. Minutes of Special Meeting of Board of Directors, Oct. 19, 1957, SCCCP, Accession 1663, box 1, pp. 1–2; Meeting held in the Office of the St. Clair Coal Company, Sept. 18, 1957, SCCCP, accession 1240, folder 18, p. 3.

78. *Pottsville Republican,* Sept. 18, 1957, p. 1; Sept. 21, 1957, p. 1.

79. Meeting held in the Office of the St. Clair Coal Company, Sept. 18, 1957, SCCCP, accession 1240, folder 18, p. 8.

80. *Pottsville Republican,* Nov. 11, 1957, pp. 1, 3.

81. Minutes of Special Meetings of Board of Directors, SCCCP, accession 1663, box 1, May 24, 1958, Oct. 31, 1959, Oct. 15, 1960.

82. Letter of Elwyn Jones to H. Gordon Smyth, May 9, 1979, SCCCP, accession 1663, box 3, file E.

83. *Pottsville Republican,* July 16, 1957, p. 1.

84. *Pottsville Republican,* July 17, 1957, p. 1; July 19, 1957, p. 1; July 20, 1957, p. 1.

85. Minutes of Special Meetings of Board of Directors, SCCCP, accession 1663, box 1, May 24, 1958.

86. A typical UMWA effort in this area is evident in "Statement of Mr. Thomas Kennedy, United Mine Workers of America, in the matter of the Scranton Spring Brook Water Service Company, Docket No. G-10,251 before the Federal Power Commission, Washington, D.C., Friday, September 14, 1956," in District 25 UMWA Records, Collection 109, series 1, subseries A, box 2, Indiana University of Pennsylvania.

87. *Anthracite Tri-District News,* 25 July 1941, 3 Aug. 1945, 11 July 1947, 11 Jan. 1952, 5 Feb. 1954; "Great Lakes St. Lawrence Seaway System: Seaway History," http://www.greatlakes-seaway.com/en/aboutus/seaway_history.html (accessed March 9, 2005).

88. *Anthracite Tri-District News,* 21 Nov. 1947.

89. These proportions are calculated from the Web site of the Energy Information Administration, "Consumption Estimates by Source, Selected Years, 1960–2000, Pennsylvania," http://www.eia.doe.gov/emeu/states/sep_use/total/use_tot_pa.html (accessed March 9, 2005). We needed to estimate the share that anthracite contributed to the proportion for coal and did so using state of Pennsylvania annual production figures for bituminous and anthracite at http://www.dep.state.pa.us/dep/deputate/minres/bmr/annualreport/2000/, tables 1 and 19 (accessed March 8, 2005). By 2000, the disparity widened with natural gas's share at 16 percent and that for anthracite at 1.7 percent.

90. To place these figures in context, in 1900, Pennsylvania anthracite had employed 144,000 workers, compared to 205,000 in bituminous coal nationwide (see table 3 in appendix 1).

91. Ivana Krajcinovic, *From Company Doctors to Managed Care: The United Mine Workers' Noble Experiment* (Ithaca: Cornell University Press, 1997), p. 27.

92. Krajcinovic, *From Company Doctors to Managed Care,* pp. 28–32; Zieger, *John L. Lewis,* p. 154.

93. Dubofsky and Van Tine, *John L. Lewis,* pp. 465–67. For a discussion that places the UMWA's approach as a response to the failure of union efforts to enact federal health insurance legislation, see Alan Derickson, "Health Security for All?: Social Unionism and Universal Health Insurance, 1935–1958," *Journal of American History* 80 (March 1994): 1345–46.

94. Local No. 1738 to John L. Lewis, March 26, 1944

and Resolution of Panther Valley General Mine Committee, March 17, 1945, both in President-District 7 Correspondence, 1932–1962, UMWA Papers, HCLA.

95. Ezra O. Koch, Local 1571 to John L. Lewis, May 14, 1947, and William Williams, Local 1738, to Thomas Kennedy, Jan. 14, 1948, both in President-District 7 Correspondence, 1932–1962, UMWA Papers, HCLA.

96. Thomas Kennedy, UMWA Vice-President, to Joseph Agor, Oct. 22, 1948, President-District 9 Correspondence, UMWA Papers, HCLA.

97. Krajcinovic, *From Company Doctors to Managed Care,* p. 147; Robert J. Myers, "Experience of the UMWA Welfare and Retirement Fund," *Industrial and Labor Relations Review* 10 (1956–57): 94–95. There is an extensive literature on the founding, growth, and decline of the UMWA Welfare and Retirement Fund for bituminous mineworkers. See, e.g., Barbara Berney, "The Rise and Fall of the UMW Fund," *Southern Exposure* 6 (1978): 95–102; Richard Mulcahy, "Partitioning the Miner's Welfare State: The Destruction of the Medical Program of the UMWA Welfare and Retirement Fund," *Mid-America* 77 (Spring/Summer 1995): 175–204; Richard J. Mulcahy, *A Social Contract for the Coal Fields: The Rise and Fall of the United Mine Workers of America Welfare and Retirement Fund* (Knoxville: University of Tennessee Press, 2000). On the broad implications of the Fund for organized labor more generally, see Jennifer Klein, *For All These Rights: Business, Labor, and the Shaping of America's Private-Public Welfare State* (Princeton, N.J.: Princeton University Press, 2003), pp. 197–200, 202, 219–21.

98. Managers of the Fund set rehabilitation as a major, initial goal of the health plan, and by 1955, some 97,000 disabled miners had received treatment. See Berney, "The Rise and Fall of the UMW Fund," 97.

99. Krajcinovic, *From Company Doctors to Managed Care,* chaps. 4 and 5, and pp. 173–75. Edwin W. Stock Jr. and William C. Gipe, "Health Care in Appalachia: The United Presbyterian Church, the United Mine Workers, and the U.S.A.," *American Presbyterians* 71 (Summer 1993): 115–26; quotation on p. 116.

100. Krajcinovic, *From Company Doctors to Managed Care,* fig. 15, p. 151.

101. On the overall UMWA-management consensus that developed in bituminous in the 1950s and 1960s, see David Brody, "Labour Relations in American Coal Mining: An Industry Perspective," in Gerald D. Feldman and Klaus Tenfelde, eds. *Workers, Owners and Politics in Coal Mining: An International Comparison of Industrial Relations* (New York: Berg, 1990), pp. 74–117, especially pp. 111–12.

102. Krajcinovic, *From Company Doctors to Managed Care,* table 13, p. 149.

103. Ibid., pp. 152–54.

104. Mulcahy, *A Social Contract for the Coal Fields,* pp. 129–30, 133–34; Krajcinovic, *From Company Doctors to Managed Care,* pp. 156–57; Stock and Gipe, "Health Care in Appalachia," pp. 117–25; *Redevelopment* 1 (May 1964): 1, 8.

105. Krajcinovic, *From Company Doctors to Managed Care,* table 13, p. 149; "Historical Data, 1931–

1977—Pennsylvania anthracite mines in the United States," table 2A, Web site of the Department of Labor, Mine Safety and Health Administration, Statistics, http://www.msha.gov/STATS/PART50/WQ/1931/wq31an 02.htm (accessed July 17, 2003).

106. For bituminous figures, see Krajcinovic, *From Company Doctors to Managed Care,* fig. 13, p. 148; for anthracite, see annual reports for the Anthracite Health and Welfare Fund, 1952–1973, accessed at the Hazleton office of the Fund, July 2003.

107. Maier B. Fox, *United We Stand: The United Mine Workers of America, 1890–1990* (n.p.: United Mine Workers of America, 1990), p. 457, provides the total of retirees; for active workers, see Commonwealth of Pennsylvania, Department of Mines, Anthracite Division, *Annual Report,* 1958. For typical communications describing the Fund's financial difficulties, see "To All Pensioners of the Anthracite Health and Welfare Fund," Sept. 18, 1958, in files of the Anthracite Health and Welfare Fund, Hazleton, Pa.; August J. Lippi to Frank B. Ellis, June 12, 1961, District 25 UMWA Records, Collection 109, series 1, subseries A, box 2, Indiana University of Pennsylvania.

108. Testimony of Joseph P. Brennan in the "Transcript of Proceedings, Pension Benefit Guaranty Corporation, 21 May 1975," p. 37, *In Re: Anthracite Health and Welfare Fund Application for Partial Termination of Benefit Payments,* in files of the Anthracite Health and Welfare Fund, Hazleton, Pa.

109. *Scranton Times,* Oct. 23, 1962, and *The Inquirer,* Feb. 4, 1962, two clippings in President's Correspondence Files, Health and Welfare Fund, 1958–1962, UMWA Papers, HCLA.

110. For a well-documented example of this logic, see the file on the Fund's application to the Pension Benefit Guaranty Corporation for a partial termination of pension benefits, 1974–1975, accessed at the Anthracite Health and Welfare Fund, Hazleton, Pa., July 2003.

111. The first federal loan to support anthracite pensions came in June 1981. See PBGC news release, June 19, 1981. This financial support was only finally repaid by the Fund and the UMWA in 1997. PBGC news release, Dec. 30, 1997; *Hazleton Standard-Speaker,* Dec. 31, 1997, clipping provided by PBGC. For recent discussion of subsequent agency difficulties as increasing numbers of employer pension plans have come under economic pressure, see Mary Williams Walsh, "An Outsider's Grim Prognosis for Pension Agency," *New York Times,* Sept. 14, 2004, pp. C1, C22.

112. Statistics computed from published annual reports of the Anthracite Health and Welfare Fund, 1952–1958.

113. For discussion of the agreement with the Jefferson Medical College to conduct research and provide limited treatment for victims of Black Lung, see Anthracite Health and Welfare Fund, *Report for the Year Ending June 30, 1952* (Hazleton, Pa., n.d.), pp. 7–11; Minutes of Trustees of the Anthracite Health and Welfare Fund, 29 Jan. 1974.

114. Memorandum, Michael H. Sheridan, Attorney for the Anthracite Health and Welfare Fund, to Thomas H. Kennedy, President, UMWA, Sept. 29, 1960, President's Correspondence Files, Health and Welfare Fund, 1958–1962, UMWA Papers, HCLA.

115. Memo from John L. Lewis to Harry M. Moses, March 12, 1951 and Moses reply memo, April 9, 1951, both in UMWA, President-District 9 Correspondence, UMWA Papers, HCLA.

116. *Anthracite Tri-District News,* Oct. 21, 1960, p. 1.

117. Dubofsky and Van Tine, *John L. Lewis,* p. 506; Krajcinovic, *From Company Doctors to Managed Care,* p. 152.

118. This overall rate of return is calculated from the published annual financial reports of the UMWA Welfare and Retirement Fund. The rate is based on the average interest and dividend income annually over the period divided by the average end-of-year reserves for the Fund.

119. Dubofsky and Van Tine, *John L. Lewis,* p. 508.

120. *Blankenship v. Boyle,* 337 F. Supp. 1089; 1971 U.S. Dist. LEXIS 13548, and 337 F. Supp. 296; 1972 U.S. Dist. LEXIS 15652. The first decision here, dated April 28, 1971 provided the court's substantive decision on the case, while the second, dated Jan. 7, 1972, addressed the damage award and the court's reasoning in limiting damages in this way. Both accessed via LexisNexis, Sept. 7, 2003. For a useful account of the decline of the UMWA Welfare and Retirement Fund as a contributor to the social welfare of bituminous miners that goes into far more detail than we are able to do here, see Berney, "The Rise and Fall of the UMW Fund."

121. Data on Fund interest income and reserves come from annual reports for 1951–1960 on file at the Fund's offices in Hazleton, Pa. The final estimate is modest, as the 3 percent hypothetical interest rate is well below the 5 percent figure employed by Judge Gesell in settling on damages in the bituminous case, *Blankenship v. Boyle,* 337 F. Supp. 296; 1972 U.S. Dist. LEXIS 15652, Jan. 7, 1972.

122. Dubofsky and Van Tine, *John L. Lewis,* p. 506.

123. Ibid., pp. 507, 509. See also A. H. Raskin, "John L. Lewis—'A Glorious Anachronism,'" *New York Times Magazine,* Feb. 13, 1955, p. 35.

124. *New York Times,* April 17, 1960, p. 7, as quoted in Dubofsky and Van Tine, *John L. Lewis,* p. 508; see also *Anthracite Tri-District News,* April 23, 1960, p. 1. For a pointed analysis of the ultimate failure of the UMWA's business unionism to secure long-term security for coal miners or the union, see Brody, "Labour Relations in American Coal Mining," pp. 112–17.

125. Klein, *For All These Rights,* p. 244.

126. Quote is from Barbara Ellen Smith, *Digging Our Own Graves: Coal Miners and the Struggle over Black Lung Disease* (Philadelphia: Temple University Press, 1987), p. 63; see also Curtis Seltzer, *Fire in the Hole: Miners and Managers in the American Coal Industry* (Lexington: University Press of Kentucky, 1985), p. 95.

127. Kenneth C. Wolensky, "Living for Reform," *Pennsylvania Heritage* 27 (Winter 2001): 17.

128. Seltzer, *Fire in the Hole,* p. 100; Alan Derickson, *Black Lung: Anatomy of a Public Health Disaster* (Ithaca: Cornell University Press, 1998), pp. 147–65.

129. As quoted in Wolensky, "Living for Reform," 18; "King's College Observes Legacy of Late Daniel J. Flood," press release, Nov. 13, 2000, http://www.kings.edu/pr/releases/releases/danfloodcent03.html (accessed March 9, 2005).

130. Derickson, *Black Lung,* pp. xi, 166–82; quote is from p. 181. See also Seltzer, *Fire in the Hole,* chap. 7.

131. Dubofsky and Van Tine, *John L. Lewis,* pp. 116–18, 450; quote on p. 450; Takamiya, *Union Organization and Militancy,* pp. 42–43, 53.

132. Brit Hume, *Death and the Mines: Rebellion and Murder in the United Mine Workers* (New York: Grossman, 1971), p. 184.

133. *Report of Mart F. Brennan, President of District No. 7, From Oct. 9, 1941 to Oct. 1, 1946,* pp. 7–9; quote on p. 7. Pamphlet in Curilla Collection.

134. Letter from David J. Stevens to John L. Lewis, May 3, 1954 with accompanying accounting of repayment of District 7's debt to the International; see also minutes, International Executive Board Meeting, May 3–6, 1954. Both in President-District 7 Correspondence, UMWA Papers, HCLA. In contrast to these communications, letters from members and officers of locals complained bitterly about the lack of democracy in the district. See sample letters on this issue: John Koller, Local Union 1572, to John L. Lewis, Nov. 15, 1953. In this letter the Lansford local's recording secretary reported two "requests" from that local. In the first his membership called on "the International Executive Board [to] set a date for election off [sic] officers and field workers, [to] be elected by popular vote." In justifying the request, Koller noted "The membership feel that they are self sustaining and now are ready for autonomy." See also Local 1572 to John L. Lewis, Oct. 12, 1954. Both letters are in President-District 7 Correspondence, UMWA Papers, HCLA.

135. Hume, *Death and the Mines,* p. 185. For an example of continuing local pleas for the restoration of democracy, see the correspondence of local 4004 in Tamaqua in December 1962, calling for the election of the district president and other officers. It had been twenty years without such elections, but the International Executive Board deferred action on the request. President-District 7 Correspondence, UMWA Papers, HCLA.

136. For biographical sketches of Kennedy, see "Golden Anniversary Testimonial Dinner for the Honorable Thomas Kennedy, International Vice-President, United Mine Workers of America," Sept. 4, 1958, program in the Curilla Collection; "Kennedy Sprang from Ranks to Lofty Niche in Miners' Union," in "Lansford Centennial Supplement," *Evening Record,* Aug. 27, 1946, as reprinted in *Valley Gazette,* no. 362 (October 2002): 9.

137. Dubofsky and Van Tine, *John L. Lewis,* pp. 473, 497, and 507–8.

138. William J. Sneed to John L. Lewis, Nov. 6, 1938, President-District 9 Correspondence, UMWA Papers, HCLA.

139. Report of Martin Brennan, President of District 9, Oct. 18, 1938, p. 3, President-District 9 Correspondence, UMWA Papers, HCLA.

140. For reminiscences noting wages in this range,

see John Bodnar, *Anthracite People: Families, Unions and Work, 1900–1940* (Harrisburg: Pennsylvania Historical and Museum Commission, 1983), pp. 52, 60.

141. Stephen Moriak Jr., to UMW Organization, June 13, 1950, President-District 1 Correspondence, UMWA Papers, HCLA.

142. John Owens to Stephen Moriak Jr., July 19, 1950, President-District 1 Correspondence, UMWA Papers, HCLA.

143. See Harry Welby to Thomas Kennedy, June 22, 1951 reporting on the misuse of local funds by officers of Local 7779 of Ashley. The allegations were upheld through an audit and hearings whose records are in this same correspondence file. For the restoration of privileges, see John Owens to Walter Baran, Nov. 25, 1953, and the accompanying digest of the case. Finally, the protest of the IEB action came signed by nine "Officers and Committee-men" of the local, Jan. 14, 1954. All of these documents are in President-District 1 Correspondence, UMWA Papers, HCLA.

144. Similar proceedings on July 29, 1953, revealed the improper handling of initiation fees by Local 466 officers over a four-year period. Owens's response is contained in John Owens to August J. Lippi, Aug. 20, 1953, President-District 1 Correspondence, UMWA Papers, HCLA.

145. Stephen Sporay to John L. Lewis, March 10, 1953, President-District 9 Correspondence, UMWA Papers, HCLA.

146. Anonymous to John L. Lewis, from "Fight for Right," typescript copy of a letter dated by hand as January 1950, President-District 1 Correspondence, UMWA Papers, HCLA.

147. Robert P. Wolensky, Kenneth C. Wolensky, and Nicole H. Wolensky, *The Knox Mine Disaster, January 22, 1959: The Final Years of the Northern Anthracite Industry and the Effort to Rebuild a Regional Economy* (Harrisburg: Pennsylvania Historical and Museum Commission, 1999) offers the fullest account of the disaster and the paragraphs that follow rely heavily on their study. Since we completed this study, the Wolenskys have edited a valuable anthology of primary sources that speaks to the issues we explore here. We wish we had been able to draw on this material. See Robert P. Wolensky, Kenneth C. Wolensky, and Nicole H. Wolensky, eds. *Voices of the Knox Mine Disaster: Stories, Remembrances, and Reflections on the Anthracite Coal Industry's Last Major Catastrophe, January 22, 1959* (n.p.: Pennsylvania Historical and Museum Commission, 2005).

148. Wolensky, Wolensky, and Wolensky, *Knox Mine Disaster,* pp. 70–71, and 77.

149. Ibid., pp. 88–98, and 101.

150. Ibid., pp. 98–100.

151. Hume, *Death and the Mines,* pp. 177–96; Wolensky, "Living for Reform," pp. 17–20.

152. Trevor Armbrister, *Act of Vengeance: The Yablonski Murders and Their Solution* (New York: E.P. Dutton, 1975), p. 150.

153. Hume, *Death and the Mines,* pp. 240–41; Joseph E. Finley, *The Corrupt Kingdom: The Rise and Fall of the*

United Mine Workers (New York: Simon and Schuster, 1972), chap. 11.

154. Hume, *Death and the Mines,* pp. 249–59; *New York Times,* July 9, 1982, June 1, 1985 (accessed via Lexis-Nexis, Sept. 7, 2003). See also Wolensky, "Living for Reform."

5. Industrial Development Efforts

1. The emphasis here will be on the efforts of the three largest and best-documented industrial development groups—those in Scranton, Wilkes-Barre, and Hazleton. However, there were more than three dozen industrial development corporations across the anthracite region. For brief accounts of some of the smaller operations that are part of the larger database developed later in the chapter but are passed over in the narrative portion of this analysis, see "Our Purposeful Pioneers," *SECDO Newsletter* (Winter 2001/2002), http://www.sed-co/newsletter_905F2002905F01/article1.asp (accessed March 9, 2005); Panther Valley Industrial Commission, *Panther Valley—Opportunities for New Industries* (Lansford, Pa., [1946]). Another excellent source on the industrial development corporations in the region is testimony presented in U.S. Senate, Committee on Labor and Public Welfare, *Area Redevelopment. Hearings before the Subcommittee on Labor of the Committee on Labor and Public Welfare, United States Senate, Eighty-fourth Congress, second session, on S. 2663, a bill to establish an effective program to alleviate conditions of excessive unemployment in certain economically depressed areas . . . ,* 2 vols. (Washington, D.C.: Government Printing Office [GPO], 1956), 1:156–61, 424–38, 464–69, 471–74, 484–86.

2. For an overview of the earliest efforts in the region, see Harold Aurand, "Diversifying the Economy of the Anthracite Region, 1880–1900," *Pennsylvania Magazine of History and Biography* 94 (Jan. 1970): 54–61. For specific examples for particular communities, see *Greater Scranton Chamber of Commerce: 100 Years of Service, 1867–1967,* pp. 12, 21; Christopher Sterba, "Family, Work, and Nation: Hazleton, Pennsylvania, and the 1934 General Strike in Textiles," *Pennsylvania Magazine of History and Biography* 120 (1996): 7–9; Harold Landau, "The Industrial Development in the Wilkes-Barre Area" (M.A. thesis, University of Scranton, 1967), pp. 26–29.

3. *Greater Scranton Chamber of Commerce,* pp. 12, 21, 22, 29, and 35.

4. Scranton Chamber of Commerce, "The Scranton Area and the Advantages It Offers as an Industrial Site," [1942,] typeset pamphlet; Scranton Chamber of Commerce, "Boost Scranton, Make Jobs," [1942,]. Both pamphlets in the research files of the Greater Scranton Chamber of Commerce, 212 Mulberry St., Scranton, Pa., courtesy of Austin Burke, President.

5. Michael J. Saada, " 'The Scranton Plan': Mining Town Licks Its Slump: New Industries Replace Vanishing Coal," *Wall Street Journal,* March 31, 1949, pp. 1, 4; George D. Wolf, *William Warren Scranton: Pennsylvania Statesman* (University Park: Pennsylvania State University Press, 1981), pp. 35–36.

6. *Greater Scranton Chamber of Commerce,* p. 37;

"Dedication: Scranton Plant, Murray Corporation of America," and "The Murray Corp. of America," [April 1944,] both in LIFE folder, Greater Scranton Chamber of Commerce files, Scranton, Pa. For parallel efforts in Wilkes-Barre during this period, see Landau, "Industrial Development in the Wilkes-Barre Area," pp. 35–40.

7. Scranton Lackawanna Industrial Building Company, *SLIBCO: Fifty Years and Building* (Scranton: SLIBCO, 1995), p. 8.

8. "Scranton Bets the Future," *Time,* April 15, 1946, p. 26.

9. *The New Scranton: A Dynamic, Diversified Industrial City* [Scranton, 1946], Scranton Public Library, unpaginated.

10. Scranton Chamber of Commerce, "The New Scranton: A Dynamic, Diversified Industrial City," [Scranton: Scranton Chamber of Commerce, 1946], unpaginated pamphlet in research files of the Greater Scranton Chamber of Commerce.

11. The employment figures for the Murray Corporation come from linkage in successive editions of the *Industrial Directory of the Commonwealth of Pennsylvania,* 1947, 1953, and 1956. Saada, "The Scranton Plan," pp. 1–2, offers a useful contemporary overview of developments in the period. Quotations from two untitled pamphlets in LIFE folder, Chamber of Commerce files, Scranton, Pa. Additional contemporary coverage of industrial development initiatives can be found in "Operation Boot Strap: How Third District Communities Are Using Their Own Capital to Build Industry," *Business Review* (Dec. 1949): 120–27. This last journal is the publication of the Third Federal Reserve District covering the eastern two-thirds of Pennsylvania.

12. "Scranton Attracts New Industry as It Fights Hard Coal Hard Times," *Wall Street Journal,* July 20, 1956, p. 1; "Tenth Annual Report of Lackawanna Industrial Fund Enterprises," 1960, pp. 9–11; "Operation Boot Strap," p. 124. See also "Statement of Ernest D. Preate," in U.S. Senate, Committee on Labor and Public Welfare, *Area Redevelopment,* 1:464–65.

13. Sterba, "Family, Work, and Nation," pp. 7–9; *The CAN DO Story: A Case History of Successful Community Industrial Development* (Hazleton: CAN DO, 1974), pp. 2–3.

14. *CAN DO Story,* p. 5.

15. Dan Rose, *Energy Transition and the Local Community: A Theory of Society Applied to Hazleton, Pennsylvania* (Philadelphia: University of Pennsylvania Press, 1981), p.112; *Upon the Shoulders of Giants: The CAN DO Story* (Hazleton, Pa.: CAN DO, 1991), pp. 1–2. We traced employment figures for the company through Pennsylvania industrial directories between 1950 and 1977, the last year for which the company was listed.

16. *CAN DO Story,* pp. 8–15, 24–26, and 38–39.

17. Sheldon Spear, *Chapters in Northeastern Pennsylvania History: Luzerne, Lackawanna, and Wyoming Counties* (Shavertown, Pa.: Jemags, 1999), pp. 2–3. See two stories about these development committees in *Wilkes-Barre Wyoming Valley Progress,* 1:1 (Sept. 1929), p. 1. Two runs of this newsletter (for 1929–1936 and

1957–1961) are available in the Chamber of Commerce files, Luzerne County Historical Society, Wilkes-Barre, Pa. (hereafter cited as WBCofC files). The Chamber of Commerce launched this newsletter with the startup of the committees, evidence that even before the stock-market crash, members of the Wilkes-Barre economic elite were concerned about the region's economic prospects.

18. *Progress*, Sept. 1929, pp. 1 and 6; *Progress*, Feb. 1930, p. 5; Lockwood Greene Engineers, Inc., "Industrial Survey of the Wyoming Valley, Pennsylvania," 1930, in WBCofC files.

19. Landau, "Industrial Development in the Wilkes-Barre Area," chap. 2; Sheldon Spear, "Life After An-thracite: Wyoming Valley's Economic Recovery in the 1950's and 1960's," *The History of Northeastern Pennsylvania* 2 (1990): 26.

20. Spear, "Life After Anthracite," 28–30. For an account of the work of the Committee of One Hundred, see "Statement of William O. Sword," in U.S. Senate, Committee on Labor and Public Welfare, *Area Redevelopment*, 1:156–61.

21. Landau, "Industrial Development in the Wilkes-Barre Area," pp. 75 and 83; Spear, "Life After Anthracite," 34–36. Spear comments that "Community involvement [in Wilkes-Barre] was never as broad as in Hazleton." See p. 42, n. 27.

22. Landau, "Industrial Development in the Wilkes-Barre Area," p. 85.

23. For UMWA support for Leader's efforts, see Thomas Kennedy to August J. Lippi, Feb. 10, 1958, District 25 UMWA Records, Collection 109, series 1, sub-series A, box 2, Indiana University of Pennsylvania.

24. For a summary of this legislation before its ultimate passage, see "Statement of Hon. George M. Leader," in U.S. Senate, Committee on Labor and Public Welfare, *Area Redevelopment*, 1:261. For the specifics of the initial passage of PIDA legislation, see Wilson, "Before the Great Society," 89–94.

25. Landau, "Industrial Development in the Wilkes-Barre Area," p. 63; Commonwealth of Pennsylvania, Office of Budget and Administration, *The Pennsylvania Industrial Development Authority: An Assessment* (Harrisburg, Pa., 1979).

26. "Loan Projects of the Pennsylvania Industrial Development Authority: May, 1956 through December, 1970," a computer report provided the authors by PIDA. These counties comprised three of sixty-seven counties in the state, so clearly their share of PIDA funds was far greater than their share of either the state's economy or its population.

27. Rose, *Energy Transition and the Local Community*, pp. 152–55; *CAN DO Story*, pp. 29–30; *Upon the Shoulders*, pp. 33, 103.

28. Pennsylvania Power & Light Company, "Report on Industrial Development, 1940—Annual Report," pp. 1–2; PP&L Industrial Development Department, 1950 Annual Report, pp. 69 and 72.

29. Joseph X. Flannery, "John Davidson: His Mission Ends," *Scranton Times*, Sept. 20, 1988, and "Area's Strug-gle Was Never Easy," *Scranton Times*, Sept. 22, 1988; John Fischer, "The Lazarus Twins in Pennsylvania: How Scranton and Wilkes-Barre Are Rising from the Dead," *Harper's Magazine* (Nov. 1968): 14.

30. Landau, "Industrial Development in the Wilkes-Barre Area," pp. 54 and 86; for PP&L interest and contributions in Wilkes-Barre at still earlier dates, see Spear, "Life After Anthracite," 40, n. 3.

31. For a fuller discussion of this source and how we linked firms recorded here to a variety of other local records, see appendix 6. Our thanks to Mr. Andrew Kelhart of the Marketing and Economic Development Department of the Pennsylvania Power & Light Company who provided access to the annual reports of the Economic Development Department that became the basis for this portion of our study.

32. See, e.g.,, Gregory Wilson, "'Our Chronic and Desperate Situation': Anthracite Communities and the Emergence of Redevelopment Policy in Pennsylvania and the United States, 1945–1965," *International Review of Social History* 47:S10 (2002): table 4, p. 152.

33. In putting together the database, we began by recording information from the PP&L annual reports between 1940 and 1987. We proceeded to link the named firms in the variety of sources described here. However, in presenting our findings in this chapter, we limit ourselves to firms that relocated into or expanded operations in the anthracite region between 1940 and 1970, thus permitting us to trace all firms in the study population for at least twenty-five years after they were first enumerated in the PP&L annual reports.

34. For discussions of this process based on oral history interviews that offer regional workers' views on developments, see Thomas Dublin and Walter Licht, "Gender and Economic Decline: The Pennsylvania Anthracite Region, 1920–1970," *Oral History Review* 27 (Winter/Spring 2000): 1–17; for fuller discussion of migration patterns that accompanied industrial decline, see Thomas Dublin, "Working-class Families Respond to Industrial Decline: Migration from the Pennsylvania Anthracite Region since 1920," *International Labor and Working Class History* 54 (Fall 1998): 40–56.

35. Rose, *Energy Transition and the Local Community*, pp. 111–15.

36. These trends in terms of recruiting firms from outside PP&L territory were consistent over the period examined. As early as 1940, the PP&L "Report on Industrial Development" (p. 3) noted that "the metropolitan area of New York was the most productive." During the previous year, some 18 prospects had relocated into PP&L territory from metropolitan New York, in comparison to 5 firms that relocated within PP&L territory, and another 3 that moved from the Philadelphia area.

37. Chi-square here was 141 for a two-by-two table with a significance of less than .0001. Relocation was without a doubt the most influential factor influencing whether or not a firm received local or PIDA assistance.

38. The data discussed in these two paragraphs come from tables provided by the Bureau of Statistics, Pennsylvania Department of Labor and Industry. The tables offer

evidence on overall employment and employment by economic sectors by county for 1955, 1960, 1965, 1970, 1975, and 1980. Our thanks to Marikay Cady for her assistance with our queries.

39. Sar A. Levitan, *Federal Aid to Depressed Areas: An Evaluation of the Area Redevelopment Administration* (Baltimore: Johns Hopkins University Press, 1964), chap. 1; Glenn Banks Fatzinger, "A Descriptive Study of the Area Redevelopment Administration (ARA)-Economic Development Administration (EDA) University Center Program, 1963–1974" (Ed.D. diss., George Washington University, 1977), pp. 41–42.

40. Charles Weissman et al., Northeast Pennsylvania Industrial Development Commission, to Sen. Edward J. Martin and others, March 24, 1954, Committee of Twelve Papers, 1930–1954, Historical Collections and Labor Archives, Pennsylvania State University.

41. Gregory Wilson, "Deindustrialization, Poverty, and Federal Area Redevelopment in the United States, 1945–1965," in Jefferson Cowie and Joseph Heathcott, eds., *Beyond the Ruins: The Meanings of Deindustrialization* (Ithaca: Cornell University Press, 2003), pp. 181–98; U.S. Senate, Committee on Labor and Public Welfare, *Area Redevelopment*, vol. 1, *passim.*

42. U.S. Senate, *Area Redevelopment Act: Hearings before a Subcommittee of the Committee on Banking and Currency, United States Senate, Eighty-Sixth Congress, First Session*, pt. I, Feb. 25, 26, and 27, 1959 (Washington, D.C.: GPO, 1959), p. 96. For other Pennsylvania testimony, see pp. 75–94, and 205–27.

43. Levitan, *Federal Aid to Depressed Areas*, pp. 30–31; Wilson, "Deindustrialization, Poverty, and Federal Area Redevelopment," 190; Fatzinger, "A Descriptive Study of the Area Redevelopment Administration," 49–54.

44. *Redevelopment* 1 (July 1964): 6–8; 1 (August 1964): 4; 1 (Nov. 1964): 8; 2 (June/July 1965): 6–7.

45. Levitan, *Federal Aid to Depressed Areas*, pp. 41–46, 60–99; Wilson, "Deindustrialization, Poverty, and Federal Area Redevelopment," 190–96. Bruce J. Shulman shares the judgment offered here about the problems of the short-lived ARA: *From Cotton Belt to Sun Belt: Federal Policy, Economic Development, and the Transformation of the South, 1938–1980* (New York: Oxford University Press, 1991), pp. 184–85.

46. Michael Harrington, *The Other America: Poverty in America* (New York: Macmillan, 1962). See also Harry M. Caudill, *Night Comes to the Cumberlands: A Biography of a Depressed Area* (Boston: Little, Brown, 1963).

47. Michael Bradshaw, *The Appalachian Regional Commission: Twenty-Five Years of Government Policy* (Lexington: University Press of Kentucky, 1991), chap. 3; quote, p. 39.

48. Monroe Newman, *The Political Economy of Appalachia: A Case Study in Regional Integration* (Lexington, Mass.: Lexington Books, 1972), pp. 100–104; Web site of the Appalachian Regional Commission, "Programs and Initiatives: Local Development Districts Program," http://www.arc.gov/index.do?nodeId=20 (accessed April 8, 2004); Bradshaw, *Appalachian Regional Commission*, p. 55.

49. For an early report on the work of the EDCNP see, Appalachian Regional Commission, *Annual Report: 1968*, p. 91.

50. Appalachian Regional Commission, *Annual Report: 1966*, p. 36, *Annual Report: 1967*, pp. 65–66, *Annual Report: 1968*, pp. 77–78, *Annual Report: 1969*, pp. 81–82.

51. See entries for Flood and McDade on the Political Graveyard Web site at http://politicalgraveyard.com (accessed March 9, 2005); Diana Evans, *Greasing the Wheels: Using Pork Barrel Projects to Build Majority Coalitions in Congress* (New York: Cambridge University Press, 2004), pp. 6–7. See also Spear, "Life after Anthracite," 33; Miller and Sharpless, *Kingdom of Coal*, pp. 325–32; Joe Skeen, "Tribute to the Honorable Joseph M. McDade," http://thomas.loc.gov/cgi-bin/query/z?r105:E08OC8-25 (accessed August 30, 2004).

52. Joseph McDade, "Reports from Washington," Winter 1992.

53. Commonwealth of Pennsylvania, Office of Budget and Administration, *The Pennsylvania Industrial Development Authority [PIDA]: An Assessment* (Harrisburg, Pa.: n.p., 1979).

54. Ibid., pp. 4–6.

55. Ibid., p. 9.

56. The turning point in terms of the shift in the distribution of state PIDA aid must have come in 1968, following passage of the new state law. Analysis of aid between 1956 and 1970, suggests that for the first eleven years, PIDA aid was directed disproportionately toward depressed areas within the state. Between 1968 and 1979, however, the distribution favored the well-heeled counties in the state, hence the critical tone of the *Assessment.*

57. *PIDA: An Assessment*, pp. i, 11–12, 14.

58. An additional way to consider the "success" of PIDA loans is to explore the extent to which state loans to expanding or relocating businesses became a commonly employed tactic in reindustrialization efforts across the country. One study concluded that "state direct loans represent a minor effort" in state and local development practices. By 1985, twenty-two states provided direct loans to businesses, but the PIDA program alone accounted for 64 percent of the total of such loans. Other states did not follow Pennsylvania's lead. Peter K. Eisinger, *The Rise of the Entrepeneurial State: State and Local Economic Development Policy in the United States* (Madison: University of Wisconsin Press, 1988), pp. 154–55.

59. *CAN DO Story; Upon the Shoulders*; Rose, *Energy Transition and the Local Community.*

60. *CAN DO Story*, p. 7.

61. *CAN DO Story*, pp. 5–6; *Upon the Shoulders*, p. 1.

62. *CAN DO Story*, pp. 7–10; *Upon the Shoulders*, p. 3.

63. "Primer for Giving," [1956], pamphlet in Scrapbook #1, CAN DO offices, Hazleton.

64. *CAN DO Story*, pp. 11–15; *Upon the Shoulders*, pp. 5–11.

65. *CAN DO Story*, pp. 16–17; *Upon the Shoulders*, pp. 21, and 29–34.

66. "Local Group Approves 'Hazleton Council of Civic Clubs' As Name," Jan. 30, 1959; "Labor Pledges Its

Support To CAN-DO Drive Next Month," March 23, 1959, both in CAN DO Scrapbook #2.

67. *Hazleton Standard Speaker* article, March 15, 1963, as reprinted in *Upon the Shoulders*, pp. 30–34.

68. *Upon the Shoulders*, pp. 21–22.

69. Ibid., p. 22. Although Dan Rose offers little concrete evidence on Hazleton workers' responses to CAN DO activities, he acknowledges the problems arising from the area's endemic low wages. See Rose, *Energy Transition and the Local Community*, p. 148.

70. Craig Thompson, "The Valley That Came Back to Life," *Saturday Evening Post*, Oct. 31, 1953, p. 139; *Coaldale Observer*, July 17, 1947, p. 1, Sept. 5, 1947, p. 1, Nov. 21, 1947, p. 1.

71. (Lansford) *Evening Record*, Sept. 14, 1945, p. 1, Sept. 23, 1945, p. 1, April 14, 1947, p. 1; *Coaldale Observer*, Feb. 8, 1946, p. 1.

72. *Coaldale Observer*, March 8, 1946, p. 1.

73. *Coaldale Observer*, March 14, 1947, p 1, Sept. 5, 1947, p. 1; *Evening Record*, April 16, 1947, p. 1, Dec. 11, 1947, p. 1, May 25, 1948, p. 1.

74. *Coaldale Observer*, March 1, 1946, p. 1.

75. *Coaldale Observer*, July 24, 1953, p. 1.

76. *Evening Record*, April 16, 1947, p. 1.

77. *Coaldale Observer*, Oct. 7, 1949, p. 1.

78. *Coaldale Observer*, Oct. 3, 1952, pp. 1–2.

79. *Evening Record*, Sept. 26, 1953, p. 1.

80. *Evening Record*, Dec. 8, 1954, p. 1.

81. *Coaldale Observer*, April 24, 1953, p. 1; *Evening Record*, Oct. 1, 1954, Oct. 12, 1954, p. 1.

82. *Evening Record*, July 20, 1955, p. 1, June 6, 1956, p. 1, Dec. 17, 1956, p. 1, Dec. 21, 1956, p. 1.

83. *Evening Record*, Dec. 21, 1956, p. 1, Jan. 11, 1957, p. 1, Jan. 14, 1957, p. 1, Jan. 16, 1957, p. 1.

84. *Evening Record*, June 6, 1958, p. 1, Sept. 17, 1958, p. 1, Sept. 19, 1958, p. 1, Sept. 26, 1958, p. 1, Sept. 30, 1958, p. 1, Nov. 6, 1958, p. 1.

85. *Evening Record*, Nov. 17, 1958, p. 1, Dec. 2, 1958, p. 1, July 22, 1959, p. 1, July 23, 1959, p. 1.

86. *Evening Record*, Jan. 27, 1961, p. 1.

87. Statistics on redevelopment initiatives in the Panther Valley come from the PP&L database examined earlier in this chapter.

88. Steve Pecha interview, July 29, 1993, p. 19; Mary Matrician interview, October 12, 1994, p. 9.

89. Pecha interview, pp. 19–20.

90. Grant Gangaware interview, May 17, 1994, p. 11.

91. Michael Mikovich interview, July 30, 1993, p. 13.

92. Irene Gangaware interview, Sept. 5, 1993, pp. 11–12.

93. Pecha interview, p. 20.

94. Mike Sabron interview, June 28, 1993, p. 48; William Wrightson interview, June 2, 1995, pp. 12–13; Mary Mogilski interview, Aug. 29, 1994, p. 12.

95. Helen Lazar and John Lazar interview, Aug. 24, 1994, pp. 31–32; Ziggie Whitecavage interview, April 8, 1995, p. 27. See also Mike Knies interview, July 19, 1994, p. 16. A similar story about a possible Ford plant in Hazleton is noted and described as "completely unfounded and untrue" in *Upon the Shoulders*, p. 79. The veracity of the

story is less important than the fact that it keeps reappearing throughout the anthracite region and is believed to be true by many residents and migrants.

96. Pennsylvania Power & Light, Area Redevelopment Division, Annual Report, 1955, p. 2 and the PP&L database; Robert Daniels interview (pseudonym), May 26, 1994, pp. 10–14; quotation is from p. 13.

97. John Bodnar, *Remaking America: Public Memory, Commemoration, and Patriotism in the Twentieth Century* (Princeton, N.J.: Princeton University Press, 1992), pp. 13–15; John Bodnar, *Our Towns: Remembering Community in Indiana* (Indianapolis: Indiana Historical Society, 2001), p. xv.

98. Melvin L. Burstein and Arthur J. Rolnick, "Congress Should End the Economic War Among the States," *The Region: Federal Reserve Bank of Minneapolis, 1994 Annual Report* 9 (March 1995), http://minneapolisfed.org/pubs/ar/ar1994.html (accessed Aug. 30, 2001).

99. Timothy Egan, "Towns Hand Out Tax Breaks, Then Cry Foul as Jobs Leave," *New York Times*, Oct. 20, 2004, pp. A1, A18.

100. Andy Zipser, "Civil War, Round Two," *Barron's* (April 3, 1995): 24.

6. Personal Responses to Decline: Fathers and Mothers, 1950–1990

1. Thomas Dublin, *When the Mines Closed: Stories of Struggles in Hard Times*, photographs by George Harvan (Ithaca: Cornell University Press, 1998), p. 56. See also Thomas Dublin, "Life After the Mines Closed," *Pennsylvania Heritage* 25 (Spring 1999): 6–15 (photographs by George Harvan). The Sabron narrative in *When the Mines Closed* is a recasting of portions of his much longer oral history interview, Summit Hill, Pa., June 28–29, 1993.

2. Dublin, *When the Mines Closed*, pp. 64–65.

3. George F. Deasy and Phyllis R. Griess, "Atlas of Pennsylvania Coal and Coal Mining, Part II: Anthracite," *Bulletin of the Mineral Industries Experiment Station* (University Park, Pa.: College of Mineral Industries, Penn State University, 1959–1963), p. 34; Thomas Dublin, "Working-Class Families Respond to Industrial Decline: Migration from the Pennsylvania Anthracite Region since 1920," *International Labor and Working-Class History* 54 (1998): 43.

4. Data discussed here are drawn from the Integrated Public Use Microdata Samples accessible at the Web site of the Minnesota Population Center at http://www.ipums.org/. The U.S. Public Use Microdata Samples have been reconfigured by the Minnesota Population Center as the Integrated Public Use Microdata Series (or IPUMS). A description of the original sample and a codebook for its variables is provided in U.S. Bureau of the Census, *Census of Population, 1950 (United States): Public Use Microdata Sample* (Ann Arbor, Mich.: Inter-university Consortium for Political and Social Research [ICPSR], 1984). The cases assembled for the 1950 census included all sampled individuals from State Economic Areas (SEAs) 365, 373, and 374, including the counties of Lackawanna, Carbon, Columbia, Monroe, Montour, Northumberland, Pike, Schuylkill, and Luzerne. Three of the counties

found in these SEAs—Columbia, Monroe, and Pike—are not dependent on mining, but there is no way to disaggregate them from anthracite counties. By focusing the analysis on families of mineworkers we minimize any complications that arise from the overinclusiveness of the SEAs sampled. For further discussion of our methods working with the IPUMS data, see appendix 5.

5. For a similar argument on the connection between occupational crowding and discrimination in employment, but in this case focused on ethnic crowding, see Oscar Handlin, *Boston Immigrants: A Study in Acculturation* (New York: Atheneum, 1972; originally published in 1941), pp. 57–60, 252. For useful evidence on the crowding of working women within a limited number of occupations in Boston in 1850 and 1900, see Thomas Dublin, *Transforming Women's Work: New England Lives in the Industrial Revolution* (Ithaca: Cornell University Press, 1994), pp. 159, 237. Claudia Goldin emphasizes the importance of the creation of distinct male and female sectors within the emerging area of clerical employment as a contributor to wage differentials between men and women; see *Understanding the Gender Gap: An Economic History of American Women* (New York: Oxford University Press, 1990), pp. 110–17.

6. The average age of miners had actually declined somewhat in the years after World War II. A survey of fifty-three anthracite companies in 1943 had found the mean age of mineworkers to be 45 with 46 percent concentrated in the 45–64 age bracket. *Anthracite Institute Bulletin*, no. 1322 (July 6, 1944): 3. According to the 1950 IPUMS sample, the mean age of mineworkers was a bit below 41. This decline no doubt resulted from the rehiring of younger miners with the return of members of the armed services after the war.

7. Statistics here are based on 532 cases in the one-in-eight random sample drawn from the employment records of Lehigh Coal & Navigation. See appendix 2 for further discussion of methods employed in this study.

8. We have explored these issues in more detail in two journal articles: Dublin, "Working-Class Families Respond to Industrial Decline" (see note 3 above), and Thomas Dublin and Walter Licht, "Gender and Economic Decline: The Pennsylvania Anthracite Region, 1920–1970," *Oral History Review* 27 (Winter/Spring 2000): 1–17.

9. "Population of Counties by Decennial Census: 1900 to 1990," compiled and edited by Richard L. Forstall, Population Division, U.S. Bureau of the Census, http://www.census.gov/population/cencounts/pa190090.txt.

10. "Population of Counties by Decennial Census: 1900 to 1990," http://www.census.gov/population/cencounts/pa190090.txt (accessed March 9, 2005).

11. U.S. Bureau of the Census, *County and City Data Book, 1962: A Statistical Abstract Supplement* (Washington, D.C.: GPO, 1962), p. 312; Lowell Eugene Galloway, "Depressed Industrial Areas: A National Problem" (Ph.D. diss., Ohio State University, 1959), p. 52. For a parallel treatment of anthracite-region outmigration for the 1930–1950 period, see Theodore Bakerman, *Anthracite Coal: A Study in Advanced Industrial Decline* (New York: Arno Press, 1979; originally completed as a University of Pennsylvania Ph.D. dissertation in Economics in 1956), chap. 4.

12. U.S. Senate, Committee on Labor and Public Welfare, *Area Redevelopment. Hearings before the Subcommittee on Labor of the Committee on Labor and Public Welfare, United States Senate, Eighty-fourth Congress, second session, on S. 2663 . . .* , 2 vols. (Washington, D.C.: GPO, 1956), 1:69, 438; Miller and Sharpless, *Kingdom of Coal*, p. 326; Frank H. Weir, "The Hard Coal Facts: Tonnage Down 70 Pct. in 36 Years," *Philadelphia Inquirer*, June ?, 1954, undated clipping courtesy of George Harvan.

13. Migrants' interviews that support the conclusion offered here include: Joseph McHugh (Wilmington and Fairless Hills), Paul Melovich (Trenton, N.J.), Jim Coon, Mary Painter, Ziggie Whitecavage (Fairless Hills and Levittown), John and Helen Mordock (Morrisville, Pa.), Andrew Andrusko (Winfield, N.J.), Don Hunsinger (Carteret, N.J.), John Valinski (Fair Lawn, N.J.), and Selina Woodring (Winfield, N.J.). For anthracite migration to Bridgeport, Conn., see *Economic Conditions in the Anthracite Coal Regions: Letter from the Federal Anthracite Coal Commission*, 77th Cong., 2d sess., House Document 709 (Washington, D.C.: GPO, 1942), pp. 5–6.

14. Interview with Joe Rodak, June 6, 1995, pp. 23–24. The hanging around of unemployed former miners was not always so purposeful, as one former Summit Hill resident recalled: "after the company started closing down the mines the men would stand on corners and in barrooms just passing the time away—'their time.'" Comments of a 1959 graduate of George Washington High School in Summit Hill submitted along with his completed questionnaire for the Panther Valley High School Survey. For more on the survey, see chapter 7 and appendix 8.

15. Interview with Paul Melovich, June 7, 1995, pp. 8, 26–29.

16. Weir, "The Hard Coal Facts"; David Popenoe, *The Suburban Environment: Sweden and the United States* (Chicago: University of Chicago Press, 1977), offers extensive comparisons of Levittown, Pa., and a new suburb of Stockholm, both established in the 1950s. Ely Chinoy, *Automobile Workers and the American Dream* (Boston: Beacon Press, 1972; originally published in 1965). For a thoughtful analysis of the General Motors plant in Linden, N.J., focusing more on the 1980s and 1990s than the 1950s, see Ruth Milkman, *Farewell to the Factory: Auto Workers in the Late Twentieth Century* (Berkeley: University of California Press, 1997), pp. 27–28, 53–55. For further discussion of migration from the anthracite region, see Dublin, "Working-class Families Respond to Industrial Decline."

17. This synopsis is based on an interview with Mary Painter, Mount Joy, Pa., Sept. 16, 1994. All cited interviews have been transcribed, and tapes and transcriptions will eventually be deposited at the Pennsylvania State Archives in Harrisburg to permit public access. Our thanks to the Ford Foundation and the New Jersey His-

torical Commission for financial support for this portion of the research.

18. David Schuyler, "Exhibit Review: Reflections on Levittown at Fifty," *Pennsylvania History* 70 (Winter 2003): 101–9.

19. Interview with Mary McHugh, May 15, 1995, p. 12; Interview with Paul Melovich, p. 33. Out-migrants to New Jersey had similar stories on the availability of tract housing, even without the presence of a Levittown development. See interview with Don Hunsinger, June 16, 1995, p. 29.

20. (Lansford) *Evening Record*, Aug. 2, 1954, pp. 5, 8.

21. *Evening Record*, Aug. 8, 1954, p. 1. See also *Evening Record*, Aug. 10, 1951, p. 1, Aug. 18, 1952, p. 1, Jan. 30, 1957, p. 1.

22. Interview with Joseph McHugh, May 15, 1994, p. 22; interview with Joe Rodak, pp. 6–7. See also interview with Henry and Dorothy Blum, June 8, 1995, p. 9. On the same pattern in a patch town in the northern anthracite field, see Donald J. Rowland, *Whites Crossing* (Carlisle, Pa.: Sunset Publications, 1995), p. 3.

23. Dublin, *When the Mines Closed*, p. 214; Interview with Helen Mordock, May 15, 1995, p. 20. See also interview with Nellie Valinski, Nov. 6, 1995, p. 16.

24. American Religion Data Archive, http://www.thearda.com/archive/CMS852CNT.html (accessed March 20, 2002); Popenoe, *Suburban Environment*, p. 124. Another contemporary survey places the proportions of Protestants and Catholics among Levittown homeowners at 49 and 31 percent respectively, further supporting the point we are making here. William H. Whyte Jr., *The Organization Man* (Garden City, N.Y.: Doubleday Anchor, 1956), p. 407.

25. The Panther Valley high school graduates' survey is discussed further in chapter 7 and appendix 8. For Levittown's class structure, see Popenoe, *Suburban Environment*, pp. 121–22, 130.

26. Harry Henderson, "Rugged American Collectivism: The Mass-Produced Suburbs, Part II," *Harper's Magazine* 207 (Dec. 1953): 81.

27. Interview with Mary Painter, pp. 11–13, 25–26; David Diamond, "The Children of Levittown, *Inquirer Magazine*, Dec. 12, 1982, p. 36.

28. Whyte, *Organization Man*, pp. 317, 330–31; Popenoe, *Suburban Environment*, pp. 137–38.

29. Interview with William Wrightson, June 2, 1995, p. 27; interview with Joe Rodak, p. 5; Dublin, *When the Mines Closed*, p. 237, quotation on p. 247.

30. Popenoe, *Suburban Environment*, p. 124. See also David R. Vásquez, "Forty Years Later, The American Dream Lives On," *Levittown Express*, Levittown 40th Anniversary Supplement, p. 21. On the Pittsburgh migration to the Fairless Hills area, see Dublin, *When the Mines Closed*, pp. 242, 246; Interview with James Coon, May 14, 1995, pp. 11–12.

31. Interviews with John Mordock, May 15, 1995, p. 1; Albert Rodzinak, April 20, 1995, p. 19; Joseph McHugh, May 15, 1995, p. 17.

32. Interview with Don Hunsinger, p. 30; interview with Paul Melovich, p. 32; interview with Joe Rodak, p. 28.

33. Interview with Mike Vitek and Mary Vitek, Nov. 24, 1995.

34. Interview with John Pavuk, April 20, 1994, p. 15–16; Interview with Mike Vitek and Mary Vitek, p. 23. Surprisingly, with all the references to long-distance commuting following the closing of the mines, interviewees rarely mentioned any resulting marital discord. For a different perspective, see the testimony of Min Matheson, of the International Ladies' Garment Workers' Union in U.S. Senate, Committee on Labor and Public Welfare, *Area Redevelopment*, 1:71–77. See also Barbara Kingsolver, *Holding the Line: Women in the Great Arizona Mine Strike of 1983* (Ithaca, N.Y.: ILR Press, 1989), p. 178.

35. Interview with Mike Vitek and Mary Vitek.

36. Interview with Elizabeth Mikovich and Bob Sabol (pseudonyms), Oct. 3, 1994, p. 14; Interview with Mike Knies, July 19, 1994, pp. 12–13; Interview with Robert Daniels (pseudonym), May 26, 1994, p. 19.

37. The oral history interviews speak strikingly to the irrelevance of the syndrome of the "happy housewife heroine" to the lives of working-class women in the anthracite region in this period. See Betty Friedan, *The Feminine Mystique* (New York: W.W. Norton, 1963), chap. 2.

38. Interview with Lillian Verona (pseudonym), Oct. 5, 1994, p. 16; Interview with Theresa Mogilski, Sept. 1, 1994, p. 10; [Washington] *Evening Star*, July 7, 1954, clipping in Committee of Twelve Papers, 1930–1954, Historical Collections and Labor Archives, Pennsylvania State University. See also interview with Henry and Dorothy Blum, p. 38.

39. Interview with Mary S. (pseudonym), Aug. 23, 1994, p. 27; Interview with Mary Daniels (pseudonym), p. 16.

40. David Glassberg, "Sense of History in Pennsylvania: Work, Craft, Ethnicity, and Place," *Pennsylvania History* 60 (Oct. 1993): 520.

41. These statistics come from analysis of the public-use microdata samples, but for an excellent discussion of these general trends, see Howard Harris, ed., *Keystone of Democracy: A History of Pennsylvania Workers* (Harrisburg: Pennsylvania Historical and Museum Commission, 1999), chaps. 5 and 6.

42. The changes in women's employment in the region were mirrored across the state as a whole. See Harris, *Keystone of Democracy*, pp. 231–32.

43. The national gap between men and women in labor force participation declined steadily after World War II. For adult men the overall proportion in the labor force fell from 87 to 75 percent between 1948 and 1994. For women the comparable figures rose from 32 to 59 percent in the same period. Cynthia Costello and Barbara Kivimae Krimgold, eds., *The American Woman, 1996–97: Where We Stand* (New York: W.W. Norton, 1996), pp. 46–47, 50.

44. This conclusion is based on analysis of the 1950 and 1990 IPUMS samples for the anthracite region.

45. We began with an occupational cohort of mineworkers in 1950 that was concentrated in the 35- to 45-year-old age group. Because of the decline of mining

over the period we could best approximate that group in future years by treating it as an age cohort. That is how we constructed the sense of change over time reported here.

46. Unfortunately, the public use sample for 1960 employed no geographical variable that would permit one to draw an anthracite region sample for that year, hence we are limited to employing data from the 1970, 1980, and 1990 censuses for this cohort analysis. For useful information on each of the decennial IPUMS samples, see Steven Ruggles and Matthew Sobek, *Integrated Public Use Microdata Series: IPUMS-95 Version 1.0*, vol. 1, *User's Guide* (Minneapolis, Minn.: Social History Research Laboratory, 1995).

47. Census statistics do not adequately account for an additional form of federal support that was important to anthracite region residents. In testimony before a Senate committee on area redevelopment, Congressman Daniel J. Flood reported that almost a sixth of residents in his congressional district in 1956 received surplus food on a weekly basis. The proportion must have been considerably higher for his elderly constituents. U.S. Senate, Committee on Labor and Public Welfare, *Area Redevelopment*, 1:69.

48. To derive these inflation-corrected figures, we used the "All Urban Consumers" consumer price index prepared by the Bureau of Labor Statistics for the period 1913–2001, using January for each of the census years. This table is accessible from the Bureau of Labor Statistics Web site at http://www.bls.gov/data (accessed March 9, 2005). Select "Overall Most Requested BLS Statistics," then retrieve "CPI for All Urban Consumers (CPI-U) 1982–84=100 (Unadjusted)." From the resulting table, choose "More Reformatting Options" and extend the beginning date to 1913, set the final year to 2001, and select January for each year. The result will be a table that permits one to adjust income figures over an 88-year period by the changing cost-of-living index.

49. In 1970 almost 50,000 former miners in Pennsylvania were receiving black-lung compensation payments that averaged $271 a month and another 47,000 widows were receiving monthly payments of $204. These monthly figures rose steadily over the next twenty years, and in 1989 the respective figures were $729 and $366 per month. See *Social Security Bulletin: Annual Statistical Supplement*, 1975:185, 1990:290.

50. Alan Derickson, *Black Lung: Anatomy of a Public Health Disaster* (Ithaca: Cornell University Press, 1998), p. 132, provides evidence for a 1958 study that estimated that three-quarters of anthracite retirees suffered from black lung.

51. Pension Benefit Guaranty Corporation (PBGC), Office of Program and Policy Development, "Sources of Income of Retired Anthracite Miners," [1975], pp. 2 and 7, a technical paper prepared by the PBGC as it considered loan assistance to the Anthracite Health and Welfare Fund. Photocopy of manuscript supplied by the PBGC, October 2003.

52. Combining of individual records required a fairly involved series of steps, and it may be helpful to describe that process here. The original 1950 sample for anthracite

region counties (and a small number of adjacent counties that could not be disaggregated) included 15,909 individuals. Variables recorded for each individual included one that noted the location of the spouse within the census household. Selecting cases on this variable identified almost 6,700 married individuals residing with their spouses. Invariably husband and wife were recorded adjacent to one another in the records, and it proved possible to combine the records for 3,348 husband-wife pairs. The following section draws on the analysis of that population. This further analysis was possible only because of the work completed by the Historical Census Project in creating the Integrated Public Use Microdata Sample (IPUMS) out of the original Public Use Microdata Sample (PUMS). For a discussion of the process involved, see Matthew Sobek and Steven Ruggles, "The IPUMS Project," *Historical Methods* 32, no. 3 (Summer 1999): 102–10.

53. There was one exception to this generalization, the Lanscoal Mining Company, a group of some twenty-five miners who operated at the water-level in the No. 9 mine of LC&N between 1960 and 1972. For more on that mine's history and the Lanscoal period, see George Harvan, *The Coal Miners of Panther Valley* (Bethlehem, Pa.: Lehigh University Art Gallery, 1995), a catalog of an exhibit of Harvan's photographs taken in the Lanscoal mine.

54. Dublin, *When the Mines Closed*, pp. 87–107.

55. Ibid., pp. 109–13.

56. Interview with Ken Ansbach and Ruth Ansbach, Sept. 6, 1993, pp. 2, 5.

57. Interview with Mary Jasinksi, Aug. 29, 1994, pp. 4, 9; Interview with Grant Gangaware, May 17, 1994, p. 14; Interview with Mike Knies, p. 14; Interview with Paul Melovich, p. 26.

58. Interview with Mary Kupec, Isabel Zickler, and Philomena Tout, Oct. 4, 1994, p. 21; Interview with Mary Jasinski, p. 7; Interview with Grace Ferrari, August 24, 1994, p. 10.

59. Interview with Sarah Fibac and Maria Bulgrin (pseudonyms), July 18, 1994, pp. 5–6; Interview with Anna Stone and Joe Orsulak, Aug. 24, 1994, pp. 18–19.

60. U.S. Senate, Committee on Labor and Public Welfare, *Area Redevelopment*, 1:71, 74–75. See further testimony along these lines at 1:438–39, 452–53.

61. Flood testimony, in U.S. Senate, *Area Redevelopment Act: Hearings before a Subcommittee of the Committee on Banking and Currency, United States Senate, Eighty-Sixth Congress, First Session*, pt. I, Feb. 25, 26, and 27, 1959 (Washington, D.C.: GPO, 1959), p. 81.

62. U.S. Senate, Committee on Labor and Public Welfare, *Area Redevelopment*, 1:450.

63. Interview with Tom Strohl, July 30, 1993, p. 19; Interview with Ella Strohl, July 30, 1993, pp. 4, 6–7; Interview with Ken Ansbach and Ruth Ansbach, p. 9; Interview with Eleanor Yelito, Aug. 29, 1994, p. 2. For an insightful discussion of the gender dimensions of work and consumption in working-class families in the interwar years, with parallels to the story here, see Susan Porter Benson, "Living on the Margin: Working-Class Marriages and Family Survival Strategies in the United States, 1919–1941," in *The Sex of Things: Gender and Consumption in*

Historical Perspective, ed. by Victoria de Grazia with Ellen Furlough (Berkeley: University of California Press, 1996), 212–43.

64. Interview with Irene Gangaware, May 5, 1993, p. 16.

65. Interview with Lillian Verona, pp. 4, 6.

66. Interview with Mike Mikovich, July 30, 1993, p. 21.

67. Dublin, *When the Mines Closed*, p. 110.

68. Interview with Paul Melovich, pp. 8–9, 11.

69. Interview with Mike Mikovich, p. 14; Interview with Ella Strohl, pp. 3–6.

70. *Rules and Regulations: Anthracite Health and Welfare Fund, Effective January 1, 1972* (Hazleton, Pa.), p. 13.

71. Interview with Mary Painter, pp. 14–15.

72. Interview with Don Hunsinger, May 16, 1995, pp. 13, 28–29.

73. The data for these two figures come from the IPUMS Sample for the 1970 U.S. Population Census. The anthracite portion of the graphs is based on the regional sample that we employed earlier in the chapter, and the national sample is based on selecting from the entire U.S. sample all men and women 50 years of age and older. Data extracted from the Web site of the Minnesota Population Center, http://www.ipums.org/usa/index.html (still accessible March 10, 2005) and subsequently analyzed with SPSS.

74. Tom Strohl interview, p. 26. For Strohl's fuller story, see Dublin, *When the Mines Closed*, pp. 87–107.

75. Dublin, *When the Mines Closed*, pp. 137, 149, 232–34; Joseph Marouchoc interview, July 13, 1994, p. 13.

76. Dublin, *When the Mines Closed*, pp. 110–11, 158, 233–34; Irene Uher Gangaware questionnaire in high school survey; Mike Mikovich interview, p. 14; Margaret Mikovich interview, July 30, 1993, p. 6; Robert Daniels interview, p. 12; Mary Daniels interview, p. 14; Helen Mordock interview, p. 17.

77. Dublin, *When the Mines Closed*, pp. 75–76, 153.

78. The results of this perspective are particularly evident in a study we have done of graduates from seven anthracite-region high schools between 1946 and 1960. Of more than 570 alumni who responded to a survey questionnaire, close to 40 percent went on to college or nursing training. Achieving advanced schooling, however, did loosen ties to the community, for we discovered that college education was the strongest predictor by far of whether respondents ultimately migrated from their hometowns. College education took these young people out of the region and revealed new possibilities to them. A college degree was a ticket to good employment elsewhere. Parents encouraged their children's education knowing full well that traditional family bonds were endangered. Still, our interviews reveal that these parents remain proud from afar of their children's educational and occupational attainments. We discuss this survey more extensively in chapter 7.

79. Interview with Theresa Pavlocak, Oct. 13, 1994, p. 33. For similar attitudes toward the education and out-migration of children, see interview with Mike Sabron, June 28 and 29, 1993, p. 35; Interview with Raymond Jones, Sept. 16, 1994, pp. 21–23.

80. Theresa Pavlocak interview, p. 15; Mary Jasinski interview, pp. 14–15.

81. Dublin, *When the Mines Closed*, pp. 154, 157; John Mordock interview, p. 33; Robert Daniels interview, p. 20.

82. Mike Sabron interview, p. 66.

7. Personal Responses to Decline: Sons and Daughters, 1950–1990

1. James Coon interview, May 14, 1995.

2. Donald L. Miller and Richard E. Sharpless, *The Kingdom of Coal: Work, Enterprise, and Ethnic Communities in the Mine Fields* (Philadelphia: University of Pennsylvania Press, 1985), pp. 122, 197; John Bodnar, "The Family Economy and Labor Protest in Industrial America: Hard Coal Miners in the 1930s," in David Salay, ed., *Hard Coal, Hard Times: Ethnicity and Labor in the Anthracite Region* (Scranton, Pa.: Anthracite Museum Press, 1984), pp. 85–86.

3. The population here consists of a 1-percent Scranton sample, a one-in-eight Lansford sample, and the entire enumerations for Coal and Mount Carmel townships from the 1920 Federal manuscript census of population, totaling 4,883 anthracite region residents overall. Children included here lived at home with one or both parents, and their fathers were employed in mining occupations. See appendix 4 for further discussion.

4. These statistics come from analysis of the 1-percent Integrated Public Use Microdata Samples (IPUMS) extracted from data accessible at the Web site of the Minnesota Population Center at http://www.ipums.org/. The samples each year consist primarily of anthracite-region residents, although the nature of aggregate geographical areas makes it impossible to select only the major anthracite counties and no others. These are the same samples employed in the analysis of fathers and mothers in chapter 6, though of course the age cohort being tracked here differs.

5. These estimates are based on multiplying the figures found in the 1950 and 1990 IPUMS samples by 100. The same procedure was used for the overall garment industry figures offered in the next paragraph.

6. Kenneth C. Wolensky, Nicole H. Wolensky, and Robert P. Wolensky, *Fighting for the Union Label: The Women's Garment Industry and the ILGWU in Pennsylvania* (University Park: Pennsylvania State University Press, 2002), p. 203.

7. The best treatment of this second wave of deindustrialization in the anthracite region is provided by Wolensky, Wolensky, and Wolensky, *Fighting for the Union Label*, chap. 7. On women workers' responses to one firm's downsizing, see an account of the 1994 Leslie Fay strike in Wilkes-Barre in Michael E. Ruane, "Tearing at the Fabric of Women's Lives: Striking Garment Workers Fight for Jobs That Have Lasted a Generation," *Philadelphia Inquirer*, June 17, 1994, p. B1.

8. Richard Corrigan, "The Shrinking Giant," *National Journal*, no. 24 (June 16, 1984): 1168–74; Bob Calandra, "The End of a Good Thing: Many Chalk Up a Decent Life to Steel Mill," *Bucks County Courier Times*, July 28, 1991, pp. 1A, 8A. This story was part of a multipart series that ran in the *Courier Times* between July 28 and August 2, 1991; Ruth Milkman, *Farewell to the Factory: Auto Workers in the Late Twentieth Century* (Berkeley: University of California Press, 1987).

9. We prepared figure 5 using aggregate national statistics for census years 1950–1990 to permit comparison of anthracite-region data with national figures. *Statistical Abstract of the United States*, 1952, 1966, 1988, and 1996. For the 1950 calculation we had to employ national age-group population figures from Bureau of the Census, *Historical Statistics of the United States: Colonial Times to 1957* (Washington, D.C.: GPO, 1960), p. 10.

10. This is one aspect of the postwar experience that fathers and sons shared. Recall from the earlier discussion in chapter 6 that in 1970 labor-force participation for anthracite-region men lagged behind that of a comparable national sample for men in the age groups between 55 and 69 (see figure 2). That finding for the generation of fathers is consistent with our evidence here for the cohort of sons in the period 1960–1980.

11. Our thanks to Larry Furey, Faye Lewis, Jean Long, Elizabeth Mikovich, and Tom Wehr for making it possible for us to send our survey to graduates of these schools. In all, some 77 percent of participants were public school graduates and 23 percent graduated from parochial schools. We sent questionnaires to graduates from 1946–1960. About 6 percent of 2,335 questionnaires mailed were undeliverable. Of 2,182 questionnaires that were likely to have been received, just over 26 percent were completed and returned. Completion rates ranged between 15 and 30 percent for individual schools. Of the 572 adequately completed forms that we got back, 13 came from respondents who graduated slightly earlier or later than the period 1946–1960. We have employed all completed forms even though 2 percent represent graduates from just outside our target period. For further discussion of our methods, see appendix 8.

12. This figure is dramatically larger than the 29 percent out-migration figure presented earlier in this chapter for 15- to 24-year-olds in Lackawanna County in the decade of the 1950s because it represents out-migration over the lifetimes of these high school graduates rather than simply a ten-year period of time.

13. By coincidence a number of those completing the questionnaires were also people we had interviewed, so for a few members of this study population we have both questionnaire and oral history evidence to draw on. In addition, twenty respondents wrote letters accompanying their questionnaires, providing additional comments on issues of growing up in the anthracite region and accommodating industrial decline. The combination of questionnaire responses, oral history, and correspondence provides multiple perspectives on this generation.

14. A similar survey of other Pennsylvania high school graduates in this period is described in Joseph E. Illick, *At Liberty: The Story of a Community and a Generation: The Bethlehem, Pennsylvania, High School Class of 1952* (Knoxville: University of Tennessee Press, 1989). Illick, a 1952 graduate of Liberty High School in Bethlehem, sent out 480 questionnaires to graduates and received more than 200 back, which he subsequently analyzed. He also conducted extensive oral history interviews with some 50 of his respondents. Bethlehem, of course, is not in the anthracite region; it was a steel town in the 1950s and one of the destinations to which migrants from the anthracite region headed when the mines closed. The insularity of Panther Valley towns is apparent in comparison to Bethlehem. Among 1952 graduates of Liberty High School, 70 percent had been born in Bethlehem, distinctly lower than the 90 percent figure for respondents of our survey. Illick, *At Liberty*, p. 68.

15. For the leading three anthracite counties in 1952, the overall denominational breakdown was similar. In Lackawanna, Luzerne, and Schuylkill counties, Roman Catholics comprised 66.5 percent of reported church membership, Protestants 31.1 percent, and Jews 2.4 percent. See the American Religion Data Archive, http://www.thearda.com/archive/CMS852CNT.html (accessed March 20, 2002) for a discussion of the religious survey conducted by the National Council of Churches on which these data are based. We downloaded the data from that site and added individual denominational figures to present the aggregated figures offered here. A published version of the study is *Churches and Church Membership in the United States: An Enumeration and Analysis by Counties, States and Regions* (New York: National Council of Churches, 1956). Comparing the self-reporting of Panther Valley high school graduates and the 1952 enumeration, two differences are apparent: (1) the Panther Valley had many fewer Jews than the more urban areas within the anthracite region; (2) it appears that Byzantine Catholic and Russian Orthodox churches did not respond to the 1952 survey because their numbers are strikingly absent from the published study. Participation in the enumeration was voluntary and somewhat less than half of religious bodies, 114 of 251, responded to the survey conducted by the National Council of Churches. The 1956 published version appeared as a series of unpaginated reports. See Series A., No. 1 for the acknowledgment of the lack of participation of "Eastern Orthodox, Old Catholic, and Polish National Catholic churches" in the survey.

16. In calculating these proportions we have excluded from the tally focused training that male graduates reported receiving within military service. Such training was open to non–high school graduates as well as high school graduates and its inclusion in the analysis would have biased the findings along lines of gender.

17. For gender, epsilon was 5.9 percent, but the chi square of 2.3 was not statistically significant at the level of .05 used throughout the study. Other variables, whose associations with children's higher education were not statis-

tically significant, included date of graduation, military service, and religion.

18. The source here is the 1950 IPUMS sample for the economic areas including anthracite-region counties. Contrary to what one might have expected, given the strength of patriarchy in so many other areas of life in the anthracite region, there was no difference between sampled men and women (overall, or in the age 35–44 cohort) in terms of having attended college. A slightly higher share of women in this age group in 1950, 2.4 compared to 2.2 percent, had attended college, but the difference was not statistically significant.

19. Hospital nursing schools were actually on the decline in the post–World War II period in which these anthracite-region women received their training. The 1948 report *Nursing for the Future,* by Esther Lucile Brown, argued that the college-educated nurses rather than those trained in hospital programs should be the basis for improving the delivery of nursing services. Still, in this transition period, miners' daughters found hospital training a sure path to a middle-class security their parents lacked. For more on this report, see Barbara Melosh, *"The Physician's Hand": Work Culture and Conflict in American Nursing* (Philadelphia: Temple University Press, 1982), pp. 45–47.

Many young women from the anthracite region did not have the funds to attend college or graduate nursing programs. While area hospitals charged only $300 for the three-year course of study, still many families of prospective student nurses could not afford that tuition. For two Panther Valley examples, see Thomas Dublin, *When the Mines Closed: Stories of Struggles in Hard Times,* photographs by George Harvan (Ithaca: Cornell University Press, 1998), pp. 75, 153. Hospital nursing programs continued despite pressure to reform nursing education because they provided hospitals with the virtually free services of student nurses. See Susan M. Reverby, *Ordered to Care: The Dilemma of American Nursing, 1850–1945* (Cambridge: Cambridge University Press, 1987), pp. 188–89.

20. To place the Panther Valley figures in perspective, see U.S. Bureau of the Census, *1990 Census of Population: Education in the United States* (Washington, D.C.: GPO, 1994), http://www.census.gov/prod/cen1990/cp3/cp-3-4.pdf (accessed March 9, 2005). Drawing on the two age groups most comparable to the Panther Valley high school graduates we surveyed, one estimates that for the United States as a whole, 25 percent of high school graduates earned college degrees and 21 percent of Pennsylvania high school graduates did so. For questionnaire respondents, more than 39 percent earned either college or nursing degrees, a distinctly higher proportion.

21. For the 2×2 contingency tables for higher education with each of these independent variables, the corresponding chi squares are: father's occupation, 10.9; mother's employment, 6.2.

22. Edward G. Herron to Walter Licht, July 4, 1994, in a letter that accompanied his completed questionnaire for the high school survey. Irene Gangaware reported the

same phenomenon in her interview: that her dad, a miner, would not let his son go into the mines. Irene Gangaware interview, Sept. 5, 1993, p. 18.

Herron's father did not permit him to work in the mines, but he himself could not escape its clutches. Lamar Herron was electrocuted by an unguarded mine trolley wire in the No. 14 mine of LNC, August 7, 1953, four years after his son graduated from high school. See Edward G. Herron, letter to the editor, *Valley Gazette,* no. 309 (May 1998): 16.

23. John Bodnar, *Anthracite People: Families, Unions and Work, 1900–1940* (Harrisburg: Pennsylvania Historical and Museum Commission, 1983), p. 99.

24. Joe Rodak interview, June 6, 1995, pp. 14–15, 19.

25. Written comments submitted by David D. DiFebo and John A. Evans accompanying their high school survey questionnaires.

26. We found this high proportion of migrants even though our methods of analyzing responses were consciously chosen to set a high bar. We did not classify as migrants individuals who left the region briefly to attend college or nursing school or serve in the military but returned to their home communities after these time-limited experiences.

The high proportion of migrants reflected the depth of the economic crisis in this part of the anthracite region between 1946 and 1960. In Bethlehem, home to the thriving Bethlehem Steel Company in this period, almost 37 percent of high school graduates remained in the city 24 years after graduation and another 27 percent resided in neighboring communities in the Lehigh Valley. There was a good deal more economic opportunity in and around Bethlehem than in the Panther Valley, a circumstance reflected in the fact that 43 respondents—almost 8 percent of Panther Valley graduates who completed the survey—lived in Bethlehem at some point after their graduation. This comparison has required reworking Illick's reported findings to exclude missing data in the calculation of proportions. See Illick, *At Liberty,* table 1, pp. 36–37.

27. Ben Marsh, "Continuity and Decline in the Anthracite Towns of Pennsylvania," *Annals of the Association of American Geographers* 77 (1987): 345.

28. Among those with no higher education, 55 percent left the Panther Valley permanently; for those with 1–2 years of higher education, 72 percent did so; and among those with 3+ years of higher education, more than 75 percent left. Chi square was 46.3 with a significance of less than .001. Contemporary survey research confirms the pattern evident here. See William H. Whyte, *The Organization Man* (Garden City, N.Y.: Doubleday Anchor, 1956), p. 298.

29. To compare the influence of a variety of independent variables, we created dichotomous versions of these variables and generated 2×2 contingency tables noting the proportions that never left or ever left the Panther Valley within the subgroups of the independent variables. The results noted below demonstrate how much more influential higher education was than any other independent variable tested.

Variable	Percentage Difference	Chi Square
Higher Education	27.8	43.3
Ever Married	24.6	7.8
Status, Current Job	12.5	9.2
Public/Catholic H.S.	10.9	5.4
Date of Graduation	10.7	6.0
Gender	9.2	5.2
Military Service	7.0	4.1
Family Size	6.4	2.6
Father's Occupation	4.1	0.9
Mother's Occupation	2.5	0.3
Religion	0.7	0.02

Using epsilon as a measure, ever married seems to have been almost as strong an influence as higher education, but chi square is not particularly high because there are so few single individuals who did not marry (31) that small changes in the distribution within the cells for that category could result in a significant decrease in the apparent differences between the subgroups here. In this case a ranking of influence by chi square offers a more reliable indicator of the association of various independent variables with migration.

30. Ann O'Connell responses on completed questionnaire in high school survey.

31. In all, 72 men in the survey population served in the military and earned a B.A. or B.S. degree. Of this number 25 went to college first and then entered the military; 5 went to college while in the military; the remaining 42 completed college after their discharges from the service, and presumably were able to make use of the G.I. Bill to help fund their education.

32. Illick found a similar association for Bethlehem high school graduates between occupational attainment and place of residence. More than 80 percent of craft workers, sales workers, operatives and clerical workers, for instance, continued to reside in Bethlehem, compared to only about 50 percent of professionals, managers, and proprietors. The pattern here parallels findings for the Panther Valley. Illick, *At Liberty*, table 8, p. 171.

33. Rather than stress how Panther Valley women differed from broader national patterns in the postwar years, we would say that it was precisely the kinds of changes evident among female high school graduates from the Panther Valley (and in thousands of other similar locales) that made possible the rising national labor force participation rate for women in this period. Between 1950 and 1984, for instance, labor force participation for adult women went from 33.9 to 53.7 percent, in precisely the period that employment for women in the anthracite-region cohort traced in census samples went from 39 to 52 to 60 percent (see figure 4). For the national figures, see Francine D. Blau and Marianne A. Ferber, *The Economics of Women, Men, and Work* (Englewood Cliffs, N.J.: Prentice-Hall, 1986), p. 70.

34. Looking for individuals among those interviewed for the project who also completed questionnaires revealed five women in both populations. In addition, one son of an interviewee and one brother of an interviewee participated in the survey of high school graduates. Exam-

ining the transcripts of the oral history interviews and the questionnaires permitted us to select two individuals, Lillian Verona and James Ferrari, whose experiences seemed to encapsulate the range of experiences among members of the younger generation who graduated from area high schools in the postwar years.

35. Lillian Verona is a pseudonym. An edited, first-person narrative based on the transcript of her interview appears in Dublin, *When the Mines Closed*, pp. 67–86.

36. James Ferrari's story draws on his completed questionnaire and the transcript of an oral history interview of his mother, Grace Ferrari, conducted by Mary Ann Landis on August 24, 1994.

37. Lillian Verona narrative in Dublin, *When the Mines Closed*, p. 75. For a second Panther Valley woman whose family could not afford sending her to hospital nursing training, see Irene Uher Gangaware narrative in Dublin, *When the Mines Closed*, p. 153.

38. For further examples of Panther Valley women working part-time in garment factories while raising young children, see the narratives of Ruth Strohl Ansbach and Irene Uher Gangaware in Dublin, *When the Mines Closed*, pp. 122, 154. See also Elizabeth Mikovich and Bob Sabol (pseudonyms) interview, Oct. 3, 1994, p. 5.

39. Dublin, *When the Mines Closed*, pp. 78–79.

40. Ibid., p. 80.

41. Ibid., p. 83.

42. Much of the information summarized here is based on a phone interview with James Ferrari, March 17, 2002. Our thanks to Ferrari for his willingness to expand on his questionnaire responses and fill us in on developments between 1994 and 2002.

43. For narratives based on the Gangawares' interviews, see Dublin, *When the Mines Closed*, pp. 137–62. This account draws on the original interviews and Irene Uher Gangaware's questionnaire in the high school survey.

44. The higher pay that Grant Gangaware received on his distant construction work was striking. Speaking of her husband's low pay in his first jobs after the mines closed, Irene Gangaware recalled, "And I swear to God, I thought everybody in this whole wide world got paid fifty dollars a week." Supervising his construction crew, by contrast, Grant earned $200 a week. Irene Gangaware interview, Sept. 5, 1993, pp. 12, 16.

45. Irene Uher Gangaware interview, Sept. 5, 1993, pp. 12–14.

46. Verona interview, p. 29.

47. Dublin, *When the Mines Closed*, pp. 153, 160; Irene Gangaware interview, pp. 18, 19.

48. Thomas Shober letter, June 29, 1994, accompanying his completed questionnaire in the high school survey.

49. Dublin, *When the Mines Closed*, pp. 115–35.

50. Ken Ansbach and Ruth Strohl lived in nearby communities in the Panther Valley, but they graduated from the same high school because in 1964 the public high schools in Nesquehoning, Lansford, Summit Hill, and Coaldale were consolidated into the Panther Valley High School. Ken and Ruth graduated together in 1966, members of the second graduating class of the new high

school. They are six years younger than the youngest members of the high school survey population, but their experiences speak to many of the generational differences survey respondents expressed.

51. Ruth Strohl Ansbach and Ken Ansbach interview, Sept. 6, 1993, p. 6.

52. Ansbach interview, p. 8.

53. Dublin, *When the Mines Closed*, pp. 101, 104.

54. Ansbach interview, pp. 19, 36.

55. The irony of Ken Ansbach's decision to work at New Jersey Zinc is that by the mid-1980s Palmerton had become an EPA Superfund site noted for extremely high levels of zinc, cadmium, and lead, which posed significant health risks to the town's residents and company employees. See Robert J. Hill, "Growing Grassroots: Environmental Conflict, Adult Education and the Quest for Cultural Authority" (Ph.D. diss., Pennsylvania State University, 1997), pp. 18–22, 52 ff.

8. Legacies

1. Edward T. Devine, "Production and Labor in United States Coal Mines: The Report of the United States Coal Commission of 1922–1923," *International Labour Review* X (Nov. 1924): 775, as quoted in Harold Kenneth Kanarek, "Progressivism in Crisis: The United Miner Workers and the Anthracite Coal Industry during the 1920's" (Ph.D. diss., University of Virginia, 1972), p. 17.

2. U.S. Department of Labor, Children's Bureau, *Child Labor and the Welfare of Children in an Anthracite Coal-Mining District,* Bureau Publication No. 106 (Washington, D.C.: GPO, 1922), p. 1, as quoted in Kanarek, "Progressivism in Crisis," p. 18.

3. Ben Marsh, "Continuity and Decline in the Anthracite Towns of Pennsylvania," *Annals of the Association of American Geographers* 77 (1987): 341.

4. George R. Leighton, "Shenandoah, Pennsylvania: The Story of an Anthracite Town," *Harper's Monthly Magazine* (Jan. 1937): 131.

5. See the Pennsylvania portion of the Abandoned Mine Land Program database, http://ismhdqa02.osmre.gov/scripts/stsweb.dll (accessed March 10, 2005).

6. *Philadelphia Inquirer*, March 4, 1990, p. A1, Jan. 17, 1994, p. B1.

7. *Philadelphia Inquirer,* Oct. 31, 1993, p. E1, Sept. 24, 1998, p. A1.

8. While the name is the same as the firm that operated mines in the Panther Valley beginning in 1822, the current Lehigh Coal & Navigation Company dates only from 1989 when James J. Curran Jr. purchased the valley's coal lands from Bethlehem Mines. He incorporated his new business under the old LC&N name, abandoned when the former company liquidated in the mid-1980s. Operating strip mines in the valley and a breaker at Greenwood, the company employed about 200 mineworkers and in 1996 produced more than 434,000 tons of anthracite coal. In January 2001 the company closed its operations and laid off 163 employees. Despite a $9 million loan from the U.S. Department of Agriculture, the firm

has not restarted its mining operations as of November 2004. A court ruling in January 2004 requires the company to pay almost $1.9 million to the Anthracite Health and Welfare Fund to cover back liability to the fund. The company also "owes hundreds of thousands of dollars in back taxes to local school districts." Its only current activity appears to be the mixing of culm wastes and fly ash for filling the Springdale stripping. In 2004, Curran denied rumors that the company would soon file for bankruptcy, but whether the company will ever mine coal again remains uncertain. See articles in the *Lehighton Times News*: James Castagnera, "Old Corporations Never Die," July 3, 2004; "Around Pennsylvania," Sept. 23, 2004; Michele Bruno, letter to the editor, Dec. 3, 2002; "$9 million USDA loan will help restart LC&N Greenwood Breaker in Tamaqua," Sept. 19, 2001; "UMW strike looming for LC&N?" June 15, 2002; "LC&N has temporary shutdown," March 5, 2003, all accessed at http://www.tnonline.com.

9. *Philadelphia Inquirer,* March 3, 2004, p. B3; *Lehighton Times News,* March 3, 2004. See also an earlier three-part story on the Springdale backfill controversy that preceded the recent proposal to use river dredge: *Lehighton Times New*, July 13, 15, 16, 2002, http://www.tnonline.com/archives/news/2002/pits/pits1.html and accompanying links (accessed March 10, 2005). On the recent controversy in Tamaqua, see *Lehighton Times News*, "Tamaqua's more than a hole in the ground," Dec. 15, 2003; "Dirty dough: Time to sever ties," June 7, 2004, http://www.tnonline.com.

10. "Influence of Prisons on Schuylkill Economy Measures in $ Millions," *SEDCO Newsletter* (Summer 2002), http://www.sed-co.com/newsletter_2002_06/article1.asp; Econo-Lodge Frackville advertisement, on the Web site for OneTravel.com, at http://hotel.onetravel.com/hothotels.aspx? Action+PRC_ByProperty& PropertyID=38998&NavHeader= . . . (accessed March 10, 2005).

11. Paul Nussbaum, "In Rural PA., Iron Bars Are the New Industry," *Philadelphia Inquirer*, June 17, 1991, p. B1. Another "winner" in the prison bidding was Indiana County, a declining bituminous mining area in western Pennsylvania. See Elizabeth Chiang, "The Great Storm that Swept Through: The Effects of Globalization on Indiana County," *Pennsylvania History* 71 (Spring 2004): 180–81.

12. Marsh, "Continuity and Decline," pp. 337, 345.

13. Thomas Dublin, *When the Mines Closed: Stories of Struggles in Hard Times* (Ithaca: Cornell University Press, 1998), p. 198.

14. Robert W. Reichard, "Coal Was King, Lansford Had Over 10,000 Residents," *Valley Gazette*, no. 304 (Dec. 1997): 25–27.

15. This inventory of Ridge Street businesses was taken on October 10, 2004.

16. For the photographs, see *Valley Gazette*, no. 243 (Sept. 1992): 25; No. 251 (May 1993): 13. Oral history interviews are full of positive recollections of the role of Bright's in the community. For representative remarks, see those of Mike Sabron, Anna Meyers, and Theresa

Pavlocak, in Dublin, *When the Mines Closed*, pp. 46, 196, 209.

17. Marsh, "Continuity and Decline," p. 349.

18. Population figures for 1880–1940 at http://www.hist.umn.edu/gardner/metro/md_Scranton.html; 2000 figures are at http://scrantonconnect.com/statistics.htm.

19. This account summarizes a much more detailed treatment available in Pennsylvania Economy League, "Revised and Updated Act 47 Recovery Plan for the City of Scranton," May 2002, available at http://www.scrantontomorrow.org/documents/RecoveryPlan_05–17–2002.pdf. Even though problems continue, the scale of the difficulty is much reduced. In December 1991, at the outset of the crisis, a state audit found the projected budget deficit to be greater than 23 percent of the city's $33 million annual budget—more than $6.6 million. (*New York Times*, Dec. 26, 1991, p. A18.) By 2002, the Plan Administrator estimated the upcoming shortfall at $1.8 million annually during the period 2003–2007, less than 4 percent of the city's projected annual budget.

In the last two years in a new effort, city officials, civic groups, and chamber of commerce leaders have launched a program to rejuvenate the center city, the Downtown Scranton Renaissance Project. While there is much optimism, and several businesses have located in downtown Scranton, the price of such moves has been guaranteed tax abatements for significant periods of time. The Scranton Lackawanna Building Company, SLIBCO, built and owns the Scranton Enterprise Center, which serves as a business incubator, "offering below market priced rent to encourage the growth of small, start-up companies and firms providing business services." The building site has qualified as a Keystone Opportunity Zone, which "eliminates state corporate and other taxes through 2013." See Marueen McGuigan, "In This Corenr [*sic*]: Neighborhood Spruce-up Vital to Downtown," *Scranton Times Tribune*, July 4, 2004. For the efforts of one civic organization promoting this effort, see http://www.scrantontomorrow.org (accessed March 10, 2005). Accessing the online archives of the *Scranton Times Tribune* revealed eight articles in the period June 2003–June 2004 dealing with new businesses locating in downtown Scranton. The area PBS affiliate, WVIA, sponsored a roundtable discussion in June 2004 of the rejuvenation efforts. It is too early to judge the prospects for these most recent economic development efforts, but they are very much in keeping with seventy years of similar programs in Scranton.

20. Commonwealth of Pennsylvania, Department of Environmental Protection, "A Brief History of the Centralia Mine Fire (Borough of Centralia, Columbia County)," Feb. 1996, http://www.dep.state.pa.us/dep/deputate/mineres/bamr/centbrf.htm (accessed March 10, 2005); *Philadelphia Inquirer,* Feb. 21, 1992, Oct. 29, 1995; *Scranton Times-Tribune*, May 26, 2004. See also Jeff Tietz, "The Great Centralia Coal Fire: How One Small Mining Town Went Up in Smoke," *Harper's Magazine* (February 2004): 47–57.

21. David DeKok, *Unseen Danger: A Tragedy of People, Government, and the Centralia Mine Fire* (Philadelphia: University of Pennsylvania Press, 1986), chap. 16.

22. J. Stephen Kroll-Smith and Stephen Robert Couch, *The Real Danger Is Above Ground: A Mine Fire and Social Conflict* (Lexington: University Press of Kentucky, 1990). p. 61; Tietz, "Great Centralia Coal Fire," 51.

23. Kroll-Smith and Couch, *Real Danger Is Above Ground*, pp. 150, 153.

24. Our focus here is necessarily on the UMWA in the anthracite region. For a fuller discussion of the revitalization of the union in bituminous after the end of the Boyle era, but also its continuing conflicts, see Daniel Marschall, "The Miners and the UMW: Crisis in the Reform Process," *Socialist Review*, no. 40–41 (July-Oct. 1978): 65–115.

25. *Philadelphia Inquirer*, Oct. 21, 1990, p. A1.

26. Joseph Kahn, "China's Coal Miners Risk Danger for a Better Wage," *New York Times*, Jan. 28, 2003, p. A3. For a view of contemporary coal mining in China, see the independent film by Li Yang, *Blind Shaft.*

27. Marie-France Bezzina, "France Marks the End of its Coal-Mining Era," *Philadelphia Inquirer*, April 24, 2004, p. A2.

28. David Waddington, Chas Critcher, Bella Dicks, and David Parry, *Out of the Ashes: The Social Impact of Industrial Contraction and Regeneration on Britain's Mining Communities* (London: The Stationery Office, 2001), pp. 188–89; Emile Boyer King, "The Last Miners," April 27, 2004, on the euro-correspondent.com Web site, http://www.euro-correspondent.com/ed48coal.htm (accessed March 10, 2005).

29. William Ashworth, *The History of the British Coal Industry*, vol. 5, *1946–1982: The Nationalized Industry* (Oxford: Clarendon Press, 1986), p. 162; Waddington, Critcher, Dicks, and Parry, *Out of the Ashes*, p. 1; Paul Stokes, "2,000 Jobs to Go at Doomed Coalfield That Lost Millions," *Daily Telegraph*, July 17, 2002, on the Millennium Environment Debate Web site, http://millennium-debate.org/tel17july022.htm (accessed March 10, 2005).

30. The political and trade union history of British coal in the second half of the twentieth century can be found in Ashworth, *History of the British Coal Industry*, vol. 5; M. J. Parker, *Thatcherism and the Fall of Coal* (Oxford: Oxford University Press, 2000).

31. Royal Geographic Society, "Regenerating Britain's Coalfields: Problems and Prospects," Nov. 1999, p. 3, http://www.rgs.org/pdf/BP1coalfields.pdf (accessed March 10, 2005).

32. The first critical study examining both the impact of mine closings and government assistance programs is Great Britain, Department of Employment and Productivity, *Ryhope: A Pit Closes: A Study in Redeployment* (London: Her Majesty's Stationery Office, 1970). For others, see E. M. Knight, *Men Leaving Mining: West Cumberland, 1966–67. Report to the Ministry of Labour* (University of New Newcastle Upon Tyne, Department of Geography, 1968); John Sewel, *Colliery Closure and Social Change: A Study of a South Wales Mining Village* (Cardiff: University of Wales Press, 1975). Three decades

of studies are reviewed in Waddington, Critcher, Dicks, and Parry, *Out of the Ashes*, chaps. 2–5.

33. Ashworth, *History of the British Coal Industry*, vol. 5, pp. 259–64, 279–80, 284–85, 412, 536; Waddington, Critcher, Dicks, and Parry, *Out of the Ashes*, pp. 18, 99–101, 165–67, 176, 181. Job training and placement and redevelopment programs are documented and evaluated in the social survey research noted above.

34. Parker, *Thatcherism and the Fall of Coal*, pp. 154–55, 206; Nicholas Jones, "Money for Nothing," *New Statesman & Society*, July 2, 1993, p. 15.

35. Ashworth, *History of the British Coal Industry*, vol. 5, pp. 166–68.

36. Waddington, Critcher, Dicks, and Parry, *Out of the Ashes*, pp.168–69. A cataloging of recent government, NGO, and EU initiatives in land reclamation and economic redevelopment in Great Britain can be found at: http://www.coalfields-regen.org.uk/index.cfm; http://www.englishpartnerships.co.uk/coalfields.htm (accessed March 10, 2005). For a mixed assessment of these initiatives, see Katy Bennett, Huw Beynon, and Raymond Hudson, *Coalfields Regeneration: Dealing with the Consequences of Industrial Decline* (Bristol: Policy Press, 2000).

37. For the decline of the French coal industry, see Donald Reid, *The Miners of Decazeville: A Genealogy of Deindustrialization* (Cambridge, Mass.: Harvard University Press, 1985), chap. 8; Waddington, Critcher, Dicks, and Parry, *Out of the Ashes*, p. 189. For a study of the German government's role in the redevelopment of the Ruhr industrial and coal-mining district and programs cushioning the displacement of coal miners, see Stefan Goch, "Betterment with Airs: Social, Cultural and Political Consequences of De-industrialization in the Ruhr," *International Review of Social History* 47, Suppl. no. 10 (2002): 87–111.

38. Waddington, Critcher, Dicks, and Parry, *Out of the Ashes*, pp. 193–95; Erik Swyngedouw, "Reconstructing Citizenship, the Re-scaling of the State and the New Authoritarianism: Closing the Belgian Mines," *Urban Studies* 33, no. 8 (1996): 1504–6, 1513–16.

39. Waddington, Critcher, Dicks, and Parry, *Out of the Ashes*, pp. 190, 196–98.

40. Historical perspectives on coal mining in Europe and the United States and respective relationships between capital, labor, and the state are provided in Gerald D. Feldman and Klaus Tenfelde, eds., *Workers, Owners and Politics in Coal Mining: An International Comparison of Industrial Relations* (New York: Berg, 1990). Japan offers another point of comparison. In 1960, Japan was the third largest producer of coal, mining 57.5 million tons that year and employing 300,000 mineworkers. The Japanese government, regulating a strong cartel of mine companies, began a concerted phase-out of coal mining in the 1950s (planning agencies opting for the importing of cheaper oil and gas). By the mid-1980s, annual production had declined to 6.3 million tons of coal and the number of miners to less than 10,000. National legislation established various programs to assist unemployed miners, with heavy emphasis on generous allowances for relocation and retraining. The Japanese example is analo-

gous to the European with even greater central government planning. See Suzanne Culter, *Managing Decline: Japan's Coal Industry Restructuring and Community Response* (Honolulu: University of Hawaii Press, 1999). For a set of essays on the recent closing of metal ore mines in Canada, Norway, Finland, Sweden, and Australia, see Cecily Neil, Markku Tykkyläinen, and John Bradbury, eds., *Coping With Closure: An International Comparison of Mine Town Experiences* (London: Routledge, 1992). The metal mines in these countries tended to be in wilderness areas. Government programs focused on relocation and retraining of displaced miners, not redevelopment, although there have been planned economic diversification initiatives in Scandinavia. Canadian trade unions have played strong roles in negotiating severance packages.

41. Barry Bluestone, "Forward," in Jefferson Cowie and Joseph Heathcott, eds., *Beyond the Ruins: The Meanings of Deindustrialization* (Ithaca: Cornell University Press, 2003), p. ix.

42. Lillian Rubin has written works for popular audiences that portray the stresses of life for workers and their families with deindustrialization and uncertain economic times. See her *Worlds of Pain: Life in the Working-Class Family* (New York: Basic Books, 1976), and *Families on the Fault Line: America's Working Class Speaks about the Family, the Economy, Race, and Ethnicity* (New York: Harper, 1995).

43. Terry F. Buss and F. Stevens Redburn with Joseph Waldron, *Mass Unemployment: Plant Closing and Community Mental Health* (Beverly Hills, Cal.: Sage, 1983), pp. 71, 85–89; Carolyn Perrucci, Robert Perrucci, Dena B. Targ, and Harry R. Targ, *Plant Closings: International Context and Social Costs* (New York: A. de Gruyter, 1988), pp. 58–59, 87, 96; Carrie R. Leana and Daniel Feldman, *Coping with Job Loss: How Individuals, Organizations and Communities Respond to Layoffs* (New York: Lexington Books, 1992), pp. 40, 64; Jean Gordus, Paul Jarky, and Louis Ferman, *Plant Closings and Economic Dislocation* (Kalamazoo, Mich.: W.E. Upjohn Institute for Employment Research, 1981), pp. 42, 80, 102–3; Paul O. Flaim and Ellen Seligal, "Reemployment and Earnings," in Paul D. Staudoher and Holly E. Brown, eds., *Deindustrialization and Plant Closure* (Lexington, Mass.: D.C. Heath, 1987), pp. 114–15; Randall W. Eberts and Joe A. Stone, *Wage and Employment Adjustment in Local Labor Markets* (Kalamazoo: W.E. Upjohn Institute for Employment Research, 1992), p. 19; Stephen High, *Industrial Sunset: The Making of North America's Rust Belt, 1969–1984* (Toronto: University of Toronto Press, 2003), p. 67; Kathryn Marie Dudley, *The End of the Line: Lost Jobs, New Lives in Postindustrial America* (Chicago: University of Chicago Press, 1994), p. 39; Ruth Milkman, *Farewell to the Factory: Auto Workers in the Late Twentieth Century* (Berkeley: University of California Press, 1997), p. 134; Steve May and Laura Morrison, "Making Sense of Restructuring: Narratives of Accommodation among Downsized Workers," in Cowie and Heathcott, *Beyond the Ruins*, pp. 261–62, 279–80; Joy L. Hart and Tracey E. K'Meyer, "Worker Memory and Narrative: Per-

sonal Stories of Deindustrialization in Louisville, Kentucky," in Cowie and Heathcott, *Beyond the Ruins*, pp. 302–3.

44. Alice Galenson, *The Migration of the Cotton Textile Industry from New England to the South* (New York: Garland, 1985); Laurence F. Gross, *The Course of Industrial Decline: The Boott Cotton Mills of Lowell, Massachusetts, 1835–1955* (Baltimore: Johns Hopkins University Press, 1993); Walter Licht, *Getting Work: Philadelphia, 1840–1950* (Cambridge, Mass.: Harvard University Press, 1992), chap. 1; Thomas Sugrue, *The Origins of the Urban Crisis: Race and Inequality in Postwar Detroit* (Princeton: Princeton University Press, 1996); Tami J. Friedman, "'A Trail of Ghost Towns across Our Land': The Decline of Manufacturing in Yonkers, New York," in Cowie and Heathcott, *Beyond the Ruins*, pp. 19–43; Paul A. Tiffany, *The Decline of American Steel: How Management, Labor and Government Went Wrong* (New York: Oxford University Press, 1988).

45. Gregory Wilson, "Deindustrialization, Poverty, and Federal Area Redevelopment in the United States, 1945–1965," in Cowie and Heathcott, *Beyond the Ruins*, pp. 181–98; Howard Gillette Jr., "The Wages of Disinvestment: How Money and Politics Aided the Decline of Camden, New Jersey," in Cowie and Heathcott, *Beyond the Ruins*, pp. 149–50; John Russo and Sherry Lee Linkon, "Collateral Damage: Deindustrialization and the Uses of Youngstown," in Cowie and Heathcott, *Beyond the Ruins*, pp. 208–10; Chiang, "The Great Storm that Swept Through," pp. 180–81. For the reliance of Youngstown on prison growth and the ensuing controversy that erupted, see Sherry Lee Linkon and John Russo, *Steeltown U.S.A.: Work and Memory in Youngstown* (Lawrence: University Press of Kansas, 2002), pp. 60–64.

46. Trenton, N.J., offers a prime example of the loss of manufacturing through corporate takeover and the liquidation of plants. See John Cumbler, *A Social History of Economic Decline: Business, Politics, and Work in Trenton* (New Brunswick: Rutgers University Press, 1989).

47. Philip Scranton and Walter Licht, *Work Sights: Industrial Philadelphia, 1890–1950* (Philadelphia: Temple University Press, 1986), esp. pp. 267–70.

48. Jefferson Cowie, *Capital Moves: RCA's Seventy-Year Quest for Cheap Labor* (Ithaca: Cornell University Press, 1999).

49. Gillette, "Wages of Disinvestment"; Sugrue, *Origins of the Urban Crisis*. For a thoughtful treatment of the interconnections of issues of race and industrial decline in the steel industry and their implications for broader state policy, see Judith Stein, *Running Steel, Running America: Race, Economic Policy, and the Decline of Liberalism* (Chapel Hill: University of North Carolina Press, 1998).

50. The proportion of African Americans in the anthracite region increased from 0.2 to 0.9 percent between 1920 and 1990. These figures are based on our 1920 anthracite region sample and IPUMS samples between 1950 and 1990. See appendixes 4 and 5.

51. Mike Knies interview, July 19, 1994, p. 6.

52. On the role of the Textile Workers Union of America in responding to plant closings in New England, see William F. Hartford, *Where Is Our Responsibility? Unions and Economic Change in the New England Textile Industry, 1870–1960* (Amherst: University of Massachusetts Press, 1996).

53. Dudley, *End of the Line*, pp. 20–26.

54. Lisa M. Fine, "The 'Fall' of Reo in Lansing, Michigan, 1955–1975," in Cowie and Heathcott, *Beyond the Ruins*, p. 44; Bruce Nissen, ed., *Fighting For Jobs: Case Studies of Labor-Community Coalitions Confronting Plant Closings* (Albany: State University of New York Press, 1995).

55. The literature on the Youngstown protests grows. For basic treatments, see Staughton Lynd, *The Fight Against Shutdowns: Youngstown's Steel Mill Closings* (San Pedro, Calif.: Singlejack Books, 1982), esp. pp. 49–61; Thomas G. Fuechtmann, *Steeple and Stacks: Religion and Steel Crisis in Youngstown* (New York: Cambridge University Press, 1989), pp. 74–80, 180–86.

56. For protests against the closing of steel mills in Pittsburgh, see William Serrin, *Homestead: The Glory and Tragedy of an American Steel Town* (New York: Times Books, 1992), pp. 333–40, 352–57; Dale A. Hathaway, *Can Workers Have a Voice?: The Politics of Deindustrialization in Pittsburgh* (University Park: Penn State Press, 1993).

57. For a sharp critique of the conservatism of U.S. trade union leaders with regard to plant closings, see Kim Moody, *An Injury to All: The Decline of American Unionism* (New York: Verso, 1988). For interpretations that see union leaders as a reflection of the resignation and passivity of U.S. workers, see Hathaway, *Can Workers Have a Voice?*, p. ix; and David Bensman and Roberta Lynch, *Rusted Dreams: Hard Times in a Steel Community* (Berkeley: University of California Press, 1988), p. 202. For the more aggressive stands of the Canadian trade union movement, see High, *Industrial Sunset*, chap. 6.

58. For key studies of the history and political economy of Central Appalachian coal mining areas, see John Gaventa, *Power and Powerlessness: Quiescence and Rebellion in an Appalachian Valley* (Urbana: University of Illinois Press, 1980); Ronald D. Eller, *Miners, Millhands, and Mountaineers: Industrialization of the Appalachian South, 1880–1930* (Knoxville: University of Tennessee Press, 1982); Joe William Trotter Jr., *Coal, Class, and Color: Blacks in Southern West Virginia, 1915–32* (Urbana: University of Illinois Press, 1990).

59. Richard Couto, *An American Challenge: A Report on Economic Trends and Social Issues in Appalachia* (Dubuque, Iowa: Kendall/Hunt, 1994), p. 42; Ada F. Haynes, *Poverty in Central Appalachia: Underdevelopment and Exploitation* (New York: Garland, 1997), p. 94.

60. For the mobilization of women's groups to open positions for women in Appalachian coal mining, see Betty Jean Hall, "Women Miners Can Dig It, Too!" in John Gaventa, Barbara Ellen Smith, and Alex Willingham, *Communities in Economic Crisis: Appalachia and the South* (Philadelphia: Temple University Press, 1990), pp. 53–60. For the short-term gains in income during the

1970s, see Couto, *An American Challenge*, p. 88; Haynes, *Poverty in Central Appalachia*, p. 65.

61. Mike Yarrow, "Voices from the Coalfields: How Miners' Families Understood the Crisis of Coal," in Gaventa, Smith, and Willingham, *Communities in Economic Crisis*, pp. 39–40, 46–48; Couto, *An American Challenge*, pp. 47–50.

62. Yarrow, "Voices from the Coalfields," pp. 38–40; Couto, *An American Challenge*, p. 47.

63. For employment figures for 1950, see Department of the Interior, Bureau of Mines, *Minerals Yearbook: 1951* (Washington, D.C.: GPO, 1954), pp. 361–73; for 1960, Department of the Interior, Bureau of Mines, *Minerals Yearbook: 1960*, vol. 2 (Washington, D.C.: GPO, 1961), p. 61; for 1968, Department of the Interior, Bureau of Mines, *Minerals Yearbook: 1969* (Washington, D.C.: GPO, 1971), p. 315; for 1978 and 1984, Cynthia Duncan, "Myths and Realities of Appalachia Poverty: Public Policy for Good People Surrounded by a Bad Economy and Bad Politics," in *The Land and Economy of Appalachia: Proceedings from the 1986 Conference on Appalachia* (Lexington: Appalachian Center, 1986), p. 28; for 2000, employment statistics of the Energy Information Administration of the Department of Energy, http://www.eia.doe.gov/cneaf/coal/cia/html/t40p01p1.html (accessed March 10, 2005). For production figures for 1960, see *Minerals Yearbook: 1960*, 2:61; for 2000, production statistics of the Office of Surface Mining of the U.S. Department of the Interior, http://www.osmre.gov/coal/2000coal.htm (accessed March 10, 2005).

64. Couto, *An American Challenge*, pp. 99–100, 127–32. 188–91; Haynes, *Poverty in Central Appalachia*, pp. 63–64; Kristin Layng Szakos, "People Power: Working for the Future in the East Kentucky Coal Fields," in Gaventa, Smith, and Willingham, *Communities in Economic Crisis*, pp. 29–30; Salim Kublawi, "The Economy of Appalachia in the National Context," in *The Land and Economy of Appalachia*, pp. 16–24; Cynthia L. Duncan, "Capital and the State in Regional Economic Development: The Case of the Coal Industry in Central Appalachia" (Ph.D. diss., University of Kentucky, 1985), pp. 173–75, 270.

65. Duncan, "Capital and the State in Regional Economic Development," pp. 21–22; Haynes, *Poverty in Central Appalachia*, pp. 59, 66; Maxine Waller, Helen M. Lewis, Clare McBrien, and Carroll L. Wessinger, " 'It Has to Come from the People': Responding to Plant Closings in Ivanhoe, Virginia," in Gaventa, Smith, and Willingham, *Communities in Economic Crisis*, pp. 19–28; John Gaventa, "From the Mountains to the *Maquiladoras*: A Case Study of Capital Flight and Its Impact on Workers," in Gaventa, Smith, and Willingham, *Communities in Economic Crisis*, pp. 85–95.

66. Duncan, "Capital and the State in Regional Economic Development," p. 225; Yarrow, "Voices from the Coalfields," pp. 38–42; Couto, *An American Challenge*, p. 47. For continuing efforts of coal mine companies to deregulate safety and environmental standards, see Christopher Drew and Richard A. Oppel Jr., "Friends in the White House Come to Coal's Aid," *New York Times*,

Aug. 9, 2004, pp. A1, A11. For recent federal bankruptcy court decisions threatening health care and retirement benefits of Appalachian mineworkers and subsequent protests, see "No Bed of Roses in the Mines," *New York Times*, Aug. 16, 2004, p. A14; Roger Alford, "Hundreds of Ky. Coal Miners Stage Protest," *The Associated Press State & Local Wire*, Aug. 31, 2004; James Dow, "Miners' Benefits Vanish with Bankruptcy Ruling," *New York Times*, Oct. 24, 2004, p. 20. These developments parallel those for deregulating airlines that recently have filed for bankruptcy, in large part because such action permits the firms to dump their employee pension plans. See Mary Williams Walsh, "An Outsider's Grim Prognosis for Pension Agency," *New York Times*, Sept. 14, 2004, pp. C1, C22; Micheline Maynard, "US Airways to Skip Pension Payment," *New York Times*, Sept. 14, 2004, p. C14.

67. Couto, *An American Challenge*, pp. 146–50. The fullest treatment of poverty in Appalachia for the last half of the twentieth century is Haynes, *Poverty in Central Appalachia*. For a contemporary study, see the following report prepared for the Appalachian Regional Commission: Roger Hammer, "Recent Trends in Poverty in the Appalachian Region: The Implications of the U.S. Census Bureau Small Area Income and Poverty Estimates on the ARC Distressed Counties Designations," http://www.arc.gov/images/reports/poverty/ARC-APLFinal.pdf (accessed March 11, 2005).

68. For a review and critique of treatments of Appalachian poverty, see Haynes, *Poverty in Central Appalachia*, pp. 12–19, 26–41.

69. Couto, *An American Challenge*, pp. 65–69; Haynes, *Poverty in Central Appalachia*, pp. 46, 60.

70. Dublin, *When the Mines Closed*, p. 217.

71. Ibid., pp. 64–65; *Lehighton Times News*, Jan. 21, 2003.

72. *New York Times*, Aug. 17, 1977; *Valley Gazette*, no. 99 (Sept. 1980): 31–32; no. 135 (Sept. 1983): 31. See also Pattie Mihalik, "The Last of the Panther Valley Miners," *Lehighton Times News*, Dec. 6, 1984, pp. 6–7. Other memorials to anthracite underground mining abound in the region. For a view of the relief sculpture of the miners' memorial in Shenandoah, see Dublin, *When the Mines Closed*, p. 236.

73. "A Tribute to the Men who Mined the Coal," *Valley Gazette*, no. 75 (Sept. 1978): 5.

74. Flood had a long history of supporting redevelopment efforts in the anthracite region as well as the federal black-lung compensation provision. The best sketch of his political contributions to the region appears in William C. Kashatus III, " 'Dapper Dan' Flood: Pennsylvania's Legendary Congressman," *Pennsylvania Heritage* 21, no. 3 (1995): 4–11. See also Miller and Sharpless, *Kingdom of Coal*, pp. 325–32.

75. *Valley Gazette*, no. 16 (Oct. 1973): 5; no. 99 (Sept. 1980): 31–32; no. 147 (Sept. 1984): 24, 26.

76. Shawn A. Hessinger, "St. Luke's Unveils Emergency Room," *Pottsville Republican & Evening Herald*, Oct. 11, 2002.

77. "Something New for Panther Valley," *Valley Gazette*, no. 1 (July 1972): 1, 22.

78. *Valley Gazette*, no. 13 (July 1973): p. 2.

79. *Valley Gazette*, no. 15 (Sept. 1973): 24–25; no. 6 (Dec. 1972): 1–2, 4–6; no. 9 (March 1973): 18–22; no. 7 (Jan. 1973): 1–3; no. 8 (Feb. 1973): 2, 4; no. 2 (Aug. 1972): 1, 2; no. 5 (Nov. 1972): 1–2, 11–12, 21–22; no. 11 (May 1973): 4; no. 12 (June 1973): 1–2.

80. *Valley Gazette*, no. 61 (July 1977): 3.

81. Jay D. Frantz's columns include at least the following items: *Valley Gazette*, no. 10 (April 1973): 21–22; no. 18 (Dec. 1973): 25; no. 20 (Feb. 1974): 10; no. 36 (June 1975): 14; no. 46 (April 1976): 6, 11; no. 53 (Nov. 1976): 20, 21; no. 59 (May 1977): 18; no. 74 (Aug. 1978); no. 76 (Nov. 1978): 13–16; no. 87 (Sept. 1979): 8; no. 126 (Dec. 1982): 9–14; no. 135 (Sept. 1983): 34; no. 170 (Aug. 1986): 21–23; no. 203 (May 1989): 9–10; no. 229 (July 1991): 26–27. For James H. Gildea's pieces, see issues of Sept. 1976, Oct. 1976, March 1977, May 1977, June 1977, and April 1983.

82. Michael O'Malley, "The Golden Miner," *Valley Gazette*, no. 20 (Feb. 1974): 16–17.

83. Tom Dublin, "The Blacklisting of His Father for Union Activity May Have Helped Forge the Fiery Character of James 'Casey' Gildea," *Valley Gazette*, no. 276 (June 1995): 11–13.

84. This treatment of Harvan is based on a critical essay on Harvan's life and work published online in Thomas Dublin and Melissa Doak, "Miner's Son, Miners' Photographer: The Life and Work of George Harvan," *Journal for MultiMedia History* 3 (March 2001), at http://www.albany.edu/jmmh (accessed March 11, 2005).

85. These individual projects are documented in the exhibits reproduced in "Miner's Son, Miners' Photographer."

86. George Harvan, *The Coal Miners of Panther Valley* (Bethlehem, Pa.: Lehigh University Art Gallery, 1995), pp. 1–2.

87. *Valley Gazette*, no. 1 (July 1972): 7. While the focus here will be on the mine museum in Lansford, similar movements to commemorate the region's mining history sprouted throughout the anthracite region. Private efforts, though with public assistance, organized mine tours at the Pioneer Tunnel in Ashland and the Lackawanna Mine in Scranton; Eckley Miners Village near Freeland and the Anthracite Heritage Museum in Scranton also preserved the region's heritage. The story of the No. 9 Mine addresses values that loom large in the region's recent cultural history.

88. Our thanks to David and Lora Tokosh for their willingness to be interviewed and for sharing minutes of meetings and correspondence for the Panther Creek Valley Foundation in its early years. The following narrative is based largely on the picture that emerges from their responses and the records they preserved and shared. See also a helpful account of the Tokoshes' work in *Valley Gazette*, no. 341 (Jan. 2001): 12–15.

89. *Valley Gazette*, no. 203 (May 1989): 31, 32, and 34; no. 259 (Jan. 1994): 36, 35; no. 272 (Feb. 1995): 36; no. 283 (March 1996): 34; no. 288 (Aug. 1996): 1, 36; no. 298 (June 1997): 36; no. 304 (Dec. 1997): 36, 35, 34. Occasionally page numbers in this newspaper are noted in descending order because some articles begin on the last page and are continued on previous pages.

90. Our thanks to Francis Karnish, former president of the Panther Creek Valley Foundation in this period, for sharing his recollections of his term in office and also newspaper clippings and the consultants' proposal for the renovation of the No. 9 Wash Shanty Museum and its future development. See also *Valley Gazette*, no. 332 (April 2000): 6; no. 355 (March 2002): 35, 34, 33; no. 356 (April 2002): 36, 35, 34; no. 360 (Aug. 2002): 36; no. 364 (Nov. 2002): 31.

91. Jack Yalch, "Backtalk," *Valley Gazette*, no. 49 (July 1976): 35.

92. The popular memory of the anthracite region in decline is very much a "vernacular" as opposed to an "official" view, as John Bodnar uses the terms in "The Memory Debate." We emphasize here the function of this perspective for ordinary residents of the region, rather than its place in an ongoing public debate. See Bodnar, *Remaking America*, pp. 13–15.

Appendixes

1. Other firms with surviving collections of early twentieth-century employment records similar to those of the Lehigh Coal & Navigation Company include the Endicott-Johnson Company, a major shoemaking firm in the Binghamton–Johnson City–Endicott, N.Y., area and the International Paper Company of Corinth, N.Y. My thanks to Professors Gerald Zahavi and Stephen Cernek for sharing their work-in-progress with these records.

2. One complicating factor during this period was the lease of the Nesquehoning colliery to the Edison Coal Company between 1938 and 1946. For employees who worked continuously at the Nesquehoning facilities in these years, employment records show transfers to the Edison employment rolls when the lease began and transfers back to LC&N at the lease's end. These were "fictive" changes in employment, and we have considered employment continuous in those instances. However, if a person transferred to the Edison Coal Company payroll in 1938 and left that company at some point before the lease ended, we lack an accurate record of the final date of employment for the individual.

3. Work on the selection, coding, and key entry of the employment records of LC&N was made possible by a major research grant from the National Endowment for the Humanities. This grant made possible the microfilming of these records and the employment of Penelope Harper for the intricate work of coding the original data. Our thanks to Dr. Harper for her fine work on this part of the study.

4. *Report on Condition of Woman and Child Wage-Earners in the United States in 19 Volumes*, vol. 4, *The Silk Industry* (Washington, D.C.: GPO, 1911).

5. Ibid., 4:13.

6. The samples and populations that compose the database for anthracite communities in 1920 were drawn from the microfilm edition of the 1920 Federal Manuscript Census of Population, T625: Scranton, reels 1578–81; Lansford, reel 1543; Mount Carmel Township, reel

1612, enumeration district 105; and Coal Township, reel 1610, enumeration district 64.

7. Our thanks to successive groups of Binghamton students in a graduate course on computers and historical research for their work on class projects that included coding several of these census samples.

8. For discussion of methodological questions involved in the use of the Public Use Microdata Samples, see James N. Gregory, "The Southern Diaspora and the Urban Dispossessed: Demonstrating the Census Public Use Microdata Samples," *Journal of American History* 82 (1995): 111–34; Rodney M. Ito II, Brian Gratton, and Joseph Wycoff, "Using the 1940 and 1950 Public Use Microdata Samples: A Cautionary Tale," *Historical Methods* 30 (1997): 137–47. Thanks to Diane Geraci of Bartle Library at SUNY Binghamton for her assistance in gaining access to original data tapes from the Inter-university Consortium for Political and Social Research at the University of Michigan.

9. See the IPUMS home page at the Minnesota Population Center Web site at http://www.ipums.umn.edu/.

10. We are grateful for the assistance we received from Leonard Carlin of the Economic Development Council of Northeastern Pennsylvania and from Andrew Kelhart of Pennsylvania Power & Light in making extensive resources available to us as we launched this portion of our research.

11. For work on Manchester, N.H., textile operatives that has been criticized for its exclusive reliance on interviews with loyalist workers who worked all their lives for the Amoskeag Company, see Tamara K. Hareven and Randolph Langenbach, *Amoskeag: Life and Work in an American Factory-City* (New York: Pantheon, 1978); Tamara K. Hareven, *Family Time and Industrial Time: The Relationship between the Family and Work in a New England Industrial Community* (New York: Cambridge University Press, 1982). Interviews with Quebecois returnees are found in Jacques Rouillard, *Ah les Etats!: les travailleurs canadiens-français dans l'industrie textile de la Nouvelle-Angleterre d'après le témoignage des derniers migrants* (Montréal: Boréal Express, 1985).

12. Here we thank Sheila Biddle, formerly of the Ford Foundation, for her support of this project from its formative stages and for the grant that made it possible to undertake the interviewing, transcribing, and indexing that this work entailed. The New Jersey Historical Commission also provided funding that supported transcription of the New Jersey migrants among the interviewees. We are also grateful to Mary Ann Landis, who conducted a good number of interviews, and to Melissa Doak and Julie Simmonds for excellent work transcribing the interviews. Melissa Doak worked very ably with the indexing capabilities of Microsoft Word and prepared the two aggregate indexes that provided us with extraordinary access to the interviews.

13. We are grateful to Larry Furey, Faye Lewis, Jean Long, Elizabeth Mikovich, and Tom Wehr for their crucial assistance in permitting use of school alumni mailing lists in this study.

Selected Bibliography

MANUSCRIPT SOURCES

ANTHRACITE HEALTH AND WELFARE FUND OFFICE,
HAZLETON, PA.
Annual Reports of the Anthracite Health and Welfare
    Fund Correspondence of Fund officials
Rules and Regulations, various years

ANTHRACITE HERITAGE MUSEUM, SCRANTON, PA.
George J. Curilla Collection, Penn Anthracite Collection

CAN DO OFFICE, HAZLETON, PA.
Clippings Scrapbooks, 1956–

FEDERAL MANUSCRIPT CENSUSES OF POPULATION
1920, T625
Scranton, Reels 1578–1581
Lansford, Reel 1543
Mount Carmel Township, Reel 1612, e.d. 105
Coal Township, Reel 1610, e.d. 64

HAGLEY MUSEUM AND LIBRARY, WILMINGTON, DEL.
St. Clair Coal Company Papers

INDIANA UNIVERSITY OF PENNSYLVANIA, INDIANA, PA.
District 25, UMWA Records, Collection 109. Districts 1,
    7, and 9 were merged into this district in 1969 and the
    collection here includes extensive materials from the
    premerger history of the anthracite districts. District
    25 lost its separate existence in 1991 and was merged
    into District 2, a bituminous district in western Penn-
    sylvania, hence the presence of District 25 records in
    the District 2 collection housed at IUP.

KING'S COLLEGE, WILKES-BARRE, PA.
Daniel Flood Papers
George Korson Collection

LIBRARY OF CONGRESS, PRINTS AND PHOTOGRAPHS
DIVISION, WASHINGTON, D.C.
Photographs of the Farm Security Administration—Office
    of War Information

LIBRARY OF CONGRESS, MANUSCRIPT DIVISION, WASH-
INGTON, D.C.
Papers of Cornelia Bryce Pinchot
Papers of Gifford Pinchot

LUZERNE COUNTY HISTORICAL SOCIETY, WILKES-
BARRE, PA.
Wilkes-Barre Chamber of Commerce files

NATIONAL ARCHIVES AND RECORDS ADMINISTRATION,
REGIONAL OFFICE, PHILADELPHIA
Records of the bankruptcy proceedings for the P&RC&I,
    Bankruptcy Case 19711

NATIONAL ARCHIVES AND RECORDS ADMINISTRATION,
WASHINGTON, D.C.
Records of the National Recovery Administration, RG 9
Records of the Bureau of Mines, RG 80
Records of the Federal Mediation and Conciliation Ser-
    vice, RG 280

NATIONAL CANAL MUSEUM, EASTON, PA.
Kynor Collection
Lehigh Coal & Navigation Company Collection
T. R. Berger Collection
W. Julian Parton Collection

PANTHER VALLEY AND MIGRANT ORAL HISTORY IN-
TERVIEWS
Cassette tapes, transcripts, and indexes of interviews of
    former coal miners and garment workers from the
    Panther Valley and of anthracite-region migrants to
    northern New Jersey and the Fairless Hills, Pennsyl-
    vania area. Ninety interviews, seventy-eight tran-
    scripts, and seventy indexes comprise this material,
    which will be deposited in the Pennsylvania State
    Archives in Harrisburg. Collection includes a tape and
    transcription of a portion of the UMWA mass meeting
    in Lansford, June 8, 1954.

PANTHER VALLEY HIGH SCHOOL GRADUATES'
SURVEY
Original questionnaires and responses of 572 graduates of
    Panther Valley public and parochial high schools be-

tween 1946 and 1960. Will be deposited in the Pennsylvania State Archives in Harrisburg.

PENNSYLVANIA POWER & LIGHT COMPANY, ALLEN-
TOWN, PA.
Annual Reports of the Industrial Development Depart-
ment, 1940–1987

PENNSYLVANIA STATE ARCHIVES, HARRISBURG, PA.
Lehigh Coal & Navigation Company records
Pennsylvania Writers Project, American Guide Series, Job
54

PENNSYLVANIA STATE UNIVERSITY, HISTORICAL COL-
LECTIONS AND LABOR ARCHIVES, PATTEE LIBRARY
Committee of Twelve Papers, 1930–1966
Papers of the United Mine Workers of America
William W. Scranton Papers

PENSION BENEFIT GUARANTY CORPORATION, WASH-
INGTON, D.C.
Documents Relating to the Current Financial Assistance
Arrangements with The Anthracite Plan, 1974
"Sources of Income of Retired Anthracite Miners [for
1974]," [1980]
Documents Relating to PBGC assistance to Anthracite
Health and Welfare Fund, 1981–1989

PUBLIC USE MICRODATA SAMPLES (PUMS) OF
ANTHRACITE-REGION CENSUSES, 1930–1990
PUMS data sets from the Inter-university Consortium for
Political and Social Research, University of Michigan,
1930–1990
IPUMS data sets from the Minnesota Population Center,
University of Minnesota, 1950–1990

SCRANTON CHAMBER OF COMMERCE OFFICE, SCRAN-
TON, PA.
Greater Scranton Chamber of Commerce files

SMITHSONIAN INSTITUTION, ARCHIVES CENTER, NA-
TIONAL MUSEUM OF AMERICAN HISTORY, WASHING-
TON, D.C.
Lehigh Coal & Navigation Company records, 1874–1954

UNIVERSITY OF VIRGINIA LIBRARY, SPECIAL COLLEC-
TIONS, CHARLOTTESVILLE, VA.
Papers of W. Jett Lauck (#4742)

UNPUBLISHED THESES AND DISSERTATIONS
Blatz, Perry K. "Ever-Shifting Ground: Work and Labor
Relations in the Anthracite Coal Industry, 1868–
1903." Ph.D. diss., Princeton University, 1987.
Cornell, Robert J. "The Anthracite Coal Strike of 1902."
Ph.D. diss., The Catholic University of America, 1957.

DeLauritus, Frank Thomas "Anthracite Coal: A Case
Study of the Social Problems of a Declining Industry."
M.A. thesis in Labor and Industrial Relations, Univer-
sity of Illinois-Urbana, 1956.
Duncan, Cynthia L. "Capital and the State in Regional
Economic Development: The Case of the Coal Indus-
try in Central Appalachia." Ph.D. diss., University of
Kentucky, 1985.
Evan, Frank J. "An Analysis of the Decline of the An-
thracite Industry with Emphasis on Wyoming Valley
from 1930 to Date with the View of Stimulating In-
dustrial Diversity." M.B.A. thesis, University of Pennsylvania, 1950.
Fatzinger, Glenn Banks. "A Descriptive Study of the Area
Redevelopment Administration (ARA)-Economic Development Administration (EDA) University Center
Program, 1963–1974." Ed.D. diss., George Washington University, 1977.
Galloway, Lowell Eugene. "Depressed Industrial Areas: A
National Problem." Ph.D. diss., Ohio State University,
1959.
Hill, Robert J. "Growing Grassroots: Environmental Conflict, Adult Education and the Quest for Cultural Au-
thority." Ph.D. diss., Pennsylvania State University,
1997.
Kanarek, Harold Kenneth. "Progressivism in Crisis: The
United Mine Workers and the Anthracite Coal Indus-
try during the 1920's." Ph.D. diss., University of Vir-
ginia, 1972.
Krause, Alfred Charles. "The Silk Industry of Pennsylva-
nia." M.A. thesis, University of Pennsylvania, 1933.
Landau, Harold. "The Industrial Development in the
Wilkes-Barre Area." M.A. thesis, University of Scran-
ton, 1967.
Lechleitner, Rodney S. "Bootlegging in Schuylkill County,
1930–1936." M.A. thesis, Bloomsburg University,
1986.
Lynch, Patrick M. "Pennsylvania Anthracite: A Forgotten
IWW Venture, 1906–1916." M.A. thesis, Bloomsburg
State College, 1974.
Mead, Richard R. "An Analysis of the Decline of the An-
thracite Industry Since 1921." Ph.D. diss., University
of Pennsylvania, 1933.
Monroe, Douglas K. "A Decade of Turmoil: John L.
Lewis and the Anthracite Miners, 1926–1936." Ph.D.
diss., Georgetown University, 1976.
Sable, Thomas F. "Lay Initiative in Greek Catholic
Parishes in Connecticut, New York, New Jersey and
Pennsylvania, 1884–1909." Ph.D. diss., Graduate
Theological Union, Berkeley, 1984.
Saye, William H. "The Development and Present Status
of the Bootleg Anthracite Industry and Its Influence
on Legitimate Producers." M.A. thesis, Temple Uni-
versity, 1941.
Wilson, Gregory S. "Before the Great Society: Liberalism,
Deindustrialization and Area Redevelopment in the
United States, 1933–1965." Ph.D. diss., Ohio State
University, 2001.

PUBLISHED SOURCES
NEWSPAPERS AND OTHER SERIALS
Anthracite Institute Bulletin, 1946–1960
Anthracite Operators' Association, Bulletins, 1943–1945
Anthracite Tri-District News, 1934–1976
Coal Age, 1911–1988
Coaldale Observer, 1910–1958, with significant gaps
The Evening Record (Lansford), 1917–1956
Glen Alden Coal Company, annual reports, 1923–1969
The History of Northeastern Pennsylvania, proceedings of annual conferences held at Luzerne County Community College, 1989–2001
Lehigh Coal & Navigation Company, annual reports, 1918–1980
Lehighton Times News, 1980–2004
New York Times, 1857–2004
Philadelphia & Reading Coal & Iron Company, annual reports, 1923–1970
Philadelphia Inquirer, 1980–2004
Pottsville Republican, 1955–1980
Pottsville Republican & Evening Herald, 1999–2004
Redevelopment, Economic Development (publications of the Area Redevelopment Administration), 1964–1965
Scranton Times, 1917–1988
Scranton Times Tribune, 1999–2004
The Truth (the newsletter of the Russian Brotherhood), 1990s
The Valley Gazette, 1972–2004
Wilkes-Barre Wyoming Valley Progress, 1929–1936 and 1957–1961, in the Wilkes-Barre Chamber of Commerce files, Luzerne County Historical Society, Wilkes-Barre, Pa.

GOVERNMENT RECORDS AND PUBLICATIONS
Commonwealth of Pennsylvania
Chance, H. M. *Second Geological Survey of Pennsylvania: 1883*, 2 vols. Harrisburg: Board of Commissioners for the Second Geological Survey, 1883.
Commonwealth of Pennsylvania, Anthracite Coal Industry Commission. *Report of Morris L. Ernst.* Harrisburg: n.p., 1937.
Commonwealth of Pennsylvania, Department of Commerce. Pennsylvania Industrial Development Authority. *35 Years of Job-Creating Loans: Summary of Loan Projects, May 1956–June 1991.* [Harrisburg, 1992.]
Commonwealth of Pennsylvania, Department of Community and Economic Development. The Pennsylvania Industrial Development Authority. *40 Years of Job-Producing Loans: Summary of Loan Projects, May 1956–June 1996.* [Harrisburg, 1997.]
Commonwealth of Pennsylvania, Department of Internal Affairs. *Annual Reports*, 1878–1910.
Commonwealth of Pennsylvania, Department of Internal Affairs, Bureau of Statistics. *Industrial Statistics for Pennsylvania, 1916 to 1956.* March 1959, Special Release S-3.
Commonwealth of Pennsylvania, Department of Internal Affairs, Bureau of Statistics, Topographic and Geologic Survey. *Pennsylvania's Mineral Heritage: The*

Commonwealth at the Economic Crossroads of Her Industrial Development. Harrisburg, 1944.
Commonwealth of Pennsylvania, Department of Mines and Mineral Industries. *Annual Reports*, 1917–1990.
Commonwealth of Pennsylvania, Office of Budget and Administration. *The Pennsylvania Industrial Development Authority: An Assessment.* Harrisburg: n.p., 1979.
Commonwealth of Pennsylvania. *Report of the Anthracite Coal Industry Commission.* Harrisburg: Murrelle, 1938.
Federal Government
Appalachian Regional Commission. *Annual Reports*, 1965–1988.
Blankenship v. Boyle, 337 F. Supp. 1089, 1971 U.S. Dist. LEXIS 13548.
Blankenship v. Boyle, 337 F. Supp. 296; 1972 U.S. Dist. LEXIS 15652 (Jan. 7, 1972).
Bureau of Mines, *Minerals Yearbook, 1937.* Washington, D.C.: GPO, 1937.
Bureau of Mines, *Minerals Yearbook, 1945.* Washington, D.C.: GPO, 1945.
Bureau of Mines, *Minerals Yearbook: 1951.* Washington, D.C.: GPO, 1954.
Bureau of Mines, *Minerals Yearbook: 1960.* Vol. 2. Washington, D.C.: GPO, 1961.
Bureau of Mines, *Minerals Yearbook: 1969.* 2 vols. Washington, D.C.: GPO, 1971.
Economic Conditions in the Anthracite Coal Regions: Letter from the Federal Anthracite Coal Commission. 77th Cong., 2d sess., House Document 709. Washington, D.C.: GPO, 1942.
Goodrich, Carter, Hugh S. Hanna, and David J. Price. "Final Report of Anthracite Committee of Secretary of Labor, April 6, 1934." Department of Labor Library, Washington, D.C.
In the District Court for the Eastern District of Pennsylvania. "In the Matter of the Philadelphia and Reading Coal and Iron Company, Debtor. In Proceedings for the Reorganization of a Corporation, No. 19711: Summary of the Report of the Examiner, Nicholas G. Roosevelt, Examiner" [December 1940].
Report on Condition of Woman and Child Wage-Earners in the United States in 19 Volumes. Vol. 4, *The Silk Industry.* Washington, D.C.: GPO, 1911.
Report to the President on the Anthracite Coal Strike of May-October 1902, by the Anthracite Coal Strike Commission. Washington, D.C.: GPO, 1903.
Report of the United States Coal Commission . . . In Five Parts. Part I, *Principal Findings and Recommendations.* Washington, D.C.: GPO, 1925.
Report of the United States Coal Commission . . . In Five Parts. Part II, *Anthracite—Detailed Studies.* Washington, D.C.: GPO, 1925.
Reports of the Immigration Commission: Immigrants in Industries, Part 19, *Anthracite Coal Mining.* Washington, D.C.: GPO, 1911.
"Summary of Unemployment Situation, Shamokin-Coal Township Area—Shamokin, Penna., June 10th 1935," in Records of the Federal Mediation and Conciliation

Service, RG 280, case 165–1812, National Archives and Records Administration.

Trapnell, W. C. "The Employment Situation in the Pennsylvania Anthracite Region." Typescript document, Works Progress Administration, October 1935.

U.S. Bureau of the Census. *Census of Manufactures: 1947.* Vol. 3, *Statistics by States.* Washington, D.C.: GPO, 1950.

U.S. Bureau of the Census. *Fourteenth Census of the United States Taken in the Year 1920: Population, 1920.* Washington, D.C.: GPO, 1921.

U.S. Bureau of the Census, *Historical Statistics of the United States: Colonial Times to 1957.* Washington, D.C.: GPO, 1960.

U.S. Bureau of the Census. *Manufactures 1939: State Series, Pennsylvania.* Washington, D.C.: GPO, 1942.

U.S. Bureau of the Census. *Religious Bodies: 1916.* Parts I and II. Washington, D.C.: GPO, 1919.

U.S. Bureau of the Census. *1990 Census of Population: Education in the United States.* Washington, D.C.: GPO, 1994.

U.S. Department of Labor, Children's Bureau. *Child Labor and the Welfare of Children in an Anthracite Coal-Mining District,* Bureau Publication No. 106. Washington, D.C.: GPO, 1922.

U.S. Senate, Committee on Labor and Public Welfare. *Area Redevelopment. Hearings before the Subcommittee on Labor of the Committee on Labor and Public Welfare, United States Senate, Eighty-fourth Congress, second session, on S. 2663, a bill to establish an effective program to alleviate conditions of excessive unemployment in certain economically depressed areas . . . ,* 2 vols. Washington, D.C.: GPO, 1956.

U.S. Senate. *Area Redevelopment Act. Hearings before a Subcommittee of the Committee on Banking and Currency, United States Senate, Eighty-Sixth Congress, First Session.* Part I, February 25, 26, and 27, 1959. Washington, D.C.: GPO, 1959.

CONTEMPORARY PUBLISHED SOURCES

Adamic, Louis. *My America: 1928–1938.* New York: Harper & Brothers, 1938.

———. "The Great 'Bootleg' Coal Industry." *The Nation* 140 (Jan. 9, 1935): 46–49.

"Among the Pennsylvania Coal-Strikers." *Harper's Weekly* XLVI (Oct. 11, 1902): 1446.

"The Anthracite Coal Trade of Pennsylvania." *North American Review,* no. 90 (January 1835): 241–56.

Archibald, James F. J. "The Striking Miners and Their Families." *Collier's Weekly* 30 (Oct. 11, 1902): 6–7.

Bigler, Tom. "The Glen Alden Story." Paper presented to the Luzerne County United Community Development Conference, Wilkes College, Wilkes-Barre, Pa., April 22, 1958.

"Brief History and Description of the Lehigh Coal and Navigation Company and of the Properties of that Company . . ." Guide prepared for the annual meeting of the American Institute of Mining and Metallurgical Engineering, Wilkes-Barre, Pa., September 12–15, 1921.

The CAN DO Story: A Case History of Successful Community Industrial Development. Hazleton, Pa.: CAN DO, 1974.

Carlson, Oliver. "Bootlegging Coal: A New Industry Appears on the Horizon." *Harper's Magazine,* April 1935, 613–22.

Crane, Stephen. "In the Depths of a Coal Mine." *McClure's Magazine* 3 (August 1894): 194–209.

Diamond, David. "The Children of Levittown." *The Inquirer Magazine* (Dec. 12, 1982): 32–36, 38–39, 41, 44.

Eastman, Crystal. *Work-Accidents and the Law.* New York: Charities Publication Committee, 1910.

Fischer, John. "The Lazarus Twins in Pennsylvania: How Scranton and Wilkes-Barre Are Rising from the Dead." *Harper's Magazine,* Nov. 1968, 13–14, 18, 20, 24, 26, 28, 30.

Flynn, John T. "Bootleg Coal." *Collier's,* December 3, 1936, 12–13, 30, 32, 34.

Henderson, Harry. "Rugged American Collectivism: The Mass-Produced Suburbs, Part II." *Harper's Magazine,* Dec. 1953, 80–86.

A History of the Lehigh Coal and Navigation Company. Philadelphia: William S. Young, 1840.

Leighton, George R. "Shenandoah, Pennsylvania: The Story of an Anthracite Town." *Harper's Monthly Magazine* (Jan. 1937): 131–47.

Lovejoy, Owen. "The Extent of Child Labor in the Anthracite Coal Industry." *Annals of the American Academy of Political and Social Science* 20 (1907): 35–49.

Macfarlane, James. *The Coal-Regions of America: Their Topography, Geology, and Development.* New York: D. Appleton, 1873.

Nichols, Francis H. "Children of the Coal Shadow." *McClure's* 20 (Feb. 1903): 435–44.

Obenauer, Marie L. "Living Conditions Among Coal Mine Workers of the United States." *The Annals* CXI (Jan. 1924): 12–23.

"Operation Boot Strap: How Third District Communities Are Using Their Own Capital to Build Industry," *Business Review* (Dec. 1949): 120–27.

Panther Valley Industrial Commission, *Panther Valley—Opportunities for New Industries.* Lansford, Pa., [1946].

Raught, John Willard. "The Tragedy of Coal Mining." *Scranton Republican,* Feb. 6, 1928, p. 8.

Roberts, Peter. *Anthracite Coal Communities: A Study of the Demography, the Social, Educational and Moral Life of the Anthracite Regions.* New York: MacMillan, 1904.

———. *The Anthracite Coal Industry: A Study of the Economic Conditions and Relations of the Cooperative Forces in the Development of the Anthracite Coal Industry of Pennsylvania.* New York: MacMillan, 1901.

Saada, Michael J. " 'The Scranton Plan': Mining Town Licks Its Slump: New Industries Replace Vanishing Coal." *Wall Street Journal,* March 31, 1949, pp. 1, 4.

Sanville, Florence Lucas. "Home Life of the Silk-Mill Workers: The Conservation of Our Young Womanhood." *Harper's Monthly Magazine* 121 (June 1910): 22–31.

———. "Silk Workers in Pennsylvania and New Jersey." *The Survey* (May 18, 1912): 307–12.

———. "A Woman in the Pennsylvania Silk-Mills: The Conservation of Our Young Womanhood." *Harper's Monthly Magazine* 120 (April 1910): 651–62.

The Story of the Old Company. Lansford, Pa.: Lehigh Navigation Coal Company, 1941.

"The Strike of the Pennsylvania Coal-Miners." *Harper's Weekly* XLIV (Sept. 29, 1900): 912–13.

Thompson, Craig. "The Valley That Came Back to Life." *Saturday Evening Post* 226 (Oct. 31, 1953): 42–43, 139–40, 142–43.

SECONDARY SOURCES

Aldrich, Mark. "The Perils of Mining Anthracite: Regulation, Technology and Safety, 1870–1945." *Pennsylvania History* 64 (Summer 1997): 361–83.

———. *Safety First: Technology, Labor, and Business in the Building of American Work Safety, 1870–1939.* Baltimore: Johns Hopkins University Press, 1997.

Armbrister, Trevor. *Act of Vengeance: The Yablonski Murders and Their Solution.* New York: E. P. Dutton, 1975.

Ashmead, Dever C. *Anthracite Losses and Reserves in Pennsylvania.* n.p.: Pennsylvania Department of Forests and Waters, 1926.

Ashworth, William. *The History of the British Coal Industry.* Vol. 5, *1946–1982: The Nationalized Industry.* Oxford: Clarendon Press, 1986.

The Atlas of Pennsylvania. Philadelphia: Temple University Press, 1989.

Aurand, Harold. "Diversifying the Economy of the Anthracite Region, 1880–1900." *Pennsylvania Magazine of History and Biography* 94 (Jan. 1970): 54–61.

Aurand, Harold W. *Coalcracker Culture: Work and Values in Pennsylvania Anthracite, 1835–1935.* Selinsgrove, Pa.: Susquehanna University Press, 2003.

———. *From the Molly Maguires to the United Mine Workers.* Philadelphia: Temple University Press, 1971.

———. *Population Change and Social Continuity: Ten Years in a Coal Town.* London: Associated University Presses, 1986.

Bakerman, Theodore. *Anthracite Coal: A Study in Advanced Industrial Decline.* New York: Arno Press, 1979 (originally completed as a University of Pennsylvania Ph.D. diss. in Economics in 1956).

Bartholomew, Craig L., and Lance E. Metz. *The Anthracite Iron Industry of the Lehigh Valley.* Easton, Pa.: Center for Canal History and Technology, 1988.

Beck, John. *Never Before In History: The Story of Scranton.* Northridge, Calif.: Windsor Publications, 1986.

Bensman, David, and Roberta Lynch. *Rusted Dreams: Hard Times in a Steel Community.* Berkeley: University of California Press, 1988.

Benson, Susan Porter. "Living on the Margin: Working-Class Marriages and Family Survival Strategies in the United States, 1919–1941." In *The Sex of Things: Gender and Consumption in Historical Perspective*, edited by Victoria de Grazia with Ellen Furlough, 212–43 (Berkeley: University of California Press, 1996).

Berney, Barbara. "The Rise and Fall of the UMW Fund." *Southern Exposure* 6 (1978): 95–102.

Bernstein, Michael A., and David E. Adler, eds. *Understanding American Economic Decline.* New York: Cambridge University Press, 1994.

Berthoff, Rowland. "The Social Order of the Anthracite Region, 1825–1902." *Pennsylvania Magazine of History and Biography* 89 (1965): 261–91.

Bezanson, Anne. "Earnings of Coal Miners." *The Annals* CXI (January 1924): 1–11.

Blatz, Perry K. *Democratic Miners: Work and Labor Relations in the Anthracite Coal Industry, 1875–1925.* Albany: State University of New York Press, 1994.

Blau, Francine D., and Marianne A. Ferber. *The Economics of Women, Men, and Work.* Englewood Cliffs, N.J.: Prentice-Hall, 1986.

Bluestone, Barry, and Bennett Harrison. *The Deindustrialization of America: Plant Closings, Community Abandonment, and the Dismantling of Basic Industry.* New York: Basic Books, 1982.

Bodnar, John. *Anthracite People: Families, Unions, and Work, 1900–1940.* Harrisburg: Pennsylvania Historical and Museum Commission, 1983.

———. *Remaking America: Public Memory, Commemoration, and Patriotism in the Twentieth Century.* Princeton, N.J.: Princeton University Press, 1992.

———. "Socialization and Adaptation: Immigrant Families in Scranton, 1880–1890." *Pennsylvania History* 43 (1976): 147–62.

Bogen, Jules. *The Anthracite Railroads: A Study in American Railroad Enterprise.* New York: Ronald Press, 1927.

Boyer, Paul S., ed. *The Oxford Companion to United States History.* New York: Oxford University Press, 2001.

Bradshaw, Michael. *The Appalachian Regional Commission: Twenty-Five Years of Government Policy.* Lexington: University Press of Kentucky, 1991.

Buss, Terry F., and F. Stevens Redburn with Joseph Waldron. *Mass Unemployment: Plant Closing and Community Mental Health.* Beverly Hills, Calif.: Sage, 1983.

Camp, Scott. *Worker Response to Plant Closings: Steelworkers in Johnstown and Youngstown.* New York: Garland, 1995.

Caudill, Harry M. *Night Comes to the Cumberlands: A Biography of a Depressed Area.* Boston: Little, Brown, 1963.

Chandler, Alfred D., Jr. "Anthracite Coal and the Beginnings of the Industrial Revolution in the United States." *Business History Review* 46 (Summer 1972): 141–81.

Chernow, Ron. *The House of Morgan: An American Banking Dynasty and the Rise of Modern Finance.* New York: Atlantic Monthly Press, 1990.

Chiang, Elizabeth. "The Great Storm that Swept Through: The Effects of Globalization on Indiana County. *Pennsylvania History* 71 (Spring 2004): 165–90.

Chinoy, Ely. *Automobile Workers and the American Dream.* Boston: Beacon Press,1972; originally published in 1965.

Churches and Church Membership in the United States: An Enumeration and Analysis by Counties, States and Regions. New York: National Council of Churches, 1956.

Coaldale: One Hundred Twenty-Five Years of Progress. Coaldale, Pa.: by the town, 1952.

Coode, Thomas H., and John F. Bauman. *People, Poverty, and Politics: Pennsylvanians during the Great Depression.* Lewisburg, Pa.: Bucknell University Press, 1981.

Couto, Richard. *An American Challenge: A Report on Economic Trends and Social Issues in Appalachia.* Dubuque, Iowa: Kendall/Hunt, 1994.

Cowie, Jefferson. *Capital Moves: RCA's Seventy-Year Quest for Cheap Labor.* Ithaca: Cornell University Press, 1999.

———, and Joseph Heathcott, eds. *Beyond the Ruins: The Meanings of Deindustrialization.* Ithaca: Cornell University Press, 2003.

Cumbler, John. *A Social History of Economic Decline: Business, Politics, and Work in Trenton.* New Brunswick: Rutgers University Press, 1989.

Curran, Daniel J. *Dead Laws for Dead Men: The Politics of Federal Coal Mine Health and Safety Legislation.* Pittsburgh: University of Pittsburgh Press, 1993.

Cyphers, Christopher J. *The National Civic Federation and the Making of a New Liberalism, 1900–1915.* Westport, Conn.: Praeger, 2002.

da Costa Nunes, "Pennsylvania's Anthracite Mines and Miners: A Portrait of the Industry in America[n] Art." *IA: The Journal of the Society for Industrial Archeology* 28, no. 1 (2002): 11–32.

Davies, Edward J., II. *The Anthracite Aristocracy: Leadership and Social Change in the Hard Coal Regions of Northeast Pennsylvania, 1800–1930.* DeKalb: Northern Illinois University Press, 1985.

Davis, Colin J. *Power at Odds: The 1922 National Railroad Shopmen's Strike.* Urbana: University of Illinois Press, 1997.

Deasy, George F., and Phyllis R. Griess. "Atlas of Pennsylvania Coal and Coal Mining, Part II: Anthracite." *Bulletin of the Mineral Industries Experiment Station.* University Park, Pa.: College of Mineral Industries, Penn State University, 1959–1963.

DeKok, David. *Unseen Danger: A Tragedy of People, Government, and the Centralia Mine Fire.* Philadelphia: University of Pennsylvania Press, 1986.

Derickson, Alan. *Black Lung: Anatomy of a Public Health Disaster.* Ithaca: Cornell University Press, 1998.

———. "Health Security for All? Social Unionism and Universal Health Insurance, 1935–1958." *Journal of American History* 80 (March 1994): 1333–56.

———. "Occupational Disease and Career Trajectory in Hard Coal, 1870–1930." *Industrial Relations* 32 (Winter 1993): 94–110.

Douglas, Paul H. *Real Wages in the United States, 1890–1926.* Boston: Houghton Mifflin, 1930.

Dublin, Thomas. Biographical Sketch of James Gildea. *Valley Gazette*, no. 276 (June 1995): 11–13.

———. "Daniel Helms Remembers the Equalization March of 1933." *Valley Gazette*, no. 275 (May 1996): 28–29.

———. "Life After the Mines Closed." *Pennsylvania Heritage* 25 (Spring 1999): 6–15 (photographs by George Harvan).

———. *Transforming Women's Work: New England Lives in the Industrial Revolution.* Ithaca: Cornell University Press, 1994.

———. *When The Mines Closed: Stories of Struggles in Hard Times.* Ithaca: Cornell University Press, 1998.

———. "Working-class Families Respond to Industrial Decline: Migration from the Pennsylvania Anthracite Region since 1920." *International Labor and Working Class History* 54 (Fall 1998): 40–56.

Dublin, Thomas, and Walter Licht. "Gender and Economic Decline: The Pennsylvania Anthracite Region, 1920–1970." *Oral History Review* 27 (Winter/Spring 2000): 1–17.

Dubofsky, Melvyn, and Warren Van Tine. *John L. Lewis: A Biography.* New York: Quadrangle/New York Times, 1977.

Dudley, Kathryn Marie. *The End of the Line: Lost Jobs, New Lives in Postindustrial America.* Chicago: University of Chicago Press, 1994.

Earley, James B. *Envisioning Faith: The Pictorial History of the Diocese of Scranton.* Devon, Pa: W.T. Cooke, 1994.

Edmunds, William E., and Edwin F. Koppe. *Coal in Pennsylvania.* n.p.: Commonwealth of Pennsylvania, Topographic and Geologic Survey, 1968.

1891–1991: A Century of Faith and Heritage. Lansford, Pa.: St. Michael the Archangel Parish, 1995.

Feldman, Gerald D., and Klaus Tenfelde, eds. *Workers, Owners, and Politics in Coal Mining: An International Comparison of Industrial Relations.* New York: Berg, 1990.

Fink, Leon. *Progressive Intellectuals and the Dilemmas of Democratic Commitment.* Cambridge, Mass.: Harvard University Press, 1997.

Finley, Joseph E. *The Corrupt Kingdom: The Rise and Fall of the United Mine Workers.* New York: Simon and Schuster, 1972.

Folsom, Burton W., Jr. *Urban Capitalists: Entrepreneurs and City Growth in Pennsylvania's Lackawanna and Lehigh Regions, 1800–1920.* Baltimore: Johns Hopkins University Press, 1981.

Foner, Eric, and John A. Garraty, eds. *Reader's Companion to American History.* Boston: Houghton Mifflin, 1991.

Fox, Maier B. *United We Stand: The United Mine Work-*

ers of America, 1890–1990. Washington, D.C.: UMWA, 1990.

Freese, Barbara. Coal: A Human History. Cambridge, Mass.: Perseus, 2003.

Friedan, Betty. The Feminine Mystique. New York: W.W. Norton, 1963.

Galbraith, John Kenneth. The Great Crash, 1929. Boston: Houghton Mifflin, 1954.

Gaventa, John. Power and Powerlessness: Quiescence and Rebellion in an Appalachian Valley. Urbana: University of Illinois Press, 1980.

Gaventa, John, Barbara Ellen Smith, and Alex Willingham. Communities in Economic Crisis: Appalachia and the South. Philadelphia: Temple University Press, 1990.

Glassberg, David. "Sense of History in Pennsylvania: Work, Craft, Ethnicity, and Place." Pennsylvania History 60 (Oct. 1993): 519–29.

Goldin, Claudia. Understanding the Gender Gap: An Economic History of American Women. New York: Oxford University Press, 1990.

Goodrich, Carter. The Miner's Freedom. Boston: Marshall Jones, 1925.

Gordus, Jean, Paul Jarky, and Louis Ferman. Plant Closing and Economic Dislocation. Kalamazoo, Mich.: W. E. Upjohn Institute for Employment Research, 1981.

Graebner, William. Coal-Mining Safety in the Progressive Period: The Political Economy of Reform. Lexington: University of Kentucky Press, 1976.

Graham, Otis L. Losing Time: The Industrial Policy Debate. Cambridge, Mass: Harvard University Press, 1992.

Great Britain, Department of Employment and Productivity. Ryhope: A Pit Closes: A Study in Redeployment. London: Her Majesty's Stationery Office, 1970.

The "Great Strike": Perspectives on the 1902 Anthracite Coal Strike. Easton, Pa.: Canal History and Technology Press, 2002.

Greater Scranton Chamber of Commerce: 100 Years of Service, 1867–1967. Scranton, Pa.: Harper & Row, honorary publisher, 1967.

Greene, Victor. The Slavic Community on Strike: Immigrant Labor in Pennsylvania Anthracite. Notre Dame: University of Notre Dame Press, 1968.

Gustafson, William F. "Bootleg Coal Mining, 1925–1953." Proceedings of the Ninth Annual Conference on the History of Northeastern Pennsylvania: The Last 100 Years, Oct. 10, 1997, 1–23.

Harrington, Michael. The Other America: Poverty in America. New York: Macmillan, 1962.

Harris, Howard, ed. Keystone of Democracy: A History of Pennsylvania Workers. Harrisburg: Pennsylvania Historical and Museum Commission, 1999.

Hartford, William F. Where Is Our Responsibility? Unions and Economic Change in the New England Textile Industry, 1870–1960. Amherst: University of Massachusetts Press, 1996.

Harvan, George. The Coal Miners of Panther Valley. Bethlehem, Pa.: Lehigh University Art Gallery, 1995.

Hathaway, Dale A. Can Workers Have a Voice?: The Politics of Deindustrialization in Pittsburgh. University Park: Pennsylvania State University Press, 1993.

Haynes, Ada F. Poverty in Central Appalachia: Underdevelopment and Exploitation. New York: Garland, 1997.

Historic Resources Study: Delaware and Lehigh Canal National Heritage Corridor and State Heritage Park. Easton, Pa.: Hugh Moore Historical Park and Museums, 1992.

Hodas, Daniel. The Business Career of Moses Taylor: Merchant, Finance Capitalist, and Industrialist. New York: New York University Press, 1976.

Holton, James L. The Reading Railroad: History of a Coal Age Empire. Volume II: The Twentieth Century. Laury's Station, Pa.: Garrigues House, 1992.

Howard, Walter T., ed. Anthracite Reds: A Documentary History of Communists in Northeastern Pennsylvania during the 1920s. New York: iUniverse, 2004.

———. Anthracite Reds. Vol. 2, A Documentary History of Communists in Northeastern Pennsylvania during the Great Depression. New York: iUniverse, 2004.

———. "Communist Activism in the Pennsylvania Anthracite during the Great Depression." Paper delivered at the 64th annual meeting of the Pennsylvania Historical Association, Bucknell University, October 13–14, 1995.

———. "The Communist Party in the Anthracite Coal Fields, 1919–1955." Paper delivered at the Fall Lecture Series of the Eckley Miners' Village, Eckley, Pa., October 2003.

———. Forgotten Radicals: Communists in the Pennsylvania Anthracite, 1919–1950. Lanham, Md.: University Press of America, 2005.

———. "The National Miners Union: Communists and Miners in the Pennsylvania Anthracite, 1928–1931." Pennsylvania Magazine of History and Biography 125 (Jan./April 2001): 91–124.

Howland, Marie. Plant Closings and Worker Displacement: The Regional Issue. Kalamazoo, Mich.: W. E. Upjohn Institute for Employment Research, 1988.

Hudson Coal Company. The Story of Anthracite. New York: Hudson Coal Company, 1932.

Hume, Brit. Death and the Mines: Rebellion and Murder in the United Mine Workers. New York: Grossman, 1971.

Hunter, Louis C. A History of Industrial Power in the United States, 1780–1930. Charlottesville: University Press of Virginia, 1985.

Hydro, Vincent, Jr. The Mauch Chunk Switchback: America's Pioneer Railroad. Easton, Pa.: Canal History and Technology Press, 2002.

Illick, Joseph E. At Liberty: The Story of a Community and a Generation: The Bethlehem, Pennsylvania, High School Class of 1952. Knoxville: University of Tennessee Press, 1989.

Jacobs, Renée. SLOW BURN: A Photodocument of Centralia, Pennsylvania. Philadelphia: University of Pennsylvania Press, 1986.

Janosov, Robert A. "The Rise, Demise, and Resurrection of 'Blue Coal.'" In *Chapters in Northeastern Pennsylvania History: Luzerne, Lackawanna, and Wyoming Counties,* edited by Sheldon Spear. Shavertown, Pa.: Jemags, 1999.

Jones, Chester Lloyd. *The Economic History of the Anthracite-Tidewater Canals.* Philadelphia: University of Pennsylvania, 1908.

Jones, Eliot. *The Anthracite Coal Combination in the United States.* Cambridge, Mass.: Harvard University Press, 1914.

Jones, William D. *Wales in America: Scranton and the Welsh, 1860–1920.* Cardiff: University of Wales Press, 1993.

Kanarek, Harold K. "Disaster for Hard Coal: The Anthracite Strike of 1925–1926." *Labor History* 15 (Winter 1974): 44–62.

———. "The Pennsylvania Anthracite Strike of 1922." *Pennsylvania Magazine of History and Biography* 99 (1975): 207–25.

Kashatus, William C., III. "'Dapper Dan' Flood: Pennsylvania's Legendary Congressman." *Pennsylvania Heritage* 21, no.3 (1995): 4–11.

Kenny, Kevin. *Making Sense of the Molly Maguires.* New York: Oxford University Press, 1998.

King, Clyde L., ed. "The Price of Coal," special issue of *The Annals* CXI (Jan. 1924).

Kinicki, Angelo. "Personal Consequences of Plant Closings: A Model and Preliminary Test." *Human Relations* 33 (1985): 197–212.

Klein, Jennifer. *For All These Rights: Business, Labor, and the Shaping of America's Private-Public Welfare State.* Princeton, N.J.: Princeton University Press, 2003.

Klein, Maury. *Rainbow's End: The Crash of 1929.* New York: Oxford University Press, 2001.

Knies, Michael. *Coal on the Lehigh, 1790–1827: Beginnings and Growth of the Anthracite Industry in Carbon County, Pennsylvania.* Easton, Pa.: Canal History and Technology Press, 2001.

Knight, E. M. *Men Leaving Mining: West Cumberland, 1966–67. Report to the Ministry of Labour.* Newcastle-upon-Tyne: University of Newcastle Upon Tyne, Department of Geography, 1968.

Kozura, Michael. "We Stood Our Ground: Anthracite Miners and the Challenge to Corporate Property, 1930–1941." In *"We Are All Leaders": The Alternative Unionism of the Early 1930s,* edited by Staughton Lynd, 199–237. Urbana: University of Illinois Press, 1996.

Krajcinovic, Ivana. *From Company Doctors to Managed Care: The United Mine Workers' Noble Experiment.* Ithaca: Cornell University Press, 1997.

Kroll-Smith, J. Stephen, and Stephen Robert Couch. *The Real Disaster Is Above Ground: A Mine Fire and Social Conflict.* Lexington: University Press of Kentucky, 1990.

Kucas, Antanas. *Shenandoah St. George Lithuanian Parish in Commemoration of the Diamond Anniversary of St. George Parish, 1891–1966.* Shenandoah: St. George Parish, 1968.

La Gorce, John Oliver. "The Industrial Titan of America." *National Geographic Magazine* 35 (May 1919): 367–406.

The Land and Economy of Appalachia: Proceedings from the 1986 Conference on Appalachia. Lexington, Ky.: Appalachian Center, 1986.

Langa, Helen. "Deep Tunnels and Burning Flues: The Unexpected Political Drama in 1930s Industrial Production Prints." *IA: The Journal of the Society for Industrial Archeology* 28, no. 1 (2002): 43–58.

Laslett, John M., ed. *The United Mine Workers of America: A Model of Industrial Solidarity?* University Park: Pennsylvania State University Press, 1996.

Leahy, Peter J., and Xiannuan Lin. "Plant Closings: A Comparison to Natural Disasters." *American Journal of Economics and Sociology* 51 (1992): 333–48.

Leana, Carrie R., and Daniel D. Feldman. "The Psychology of Job Loss." *Research in Personnel and Human Resources Management* 12 (1994): 271–302.

———. *Coping with Job Loss: How Individuals, Organizations, and Communities Respond to Layoffs.* New York: Lexington Books, 1992.

Leboutte, René. "Les bassins industriels en Europe: Production et mutation d'un espace, 1750–1992." *EUI Working Papers in History.* San Domenico, Italy: European University Institute, 1993.

LeDuc, Thomas. "Carriers, Courts, and the Commodities Clause." *Business History Review* 39 (1965): 57–73.

Leuchtenberg, William E. *Franklin D. Roosevelt and the New Deal, 1932–1940.* New York: Harper & Row, 1963.

Lewis, W. David. "The Early History of the Lackawanna Iron and Coal Company: A Study in Technological Adaptation." *Pennsylvania Magazine of History and Biography* 96 (1972): 424–68.

Licht, Walter. *Getting Work: Philadelphia, 1840–1950.* Cambridge, Mass.: Harvard University Press, 1992.

———. *Industrializing America: The Nineteenth Century.* Baltimore: Johns Hopkins University Press, 1995.

Lichtenstein, Nelson. *Labor's War at Home: The CIO in World War II.* Cambridge: Cambridge University Press, 1982.

Lockard, Duane. *Coal: A Memoir and Critique.* Charlottesville: University Press of Virginia, 1998.

Long, Priscilla. *Where the Sun Never Shines: A History of America's Bloody Coal Industry.* New York: Paragon House, 1989.

Lynd, Staughton, ed. *"We Are All Leaders": The Alternative Unionism of the Early 1930s.* Urbana: University of Illinois Press, 1996.

———. *The Fight Against Shutdowns: Youngstown's Steel Mill Closings.* San Pedro, Calif.: Singlejack Books, 1982.

Margrave, Richard Dobson. *The Emigration of Silk Workers from England to the United States in the Nineteenth Century: With Special Reference to Coventry, Macclesfield, Paterson, New Jersey and South Manchester.* New York: Garland, 1986.

Marschall, Daniel. "The Miners and the UMW: Crisis in

the Reform Process." *Socialist Review*, no. 40–41 (July–Oct. 1978): 65–115.

Marsh, Ben. "Continuity and Decline in the Anthracite Towns of Pennsylvania." *Annals of the Association of American Geographers* 77 (1987): 337–51.

McKerns, Joseph P. "The 'Faces' of John Mitchell: News Coverage of the Great Anthracite Strike of 1902 in the Regional and National Press." In *The "Great Strike": Perspectives on the 1902 Anthracite Coal Strike*. Easton, Pa.: Canal History and Technology Press, 2002.

Melosh, Barbara. *"The Physician's Hand": Work Culture and Conflict in American Nursing*. Philadelphia: Temple University Press, 1982.

Milkman, Ruth. *Farewell to the Factory: Auto Workers in the Late Twentieth Century*. Berkeley: University of California Press, 1997.

Miller, Donald L., and Richard E. Sharpless. *Kingdom of Coal: Work, Enterprise, and Ethnic Communities in the Mine Fields*. Philadelphia: University of Pennsylvania Press, 1985.

Miller, Randall M., and William Pencak, eds. *Pennsylvania: A History of the Commonwealth*. University Park: Pennsylvania State University Press, 2002.

Moody, Kim. *An Injury To All: The Decline of American Unionism*. New York: Verso, 1988.

Mowry, George E. *The Era of Theodore Roosevelt and the Birth of Modern America, 1900–1912*. New York: Harper & Row, 1962.

Mulcahy, Richard. "Partitioning the Miner's Welfare State: The Destruction of the Medical Program of the UMWA Welfare and Retirement Fund." *Mid-America* 77 (Spring/Summer 1995): 175–204.

———. *A Social Contract for the Coal Fields: The Rise and Fall of the United Mine Workers of America Welfare and Retirement Fund*. Knoxville: University of Tennessee Press, 2000.

Myers, Robert J. "Experience of the UMWA Welfare and Retirement Fund." *Industrial and Labor Relations Review* 10 (1956–57): 93–100.

Nash, Michael. *Conflict and Accommodation: Coal Miners, Steel Workers, and Socialism, 1890–1920*. Westport, Conn.: Greenwood, 1982.

Nearing, Scott. *Anthracite: An Instance of Natural Resource Monopoly*. Philadelphia: John C. Winston, 1915.

Neil, Cecily, Markku Tykkyläinen, and John Bradbury, eds. *Coping With Closure: An International Comparison of Mine Town Experiences*. London: Routledge, 1992.

Nelson, Steve, James R. Barrett, and Rob Ruck. *Steve Nelson: American Radical*. Pittsburgh: University of Pittsburgh Press, 1981.

Newman, Monroe. *The Political Economy of Appalachia: A Case Study in Regional Integration*. Lexington, Mass.: Lexington Books, 1972.

The New York Times. *The Downsizing of America*. New York: Times Books, 1996.

Nissen, Bruce, ed. *Fighting For Jobs: Case Studies of Labor-Community Coalitions Confronting Plant Closings*. Albany: State University of New York Press, 1995.

Novak, Michael. *The Guns of Lattimer: The True Story of a Massacre and a Trial, August 1897–March 1898*. New York: Basic Books, 1978.

Opportunity Realized: The Greek Catholic Union's First One Hundred Years, 1892–1992. Beaver, Pa.: Greek Catholic Union of the U.S.A., 1994.

Palladino, Grace. *Another Civil War: Labor, Capital and the State in the Anthracite Regions of Pennsylvania, 1840–68*. Urbana: University of Illinois Press, 1990.

Parker, M. J. *Thatcherism and the Fall of Coal*. Oxford: Oxford University Press, 2000.

Parker, Mike. *The Politics of Coal's Decline: The Industry in Western Europe*. London: Earthscan Publications, 1994.

Parton, W. Julian. *The Death of a Great Company: Reflections on the Decline and Fall of the Lehigh Coal and Navigation Company*. Easton, Pa.: Canal History and Technology Press, 1998; originally published in 1986.

Perucci, Carolyn C., Robert Perucci, Dena B. Targ, and Harry R. Targ. *Plant Closings: International Context and Social Costs*. New York: A. de Gruyter, 1988.

Phelan, Craig. *Divided Loyalties: The Public and Private Life of Labor Leader John Mitchell*. Albany: State University of New York Press, 1994.

Piore, Michael J., and Charles F. Sabel, *The Second Industrial Divide: Possibilities for Prosperity*. New York: Basic Books, 1984.

Popenoe, David. *The Suburban Environment: Sweden and the United States*. Chicago: University of Chicago Press, 1977.

Powell, H. Benjamin. *Philadelphia's First Fuel Crisis: Jacob Cist and the Developing Market for Pennsylvania Anthracite*. University Park: Pennsylvania State University Press, 1978.

Reid, Donald. *The Miners of Decazeville: A Genealogy of Deindustrialization*. Cambridge, Mass.: Harvard University Press, 1985.

Reverby, Susan M. *Ordered to Care: The Dilemma of American Nursing, 1850–1945*. Cambridge: Cambridge University Press, 1987.

Roberts, Ellis W. *Journey Through Welsh Hills and American Valley*. Wilkes-Barre, Pa.: Wyoming Historical and Geological Society, 1986.

Rose, Dan. *Energy Transition and the Local Community: A Theory of Society Applied to Hazleton, Pennsylvania*. Philadelphia: University of Pennsylvania Press, 1981.

Rowland, Donald J. *Whites Crossing*. Carlisle, Pa.: Sunset Publications, 1995.

Salay, David, ed. *Hard Coal, Hard Times: Ethnicity and Labor in the Anthracite Region*. Scranton, Pa.: Anthracite Museum Press, 1984.

Scargill, D. Ian. "French energy: the end of an era for coal." *Geography* 76 (April 1991): 172–75.

Schlegel, Marvin W. *Ruler of the Reading: The Life of Franklin B. Gowen, 1836–1889*. Harrisburg, Pa.: Archives Publishing Company of Pennsylvania, 1947.

Schuyler, David. "Exhibit Review: Reflections on Levittown at Fifty." *Pennsylvania History* 70 (Winter 2003): 101–9.

Scranton, Philip, and Walter Licht. *Work Sights: Industrial Philadelphia, 1890–1950.* Philadelphia: Temple University Press, 1986.

Scranton, Philip, ed. *Silk City: Studies on the Paterson Silk Industry, 1860–1940.* Newark: New Jersey Historical Society, 1985.

———. *Proprietary Capitalism: The Textile Manufacture at Philadelphia, 1800–1885.* New York: Cambridge University Press, 1983.

Seitchik, Adam, and Jeffrey Zumitsky. *From One Job to the Next: Worker Adjustment in a Changing Labor Market.* Kalamazoo, Mich.: W. E. Upjohn Institute for Employment Research, 1989.

Seltzer, Curtis. *Fire in the Hole: Miners and Managers in the American Coal Industry.* Lexington: University Press of Kentucky, 1985.

Serrin, William. *Homestead: The Glory and Tragedy of an American Steel Town.* New York: Times Books, 1992.

Sewel, John. *Colliery Closure and Social Change: A Study of a South Wales Mining Village.* Cardiff: University of Wales Press, 1975.

The Shamokin Area Centennial: 1864–1964. [Shamokin, Pa., 1964].

Siegelbaum, Lewis, and Daniel J. Walkowitz. *Workers of the Donbass Speak: Survival and Identity in the New Ukraine, 1989–1992.* Albany: State University of New York Press, 1995.

Singer, Alan. "Communists and Coal Miners: Rank-and-File Organizing in the United Mine Workers of America during the 1920s." *Science and Society* 55 (Summer 1991): 132–57.

Sipe, C. Hale. *The Indian Wars of Pennsylvania.* 2nd ed. Lewisburg, Pa.: Wennawoods, 1995; originally published in 1931.

Smith, Barbara Ellen. *Digging Our Own Graves: Coal Miners and the Struggle over Black Lung Disease.* Philadelphia: Temple University Press, 1987.

Spear, Sheldon. *Chapters in Northeastern Pennsylvania History: Luzerne, Lackawanna, and Wyoming Counties.* Shavertown, Pa.: Jemags, 1999.

———. "Life After Anthracite: Wyoming Valley's Economic Recovery in the 1950's and 1960's." *History of Northeastern Pennsylvania* 2 (1990): 26–42.

Sperry, J. R. "Rebellion Within the Ranks: Pennsylvania Anthracite, John L. Lewis, and the Coal Strikes of 1943." *Pennsylvania History* 40 (1973): 293–312.

Staudoher, Paul D., and Holly E. Brown, eds. *Deindustrialization and Plant Closure.* Lexington, Mass.: D.C. Heath, 1987.

Stepenoff, Bonnie. *Their Fathers' Daughters: Silk Mill Workers in Northeastern Pennsylvania, 1880–1960.* Selinsgrove, Pa.: Susquehanna University Press, 1999.

Sterba, Christopher. "Family, Work, and Nation: Hazleton, Pennsylvania, and the 1934 General Strike in Textiles." *Pennsylvania Magazine of History and Biography* 120 (1996): 3–35.

Strouse, Jean. *Morgan: American Financier.* New York: Random House, 1999.

Sutherland, Anthony X. *The Pennsylvania Slovak Catholic Union: A Century of Brotherhood, 1893–1993.* Wilkes-Barre, Pa.: PSCU, n.d.

Swyngedouw, Erik. "Reconstructing Citizenship, the Re-Scaling of the State and the New Authoritarianism: Closing the Belgian Mines." *Urban Studies* 33 (1996): 1499–1521.

Takamiya, Makoto. *Union Organization and Militancy: Conclusions from a Study of the United Mine Workers of America, 1940–1974.* Masenheim am Glan: Verlag Anton Hain, 1978.

Tenfelde, Klaus, ed. *Towards a Social History of Mining in the 19th and 20th Centuries.* Munich: Verlag C. H. Beck, 1992.

Tietz, Jeff. "The Great Centralia Coal Fire: How One Small Mining Town Went Up in Smoke." *Harper's Magazine,* February 2004, pp. 47–57.

Trachtenberg, Alexander. *The History of Legislation for the Protection of Coal Miners in Pennsylvania, 1824–1915.* New York: International Publishers, 1942.

Trattner, Walter N. *Crusade for the Children: A History of the National Labor Committee and Child Labor Reform in America.* Chicago: Quadrangle, 1970.

Upon the Shoulders of Giants: The CAN DO Story. Hazleton, Pa.: CAN DO, 1991.

Vogel, Robert M. *Roebling's Delaware & Hudson Canal Aqueducts.* Washington, D.C.: Smithsonian Institution Press, 1971.

Vosler, Nancy R. "Displaced Manufacturing Workers and Their Families: A Research-Based Practice Model." *Families in Society: The Journal of Contemporary Human Services* 39 (1994): 105–17.

Yarrow, Mike. "Voices from the Coalfields: How Miners' Families Understood the Crisis of Coal," In *Communities in Economic Crisis: Appalachia and the South,* edited by John Gaventa, Barbara Ellen Smith, and Alex Willingham. Philadelphia: Temple University Press, 1990.

Waddington, David, Chas Critcher, Bella Dicks, and David Parry. *Out of the Ashes: The Social Impact of Industrial Contraction and Regeneration on Britain's Mining Communities.* London: The Stationery Office, 2001.

Wallace, Anthony F. C. *St. Clair: A Nineteenth Century Coal Town's Experience with A Disaster-Prone Industry.* Ithaca: Cornell University Press, 1981.

Wallace, Michael. "Dying for Coal: The Struggle for Health and Safety in American Coal Mining, 1930–1982." *Social Forces* 66 (1987): 336–64.

Wallace, Paul A. W. *Indians in Pennsylvania.* 2nd ed.. Harrisburg, Pa.: Pennsylvania Historical and Museum Commission, 1991.

Warfel, Stephen G. *A Patch of Land Owned by the Company.* Harrisburg: Pennsylvania Historical and Museum Commission, 1993.

Weinstein, James. *The Corporate Ideal in the Liberal State, 1900–1918.* Boston: Beacon Press, 1968.

Whyte, William H., Jr., *The Organization Man.* Garden City, N.Y.: Doubleday Anchor, 1956.

Wiebe, Robert. "The Anthracite Strike of 1902: A Record

of Confusion." *Mississippi Valley Historical Review* 48 (1961): 229–51.

Wilson, Gregory. "'Our Chronic and Desperate Situation': Anthracite Communities and the Emergence of Redevelopment Policy in Pennsylvania and the United States, 1945–1965." *International Review of Social History* 47:S10 (2002): 137–58.

Wolensky, Kenneth C. "Living for Reform." *Pennsylvania Heritage* 27 (Winter 2001): 14–23.

Wolensky, Kenneth C., and Judith Rich. *Child Labor in Pennsylvania.* Harrisburg: Pennsylvania Historical and Museum Commission, 1998.

Wolensky, Kenneth C., Nicole H. Wolensky, and Robert P. Wolensky. *Fighting for the Union Label: The Women's Garment Industry and the ILGWU in Pennsylvania.* University Park: Pennsylvania State University Press, 2002.

Wolensky, Robert P. "The Subcontracting System and Industrial Conflict in the Northern Anthracite Coal Field." In *"The Great Strike": Perspectives on the 1902 Anthracite Coal Strike*, 67–94. Easton, Pa.: Canal History and Technology Press, 2002.

Wolensky, Robert P., Kenneth C. Wolensky, and Nicole H. Wolensky. *The Knox Mine Disaster, January 22, 1959: The Final Years of the Northern Anthracite Industry and the Effort to Rebuild a Regional Economy.* Harrisburg: Pennsylvania Historical and Museum Commission, 1999.

———. *Voices of the Knox Mine Disaster: Stories, Remembrances, and Reflections on the Anthracite Coal Industry's Last Major Catastrophe, January 22, 1959.* N.p.: Pennsylvania Historical and Museum Commission, 2005.

Wolf, George D. *William Warren Scranton: Pennsylvania Statesman.* University Park: Pennsylvania State University Press, 1981.

Yates, W. Ross. *Lehigh University: A History of Education in Engineering, Business, and the Human Condition.* Bethlehem: Lehigh University Press, 1992.

Yearley, C. K., Jr. *Enterprise and Anthracite: Economics and Democracy in Schuylkill County, 1820–1875.* Baltimore: Johns Hopkins University Press, 1961.

Zerbey, Joseph H. *History of Coaldale.* Pottsville, Pa.: J. H. Zerbey, 1934.

Zieger, Robert H. *The CIO, 1935–1955.* Chapel Hill: University of North Carolina Press, 1995.

———. *John L. Lewis: Labor Leader.* Boston: Twayne, 1988.

———. "Pennsylvania Coal and Politics: The Anthracite Strike of 1925–1926." *Pennsylvania Magazine of History and Biography* 92 (April 1969): 244–62.

———. "Pinchot and Coolidge: The Politics of the 1923 Anthracite Crisis." *Journal of American History* 52 (1965): 566–81.

Acknowledgments

This book has been more than a decade in the making, and over the years we have benefited enormously from the backing of others. Quantitative evidence and oral history interviews are critical ingredients of the book, and we could not have conducted our research without extensive grant support. The National Endowment for the Humanities provided two collaborative research grants, and the Ford Foundation and the New Jersey Historical Commission funded the oral history component of the project. The University of Pennsylvania and the State University of New York at Binghamton afforded us vital research funding as well.

During the project, Thomas Dublin received grants from the Pennsylvania Historical and Museum Commission (for an oral history pilot) and the Balch Institute for Ethnic Studies (for research on ethnic fraternal organizations). He was also the beneficiary of fellowships from the John Simon Guggenheim Memorial Foundation and the Institute for the Advanced Study of Religion at Yale and benefited from a semester in residence at the National Humanities Center in Research Triangle Park, N.C. Walter Licht received fellowship support from the Hagley Museum and Library and from the British Academy, which allowed him to study the closing of mines in Great Britain and internationally. We are extremely grateful for this support.

We relied heavily on a strong and resourceful group of student research assistants, both graduate and undergraduate. At SUNY–Binghamton, they included Penelope Harper, Melissa Doak, Julie Simmonds, Halle Lewis, Laura Murphy, Linda Janke, Eric Contreras, Christa Griswold, and Rebecca Reisen. At the University of Pennsylvania, they included Lynne Snyder, Judd Goldberg, Steve Colella, Dafna Gold, Aaron Shapiro, David Spinks, Dain Landon, Aruna Bewtra, and Seth Schreiberg. Also at the University of Pennsylvania, great thanks are due to Shira Fix, who enabled Walter Licht to be university administrator, professor, and historian at the same time.

The staff at the computer center, Research Foundation, and library at SUNY–Binghamton have been remarkable in responding to our innumerable requests for assistance. We are especially appreciative of the services of Deanna France, Stephen Gilje, Kim Riley, Rose Sherick, Lary Jones, Ralph Hansen, Diane Geraci, Helen Insinger, David Vose, Sara Maximiek, Chris Focht, and Melissa Doak.

At the numerous libraries, archives, and repositories that provided primary documents for this research, archivists and curators have shared their expertise and made our work possible. Special thanks go out to Mike Knies, Lance Metz, and Tom Smith at the National Canal Museum; Jesse Teitelbaum at the Luzerne County Historical Society; Chester Kulesa and Richard Stanislaus at the Anthracite Heritage Museum; Vance Packard Jr., Dave Dubick, and Keith Parrish at Eckley Miners Village; Denise Conklin at Historical Collections and Labor Archives at Pennsylvania State University; Linda Shopes, Bob Weible, and Kenneth Wolensky at the Pennsylvania Historical and Museum Commission; Mike Sherbon, Jonathan Stayer, and Diane Wallace of the Pennsylvania State Archives; and Irwin Marcus, Elizabeth Ricketts, and Phillip Zorich of Indiana University of Pennsylvania.

On the oral history portion of this book, we are particularly grateful to Dorothy Pennachio, James Coon, and George Harvan for recommendations for interviewees. Mary Ann Landis proved a wonderful interviewer for a third of the Panther Valley residents included in the study. George Harvan took photographs of many of the interviewees, a number of which are included in this book, and he proved a wonderful guide to the Panther Valley's history and people. We also thank Ed Gildea, Christine Harvan, and Ed Dougert for permitting the use of their photographs. Ron Grele and Michael Frisch proved excellent guides to newcomers in the oral history field.

Paul Hackash of Lansford, Sister Mercedes of the Jankola Library in Danville, Joseph Yenchko of Hazleton, Charles Petrillo in Wilkes-Barre, and Bob Urban of the *Lehighton Times News* all shared rich resources that contributed to the research. We extend our thanks also to Kevin Kenny and Susan Campbell Bartoletti for assistance with illustrations for the book, and to Joseph Gilmore for his skillful rendering of the book's map and graphs.

We presented our developing ideas for this project at numerous colloquia, conferences, and informal meetings and would like to acknowledge the questions and comments that helped to shape our analysis. Our thanks go to the following colleagues: Giovanni Arrighi, Fitzhugh Brundage, Jon Butler, Ardis Cameron, John Cumbler, Cathy Davidson, David Elesh, Mary Frederickson, Michael Frisch, Jacquelyn Hall, Michael Honey, Molly Nolan, Karen Brodkin Sacks, Harry Stout, and Louise Tilly. Thomas Dublin is particularly grateful to Jean Heffer, Francois Weil, and Pap Ndiaye of the Ecole des Hautes Etudes en Sciences Sociales in Paris for the opportunity to present a series of lectures to American Studies students and faculty in January 1999.

Earlier versions of portions of this book have appeared in print, and we acknowledge permission to reprint that material and thank the journal editors and anonymous readers for their helpful criticism. Chapter 3 took its initial form as "The Equalization of Work: An Alternative Vision of Industrial Capitalism in the Anthracite Region of Pennsylvania in the 1930s," *Canal History and Technology Proceedings* 13 (1994): 81–98. "Working-class Families Respond to Industrial Decline: Migration from the Pennsylvania Anthracite Region since 1920," *International Labor and Working Class History,* 54 (Fall 1998): 40–56 permitted development of our ideas about migration explored further in chapters 6 and 7. Our interpretations of oral history took shape in two articles: "Género y Decadencia Económica: La Región de las Minas de Antracita de Pennsylvania," *Historia, Antropologia y Fuentes Orales,* no. 17 (October 1997): 59–72, and "Gender and Economic Decline: The Penn-sylvania Anthracite Region, 1920–1970," *Oral History Review* 27 (Winter/Spring 2000), 1–17.

Over the course of this research Thomas Dublin had a rich collaboration with the anthracite photographer, George Harvan, and learned a great deal about the region in this shared work. The book draws repeatedly on that joint work and three publications in particular: "Miner's Son, Miners' Photographer: The Life and Work of George Harvan," co-authored with Melissa Doak, in *The Journal for MultiMedia History* 3 (March 2001), online at http://www.albany.edu/jmmh; "Life After the Mines Closed," *Pennsylvania Heritage,* 25 (Spring 1999): 6–15; and "When the Mines Closed: One Worker's Oral History," *Labor's Heritage* 9, no. 4 (Spring 1998): 46–59.

Our special thanks to Kathryn Kish Sklar; who read and commented on two versions of the manuscript and whose editorial suggestions have greatly improved this book. Thanks also to fellow historians who read the manuscript and whose comments have contributed to our thinking about coal's decline—John Bodnar, Ron Grele, Bob Janosov, Lance Metz, Ken Wolensky, Bob Wolensky, and anonymous readers for Cornell University Press. At Cornell University Press we benefited from the support of our editors, Peter Agree, Sherri England, and Alison Kalett, and the talents of Susan Barrett, Karen Hwa, Cathi Reinfelder, and George Whipple. It takes a village, and this book is much better for this collaboration.

In the Panther Valley numerous people shared their history. Ed Gildea's generosity knew no bounds, and Jack Yalch, Mike Sabron, Francis Karnish, David Kuchta, and David and Lora Tokosh repeatedly went out of their ways to help. We interviewed 70 valley residents in the course of the research, and another 570 responded to our survey questionnaire of area high school graduates. We are grateful for the openness with which residents and former residents responded to our inquiries and offer them a collective thank you here. Many are acknowledged individually in appropriate places in the notes. Our greatest debt is to George Harvan, who supported our efforts from the outset, and whose generosity has made this study and our lives much richer.

Index

Thomas Dublin is Professor of History at Binghamton University, State University of New York. He is the author of many books, including *When the Mines Closed: Stories of Struggles in Hard Times* and *Transforming Women's Work: New England Lives in the Industrial Revolution*, both from Cornell, and *Women at Work: The Transformation of Work and Community in Lowell, Massachusetts, 1826–1860*, winner of the Bancroft Prize and the Merle Curti Award. He is also coeditor of the Web site, Women and Social Movements in the United States, 1600–2000.

Walter Licht is Professor of History at the University of Pennsylvania. He is the author of several books, including *Working for the Railroad: The Organization of Work in the Nineteenth Century*, winner of the Philip Taft Labor History Prize, *Work Sights: Industrial Philadelphia, 1890–1950*, *Getting Work: Philadelphia, 1840–1950*, and *Industrializing America: The Nineteenth Century*.